Intelligent Systems Reference Library

Volume 137

Series editors

Janusz Kacprzyk, Polish Academy of Sciences, Warsaw, Poland
e-mail: kacprzyk@ibspan.waw.pl

Lakhmi C. Jain, University of Canberra, Canberra, Australia;
Bournemouth University, UK;
KES International, UK
e-mail: jainlc2002@yahoo.co.uk; jainlakhmi@gmail.com
URL: http://www.kesinternational.org/organisation.php

The aim of this series is to publish a Reference Library, including novel advances and developments in all aspects of Intelligent Systems in an easily accessible and well structured form. The series includes reference works, handbooks, compendia, textbooks, well-structured monographs, dictionaries, and encyclopedias. It contains well integrated knowledge and current information in the field of Intelligent Systems. The series covers the theory, applications, and design methods of Intelligent Systems. Virtually all disciplines such as engineering, computer science, avionics, business, e-commerce, environment, healthcare, physics and life science are included.

More information about this series at http://www.springer.com/series/8578

Dawn E. Holmes · Lakhmi C. Jain
Editors

Advances in Biomedical Informatics

 Springer

Editors
Dawn E. Holmes
Department of Statistics and Applied
  Probability
University of California Santa Barbara
Santa Barbara, CA
USA

Lakhmi C. Jain
University of Canberra
Canberra
Australia

and

Bournemouth University
Poole
UK

and

KES International
Shoreham-by-Sea
UK

ISSN 1868-4394   ISSN 1868-4408   (electronic)
Intelligent Systems Reference Library
ISBN 978-3-319-88442-4   ISBN 978-3-319-67513-8   (eBook)
https://doi.org/10.1007/978-3-319-67513-8

Printed on acid-free paper

This Springer imprint is published by Springer Nature
The registered company is Springer International Publishing AG
The registered company address is: Gewerbestrasse 11, 6330 Cham, Switzerland

Preface

Recent advances in computer hardware and software have generated tremendous interest among researchers in every field of discipline including engineering, science, aviation, business, and commerce. Biomedical informatics is no exception.

Biomedical informatics represents a growing area of interest and innovation in the management of health-related data and the development of focused computational models. In compiling this volume, we have brought together contributions from some of the most prestigious researchers in this field. Each chapter is self-contained. Both theoreticians and application scientists will find this volume of interest. It also provides a useful sourcebook for graduate student since it shows the direction of current research.

We are grateful to the authors and reviewers for their excellent contributions for making this book possible. Thanks are due to the Springer-Verlag for their assistance during the development phase of this book.

Santa Barbara, USA   Dawn E. Holmes
Canberra, Australia/Poole, UK   Lakhmi C. Jain

Contents

About the Editors

Dawn E. Holmes holds BA(Hons), MA, M.Sc., Ph.D., is a Senior Member IEEE, and serves as a faculty member in the Department of Statistics and Applied Probability, University of California, Santa Barbara, USA.

Dr. Holmes has research interests in Bayesian Networks, Maximum Entropy, Data Mining, and Machine Learning and has published extensively in these areas. As a recipient of the University of California Academic Senate Distinguished Teaching Award, she continues to enjoy teaching at all levels. She is an Associate Editor for the KES Journal and also serves on the international program committees of several KES conferences.

Lakhmi C. Jain holds BE(Hons), ME, Ph.D., is a Fellow (Engineers Australia), and serves as a Visiting Professor in Bournemouth University, UK, and Adjunct Professor in the Faculty of Education, Science, Technology & Mathematics In the University of Canberra, Australia.

Dr. Jain founded the KES International for providing a professional community the opportunities for publications, knowledge exchange, cooperation, and teaming. Involving around 5000 researchers drawn from universities and companies worldwide, KES facilitates international cooperation and generates synergy in teaching and research. KES regularly provides networking opportunities for professional community through one of the largest conferences of its kind in the area of KES.

www.kesinternational.org

His interests focus on the artificial intelligence paradigms and their applications in complex systems, security, e-education, e-healthcare, unmanned air vehicles, and intelligent systems.

Chapter 1
Advances in Biomedical Informatics: An Introduction

Dawn E. Holmes and Lakhmi C. Jain

Abstract This chapter presents a summary of a sample of research in the field of biomedical informatics. The topics include digital health research, medical decision support systems, Bayesian networks, tele-monitoring, preprocessing in high dimensional datasets.

Keywords Biomedical informatics · Big data · Telemonitoring · Decision support systems · Medical diagnosis · Machine learning · Bayesian network · Digital health

1.1 Introduction

Recent advances in computing hardware and software and their easy availability have resulted in the burgeoning field of biomedical informatics.

Biomedical informatics is an interdisciplinary research area, combining knowledge from the fields of data science and medicine. In broader terms that includes statistics, epidemiology and computer science as well as clinical medicine. In essence, biomedical informatics is big data restricted to the medical field. There is now a vast amount of medical data collected electronically which is pooled and made available to researchers internationally through vast databases. New data analysis techniques are continually under development in order to cope with this data explosion. Some of the data is structured in the traditional manner and stored in

D.E. Holmes
Department of Statistics and Applied Probability, University of California Santa Barbara, Santa Barbara, CA 93106-3110, USA

L.C. Jain (✉)
Faculty of Education, Science, Technology and Mathematics, University of Canberra, Canberra, ACT 2601, Australia
e-mail: jainlakhmi@gmail.com

L.C. Jain
Bournemouth University, Poole, UK

relational databases but much is unstructured and not amenable to classical statistical analysis. For example, the data gathered by Google to aid in the prediction of influenza included information from casual online searches, which are clearly lacking in the general structure associated with relational databases.

Other issues included how this data is to be stored and accessed, and how it is to be shared without violating patient privacy. But the ultimate aim is to extract useful knowledge from all this data, and so enhance medical decision making and improve health care.

The chapters in this volume, described below, range across a number of important topics, starting with an overview of digital health research. Machine learning is discussed in several chapters. Medical decision support systems, including Bayesian networks are followed by papers on tele-monitoring and preprocessing in high dimensional datasets. The book concludes with a chapter discussing in detail a questionnaire design and analysis.

1.2 Chapters of the Book

Chapter 2 by Gray and Gilbert on **Digital Health Research Methods and Tools: Suggestions and Selected Resources for Researchers**, presents an overview of digital health research, aimed at people who are new to conducting investigations in this area. This chapter discusses research methods with health apps, patient generated health data, social media and wearable self-tracking devices. It outlines the essential theoretical and conceptual frameworks, ethical considerations and research methods necessary for work in digital health. Examples of tools applicable for studies of digital health interventions are presented and practical advice is given on techniques such as critically appraising digital health research literature, primary data collection from devices and services, study reporting and publishing results. A key feature of this chapter is the extensive bibliographical notes. Given the interdisciplinary nature of this area, the authors have presented considerable suggestions for further reading within each section, rather than at the end of the chapter.

In Chap. 3 Brett Beaulieu Jones discusses **Machine Learning for Structured Clinical Data**. The data produced from electronic health records (EHR's) and billing codes focuses this chapter. Areas in which machine learning for disease stratification could be useful are introduced and related to billing codes. Examples are given. The challenges associated with using machine learning in the structured HER environment are discussed in detail including, among other issues, the need to address missing data. Data ethics are also raised and finally, future potential is considered.

Chapter 4 **Defining and Learning Interactive Causes** by Xia Jiang and Richard Neapolitan considers the problem of learning causal influences from passive data. The authors point out that the techniques developed and tested hitherto entail that they cannot in general learn interactive causes with little marginal effects.

Following examples of where this does not hold, Jiang and Neapolitan have developed an exhaustive search algorithm to address the problem. A brief introduction to Bayesian networks and information gain is given as essential background to their current results. Details of the algorithm are given together with examples of its successful application to real clinical breast cancer datasets.

In Chap. 5 Yoshiaki Miyauchi and Haruhiko Nishimura also consider the use of Bayesian networks in their paper entitled **Bayesian Network Modeling for Specific Health Checkups on Metabolic Syndrome**. Health check-up data from Japan, determined from medical examination and answers to a questionnaire are analyzed with the intention of providing lifestyle advice to patients characterized as high-risk according to certain metabolic factors. The methods used to construct a Bayesian network using this data are described and evaluated in detail.

Lifestyle changes are again under consideration in Chap. 6 where Gina Sprint, Diane J. Cook, and Maureen Schmitter-Edgecombe discuss **Unsupervised Detection and Analysis of Changes in Everyday Physical Activity** Data. Wearable fitness monitoring devices are now very popular and provide valuable data, often leading to changes physical activity goals. The authors present their Physical Activity Change Detection (PACD) approach, which provides a framework for detecting time-based changes and determining their significance as well as providing an analysis of these changes. Three existing change-detention algorithms are evaluated together with one proposed algorithm. PACD is evaluated using Fitbit data.

Chapter 7, co-authored by Beatriz Remeseiro, Noelia Barreira, Luisa Sanchez-Brea, Lucia Ramos and Antonio Mosquera considers **Machine Learning Applied to Optometry Data**. In particular, the paper explains how machine learning techniques can be applied in some medical tests for dry eye syndrome (DES). After setting the scene by providing background on the nature of DES, the authors examine the main steps for the automatic classification of the lipid layer patterns with image analysis and feature selection being considered. Detailed analyses are provided with the conclusion that the best performance was achieved by the decision tree classifier described in the paper.

Intelligent Decision Support Systems in Automated Medical Diagnosis are discussed by Florin Gorunescu and Smaranda Belciug in Chap. 8. In recent years the data explosion, leading to the growth in medical databases, has enabled the development of efficient intelligent decision support systems (IDSS's). In this chapter, the authors review some of the IDSS's used in computer-aided medical diagnosis (CAMD) and draw to the attention of a wide-audience, the challenges and potential of such systems. A thorough overview of neural-networks and their application in CAMD is given, aimed at medical professionals. Support vector machines and their applications are among the other systems discussed.

Chapter 9 continues the theme of medical support systems in the contribution by Vicente Moret-Bonillo, Isaac Fernández-Varela, Elena Hernández-Pereira, Diego Alvarez-Estévez and Volker Perlitz, On **the Automation of Medical Knowledge and Medical Decision Support Systems**. The authors review and build on the pioneering work of Ledley and Lusted in reasoning for medical diagnosis and discuss first categorical logic models and Bayesian methods. A fully worked clinical case based on the medical issues surrounding sleep-apnea is given.

In Chap. 10, Hariton Costin, Cristian Rotariu discuss **Vital Signs Telemonitoring by using Smart Body Area Networks, Mobile Devices and Advanced Signal Processing**. Wearable body area networks are introduced and their key features reviewed. The authors then discuss the design and implementation issues of their own integrated system for remote monitoring of vital signs. The hardware and software solutions, the power management of wireless devices integrated in WBAN, and the on-chip bio-signal processing software are also discussed in detail. The idea of a regional telecenter is presented and evaluated.

In Chap. 11, **Preprocessing in High Dimensional Datasets** Amparo Alonso-Betanzos, Veronica Bolon-Canedo, Carlos Eiras-Franco, Laura Moran-Fernandez and Borja Seijo-Pardo explore the problems associated with the Big Dimensionality concomitant with Big Data. In today's ever-increasingly large datasets. The more traditional methods of machine learning cease to be effective. The authors explain how preprocessing techniques help reduce data dimensionality and illustrate their methods with detailed examples. Pointers toward further research are given.

Analysis of Questionnaire Survey on Psychic Characteristics in the Elderly Using Quantification Theory II is the topic discussed by Takayuki Kawaura, Yasuyuki Sugatani in 12. Changing population characteristics and family dynamics in Japan prompted the authors to conduct a survey on how the over 65 age group viewed these changes. The data collected were analyzed using quantification theory II.

1.3 Conclusion

This book presents a collection of selected contributions of leading subject-matter expert in the field of biomedical informatics. The book is intended for students, professionals and academics from all related disciplines.

Chapter 2
Digital Health Research Methods and Tools: Suggestions and Selected Resources for Researchers

Kathleen Gray and Cecily Gilbert

Abstract This chapter provides an overview of digital health research, aimed at people new to conducting investigations in this field who seek to engage seriously with patients, clients and consumers. Digital health is not a scientific discipline. This chapter argues that health and biomedical informatics offers a strong scholarly basis for research in this field, and it outlines the theoretical and conceptual frameworks, ethical considerations, research methods, and examples of tools applicable for studies of digital health interventions. Researchers from clinical, IT, engineering and similar domains who plan to undertake studies involving digital health applications will be introduced to methodologies such as using guidelines and standards, performance indicators, validated input models and outcome measures, and evaluation resources. In the specific area of consumer health informatics research, an increasing array of tools and methods exist to investigate the interaction between consumers and their health data. In addition this chapter discusses research methods with health apps, patient-generated health data, social media and wearable self-tracking devices. Practical advice is given on techniques such as critically appraising digital health research literature, primary data collection from devices and services, study reporting and publishing results.

Keywords Biomedical informatics · Consumer health informatics · Digital health · Health informatics · Research methods

K. Gray (✉) · C. Gilbert
Health and Biomedical Informatics Centre, The University of Melbourne,
Parkville VIC 3010, Australia
e-mail: kgray@unimelb.edu.au

C. Gilbert
e-mail: cecilyg@unimelb.edu.au

© Springer International Publishing AG 2018
D.E. Holmes and L.C. Jain (eds.), *Advances in Biomedical Informatics*,
Intelligent Systems Reference Library 137,
https://doi.org/10.1007/978-3-319-67513-8_2

2.1 Introduction

2.1.1 Understanding What Digital Health Means

This chapter provides an overview of research methods and tools that are useful in human research settings in clinical care and public health, specifically those where the research design depends on patients, clients or consumers using internet-connected Information and Communication Technology (ICT) as part of a formal health service.

Internet-connected ICT has many names and nuances in health: health IT, health innovation, health social media, health or bio- or med- tech, health 2.0 or 3.0, connected health, ehealth, mhealth or mobile health, online health, P4 medicine, smart health, telehealth or telemedicine, wearable health, and so on [1]. For the purpose of this paper the term 'digital health' is used to describe this phenomenon, with a deliberate emphasis on the ways that these technologies enable patients, clients or consumers to participate actively in clinical care and health research.

This paper selects and synthesises reports on existing work in order to provide a sound foundation for anyone embarking on their own digital health research project of this kind. In these digital health research projects, technology is not used simply as a novel way for expert researchers to process participants' health data. Rather, the choice to use a particular technology is deliberately inclusive of participants and is intended to support them to take an interest in their own health data. This choice must be understood as a health intervention, with the ultimate objective of improving participants' health behaviours or health outcomes. The technology may also play combined or extended roles, for instance in projects that explore the potential for a learning health system [2].

The main aim of this paper is to serve as an introduction and resource for digital health researchers—who may be specialists in their field of health but at the same time novices in the foundation discipline of health and biomedical informatics. Digital health is a relatively new focus for research and is attracting wide interest from many people with diverse theory and practice backgrounds. These researchers may be unaware of the existence of relevant methods and tools and of the value these can have in strengthening their research.

To expand our understanding of digital health inevitably entails work to adapt existing research methods and tools and to develop new ones. It is important that such work builds on a solid foundation of knowledge about the existing contributions to research quality in this field. So this paper has a secondary aim, that is, where existing approaches to digital health research may be lacking, it aims to provide a base for informed innovation.

2.1.2 Defining Your Research as Digital Health Research

The term digital health entered the research literature in the 1990s to broadly characterise the impact of Internet-connected ICTs on health care [3, 4]. It has many synonyms, as noted; it includes an array of technologies, particularly when they are implemented at scale and integrated to work within or across health service provider organisations (for example, electronic health records, mobile telehealth, electronic referral and prescribing systems, automated clinical decision support, registry databases, direct-to-consumer online health services, smart biomedical devices; and also health-information-related aspects of apps and social technologies, analytics, ontologies, machine learning, sensors and robotics). Major digital health initiatives may bundle some of these technologies and can be categorised by the scope of the vision and the size of the investment in systems [5].

Progress to advance digital health has occurred in countries around the world [6]. For example, the Australian government established the Australian Digital Health Agency in 2016 to continue previous work by the National EHealth Transition Authority on consumer-controlled electronic health records, telehealth and related infrastructure [7]. The 2015–2016 federal budget allocated $485 million to redevelop the My Health Record system and to strengthen national digital health governance through an Australian Commission for eHealth. Digital health is not compatible with 'business as usual'; at levels from whole-of-clinic to whole-of-health system and whole-of-health-profession, the effects are expected to be transformative or disruptive [8–10].

A major element of these transformations is the increasing autonomy of patients, clients and consumers of health services and citizens with an informal interest in health. Since the 1990s the Internet has democratised access to resources, including access to medical literature, connection with patient social networks and co-creation of open health data sets. Some of these resources are purpose-built, such as health information literacy aids, personal health records and patient portals, while others have been appropriated and repurposed, for instance when people use Facebook, YouTube or Twitter for health self-management. In health as in every other respect, the Internet is also a massive source of fake data, misinformation and pseudoscience.

Another major element of the transformations occurring as a result of digital health is the availability of big data sets and associated research into storage, integration, analytics, machine learning and related fields. These topics are inter-related closely with the use of consumer technologies in clinical care and public health. However often the research problems within these topics do not involve working directly with patients or citizens, once their data has been captured for study. So these topics are not treated in any detail in this paper.

2.1.3 Drawing from Health Informatics
for Digital Health Research

In contrast to the definition of digital health, health and biomedical informatics is an established field that advances the effective use of data, information and knowledge in scientific inquiry, problem solving, behaviour change, decision making and service design so as to improve human health. Across the spectrum from molecular medicine to population health, health and biomedical informatics provides the scientific and scholarly foundations for managing raw health data, organising it into meaningful health information and systematising it as health knowledge.

The field traces its origins from the development of the Index Medicus in 1879, the establishment of the American Association of Medical Records Librarians in 1928 and the Deutsche Gesellschaft fur Medizinische Dokumentation, Informatik und Statistik in 1949, and the formation of the UNESCO International Federation for Information Processing, Technical Committee 4 on Health Care and Biomedical Research in 1967. Health informatics research work with the Internet began in the early 1970s. In other words, this is a longstanding interdisciplinary field of scholarly research and professional practice internationally, with its own peer-reviewed journals, scientific conferences and learned societies. Its principles, methods and tools can add rigor and relevance to any health research project that involves planning, development, implementation, operationalisation or evaluation of information and communication services, systems or technologies.

Some examples of health informatics methods and tools that can be leveraged for digital health research are: guidelines and standards for health apps and health data-sharing; systematic reviews in related areas, including search strategies and sources; health system performance indicators to contextualise health ICT research findings; validated instruments to measure technical performance and health outcomes of health innovations; consumer and community input models for public-facing ehealth systems; health social media practices to recruit participants, source data, share findings, and crowd-source support; online survey methods and tools; research protocols for clinical studies of ICT interventions; data integration frames for self-tracking health data; specifications for health ICT tool development within a research project.

The technologies of digital health are increasing convergence between the informatics tools and methods that are designed for use by clinicians and public health workers, and those for patients, clients, consumers and citizens. New health and biomedical informatics methods are being used to facilitate 'systems medicine', that is, the evolution from reactive disease care to care that offers personalisation and precision, and services that emphasise participation, prediction and prevention [11, 12].

Digital health research can benefit particularly from consumer health informatics research, which is concerned with information structures and processes used by people who are not clinically trained (so-called "consumers") to take an active interest in health. Consumer health informatics focuses on the types of tools or methods that make valid data, information and knowledge resources available to consumers, and aims to understand and improve the ways that these tools or methods work, for example: to enable health access materially and intellectually by consumers; to address the health needs, interests and contexts of consumers; to allow consumers to interact directly with resources without a healthcare professional's facilitation; to personalise and/or socialise consumers' interactions about their health needs and interests; to aid consumers' self-management and self-monitoring of health care plans; and to deepen consumers' engagement in clinical diagnosis, treatment and research [13, 14].

2.1.4   Respecting Participation in Digital Health Research by Patients, Clients and Citizens

Although digital health research occurs in the health sector context, where technological change has been slow, it also occurs in the context of a digital society and a digital economy, where the Internet has changed many aspects of how we live and a second generation of digitally native citizens is rising. The involvement of patients, clients and other citizens as full participants in digital health research responds to a social movement that is over a decade old. If it is new to you there are a variety of introductions to the concept [15, 16] and a range of resources to help you do it well. Some examples follow.

Does your digital health research describe patients, clients or other citizens with a blanket term such as 'participants' or 'users'? If so, what terms do you employ to reflect their equal status with clinicians, public health workers and health service managers who may be considered 'participants' in the service innovation or 'users' of the digital tool that is under investigation [17]? How fully does your research capture the interactions facilitated by the technology as a whole system or, if it does not, how fully is this acknowledged as a limitation of the study [18]?

At what point should you involve patients, clients or other citizens in your digital health research? The greatest benefit for you and for them may come from engaging them in defining and scoping the research at early stages. Strongly participant-driven digital health research approaches may be described as a form of 'citizen science' [19–22].

Are the methods well-founded that your digital health research employs to secure full participation by patients, clients or other citizens? Your strategy will be strengthened by building on the considerable experience of recruitment that has been reported in the literature [23, 24].

Crucially, engagement with your target participants may be affected their levels of interrelated literacies – basic literacy, health literacy, information literacy and computer literacy. Bear in mind too that levels of information and computer literacies not only among lay people but also among health professionals or administrators may be sub-optimal for full participation; and in fact some lay people may be at more advanced levels than some health workers. There is a range of integrated approaches to these literacies from which to choose in screening your participants or collecting their baseline data [25, 26].

Consumer participation in digital health design projects is not uncommon, but generic models and methods that enable this collaborative activity to be described as a research process are not well-documented overall; a selection of potentially relevant approaches is provided in [27–35].

2.2 Methodological Considerations

2.2.1 Framing Your Digital Health Research

Since digital health research is an immature field it is possible to make a substantial contribution to its body of knowledge if you think expansively. Determine how your research will address the opportunities and challenges that have been acknowledged in this field [36–39]. Consider the scope and the calibre of the contribution that your research will make to theory and to practice.

Whether you are a health researcher unfamiliar with the technology sector or an ICT researcher unfamiliar with the health sector, it is important to be explicit about the thinking that frames your digital health research project [40]. Start by exploring the rationale behind your research project. In what ways do you want to advance knowledge and practice in the health sciences and in the ICT disciplines? What particular assumptions and worldviews form the basis for these aspirations? [41–44].

2.2.2 Contributing to the Clinical Evidence Base Through Digital Health Research

It is a fundamental expectation in health science that good professional practice is based on best available research evidence, preferably evidence that arises from synthesis of findings from randomised controlled trials.

In the main, digital health does not yet have an extensive evidence base or well-established research protocols [45–47]. Reviews (such as [48–50]) reveal not only that more research is needed but also that the research methods used—whether quantitative or qualitative—need more rigor.

Some digital health research proceeds along conventional lines, following established protocols for the conduct of clinical trials, controlled case studies and cohort studies. Wherever possible, digital health researchers should adopt these strong methods for generating data and making sense of—these data in their studies. Selected recent examples of meta-analyses of randomised controlled trials show that this level of rigor is being applied in many fields of research, such as cancer, diabetes and other chronic diseases, exercise and nutrition, mental health and sexual health as shown in Table 2.1.

An issue with randomised controlled trials for digital health research is that in many cases they are not controlled for all of the ICT factors that may influence the

Table 2.1 Examples of meta-analyses of randomised controlled trials used in digital health research	Agboola SO, Ju W, Elfiky A, Kvedar JC, Jethwani K (2015) The effect of technology-based interventions on pain, depression, and quality of life in patients with cancer: A systematic review of randomized controlled trials. J Med Internet Res 17 (3):e65. doi:10.2196/jmir.4009
	Direito A, Carraca E, Rawstorn J, Whittaker R, Maddison R (2016) mHealth technologies to influence physical activity and sedentary behaviors: Behavior change techniques, systematic review and meta-analysis of randomized controlled trials. Ann Behav Med doi:10.1007/s12160-016-9846-0
	Mita G, Ni Mhurchu C, Jull A (2016) Effectiveness of social media in reducing risk factors for noncommunicable diseases: A systematic review and meta-analysis of randomized controlled trials. Nutr Rev 74 (4):237–247. doi:10.1093/nutrit/nuv106
	Or CK, Tao D (2014) Does the use of consumer health information technology improve outcomes in the patient self-management of diabetes? A meta-analysis and narrative review of randomized controlled trials. Int J Med Inform 83 (5):320–329. doi:10.1016/j.ijmedinf.2014.01.009
	Rasekaba TM, Furler J, Blackberry I, Tacey M, Gray K, Lim K (2015) Telemedicine interventions for gestational diabetes mellitus: A systematic review and meta-analysis. Diabetes Res Clin Pract 110 (1):1–9. doi:10.1016/j.diabres.2015.07.007
	Spijkerman MP, Pots WT, Bohlmeijer ET (2016) Effectiveness of online mindfulness-based interventions in improving mental health: A review and meta-analysis of randomised controlled trials. Clin Psychol Rev 45:102–114. doi:10.1016/j.cpr.2016.03.009
	Wayal S, Bailey JV, Murray E, Rait G, Morris RW, Peacock R, Nazareth I (2014) Interactive digital interventions for sexual health promotion: A systematic review and meta-analysis of randomised controlled trials. Lancet 384 (Spec Issue):S85. doi:10.1016/S0140-6736(14)62211-X
	Williams G, Hamm MP, Shulhan J, Vandermeer B, Hartling L (2014) Social media interventions for diet and exercise behaviours: A systematic review and meta-analysis of randomised controlled trials. BMJ Open 4 (2): e003926-002013-003926. doi:10.1136/bmjopen-2013-003926
	Zachariae R, Lyby MS, Ritterband LM, O'Toole MS (2016) Efficacy of internet-delivered cognitive-behavioral therapy for insomnia - a systematic review and meta-analysis of randomized controlled trials. Sleep Med Rev 30:1–10. doi:10.1016/j.smrv.2015.10.004

results. This is especially likely if researchers who are specialists in a clinical area but naïve about health and biomedical informatics make 'common sense' assumptions about the technology to be used in the trial. Digital health interventions are complex interventions and such trials need to use appropriate methods [51–54].

Two key steps to avoid oversimplifying the organisational and cultural factors inherent in the introduction of digital health interventions into a health care environment are firstly, to incorporate socio-technical theory in the research design [55, 56] and secondly, to conduct a suitably detailed feasibility study [57, 58].

2.2.3 Positioning Digital Health Research as Health Services and Systems Research

When one looks closely at the health improvement or advancement aim that is expressed in a digital health research project, often one finds a question or problem that is not strictly clinical, in other words, a project where treatment interventions and outcomes are not the only focus or even the main focus of research. Such projects are strengthened if they are broadly informed by constructs from health services and systems research [59–61].

Digital health researchers may wish to connect their aims with performance indicators and criteria selected from those commonly applied in public policy to monitor the operation of health care systems [62]. The variety of these indicators ranges from accessibility and appropriateness to safety and trust. The next paragraph shows a worked example, namely the indicators that could be used to contextualise the aim of a digital health research project in Australian policy. By making reference to the equivalent high level indicators, a similar approach can be taken in other jurisdictions. Using such a method to specify, categorise and evaluate research findings in terms of impact on the overall performance of a health system is a stronger research approach than selecting random criteria or applying unconventional measures to make sense of digital health research findings.

In the case of Australia, a National Health Performance Framework sets out performance indicators for hospitals and health networks. These indicators are underpinned by the Review of Government Service Provision Framework, which distinguishes between outputs (the actual service delivered), and outcomes (the impact of a service on the status of an individual or a group and on the success of the service achieving its objectives). This framework emphasises three top priorities —equity, effectiveness and efficiency—and unpacks them into six aspects of performance: accessibility; continuity of care; effectiveness; efficiency and sustainability; responsiveness; safety [63, 64]. We can investigate digital health by locating specific research projects in relation to these six aspects of performance. We can go further, by augmenting them with predefined indicators of performance (for example, in Table 2.2), to understand the impact of digital health in four dimensions: its contribution to consumer centred care; to clinical safety and quality of care; to service sustainability; and to infrastructure utility.

2.2.4   Recognising Computer Science and Information Systems Research in Digital Health Research

To someone who is unfamiliar with ICT research, it may appear that a digital health study can proceed simply on the basis of a simple licensing agreement with a software vendor or a short-term contract with a software developer, plus some 'common sense' survey questions to deal with attitudinal and behavioural aspects of the project. This may not suffice.

Such an approach in your research design may imbue your work with potentially serious oversights or shortcomings. To address these may entail revising your expectations about what your project can achieve within your timeline and resources. You may need to do further background research, invite an expert co-investigator to join your project team, and/or add detailed specifications to your agreement with a vendor or contractor.

Even though the aim of digital health research is to improve or advance health, this research draws heavily on research methods and tools that have been refined in ICT disciplines such as computer science and information systems. A range of well-founded ICT research methods exists already that is suited to explore many aspects of digital health. It is preferable to use these where possible rather than inventing idiosyncratic methods. It is possible to gain some insights into the methods in this field by scanning selected summary papers (such as [69–71]).

Your digital health research will benefit even more if you have a basic understanding not just of computer science and information systems concepts, but also of the special modifications to these in health settings. Even skimming the contents of introductory health informatics textbooks can be useful to help you reflect on aspects of the ICT body of knowledge where your project may need more attention to detail (for example [72–75]).

Three examples serve to illustrate the range of methodological issues, in areas essential for digital health researchers to consider, that are being addressed in current ICT research: information retrieval [76, 77]; privacy and security [78, 79]; and human-computer interaction [80, 81].

Table 2.2 High-level indicators of health system performance to guide digital health research: an Australian example

Dimension and Source	Indicators
1 Consumer centred care Consumers Health Forum of Australia [65]	Accessible and affordable care Appropriate care that meets the needs and preferences of individuals, that is evidence based, high quality and safe Whole of person care that takes into account people's lives and personal and cultural values; that is inclusive of carers and family Coordinated and comprehensive care that provides multidisciplinary care and facilitates continuity across the different levels of the healthcare system Trust and respect at all times, including timely and efficient complaint resolution processes Support to enable informed decision-making including access to clear and understandable information about treatment options, risks and costs Meaningful involvement of people at all levels of planning, system design, service development and in key governance structures to ensure sustainability
2 Clinical safety and quality of care Australian Commission on Safety and Quality in Health Care [66]	Appropriateness: health summary; timely initial needs identification; client assessment; complete care plan and timely review; recalls and reminders; adherence to clinical guidelines; medication review Effectiveness: client improvement/stabilization; attainment of goals of care Coordination: referral process and content; allocation of care coordinator; timely communication with care team Safety: adverse drug reactions and medication allergies; documented near misses or adverse events investigated and followed up; infection control
3 Service sustainability [Dimensions covered by 1 or 2 above are bracketed] Australian Council on Healthcare Standards [67]	[Service delivery] [Provision of care] Information management: health records management systems support the collection of information and meet the consumer/patient and organisation's needs; corporate records management systems support the collection of information and meet the organisation's needs; data and information are collected, stored and used for strategic, operational and service improvement purposes; the organisation has an integrated approach to the planning, use and management of information and communication technology (ICT) Workforce planning and management: workforce planning supports the organisation's current and future ability to address needs; the recruitment, selection and appointment system ensures that the skill mix and competence of staff, and mix of volunteers, meets the

(continued)

Table 2.2 (continued)

Dimension and Source	Indicators
	needs of the organisation; the continuing employment and development system ensures the competence of staff and volunteers; employee support systems and workplace relations assist the organisation to achieve its goals Corporate systems and safety: the organisation provides quality, safe health care and services through strategic and operational planning and development; governance is assisted by formal structures and delegation practices within the organisation; external service providers are managed to maximise quality, safe health care and service delivery; the organisation's research program develops the body of knowledge, protects staff and consumers/patients and has processes to appropriately manage the organisational risk; safety management systems ensure the safety and wellbeing of consumers/patients, staff, visitors and contractors; buildings, signage, plant, medical devices, equipment, supplies, utilities and consumables are managed safely and used efficiently and effectively; emergency and disaster management/security management/waste and environmental management supports safe practice and a safe environment
4 Infrastructure utility and warranty Australia College of Rural & Remote Medicine [68] *	Adequate performance of IT equipment and infrastructure: The equipment works reliably and well over the locally available network and bandwidth./The equipment is compatible with the equipment used at the other health sites and in the home./All the participating healthcare organisations meet the standards required for security of storage and transmission of health information./Peripheral devices are used in a fit for purpose manner jointly determined by the patient and clinician./The equipment is installed according to producer's guidelines, where possible in collaboration with other organisations/clinicians using the system./The equipment and connectivity are tested jointly by the participating healthcare organisation to ensure that they do what the producer claims they will. IT risk management: Risk analysis is performed to determine the likelihood and magnitude of foreseeable problems./There are procedures for detecting, diagnosing, and fixing equipment problems./Technical support services are available during the time that equipment is operating./There is back-up to cope with equipment or connectivity failure, proportionate to the consequences of failure.

*Telehealth is used as an example here. Equally it is possible to substitute technology management and governance frameworks applicable to other kinds of healthcare infrastructure, for example, digital hospitals (https://infostore.saiglobal.com/en-us/Standards/SA-HB-163-2017-1919534/) or smart medical devices (https://www.fda.gov/downloads/AboutFDA/CentersOffices/OfficeofMedical ProductsandTobacco/CDRH/CDRHReports/UCM459368.pdf)

2.3   Research Techniques

2.3.1   Reviewing Scientific and Technical Literature Related to Digital Health Research

Conducting reviews both of existing research literature and of the state of current technology is advisable as a preliminary to designing a digital health research project. Because of the complexity that is characteristic of digital health interventions, neither type of review is straightforward. Sources of information are more widely scattered and more difficult to synthesise, and methods of reviewing are more variable, than you would expect in a narrower field of research [82, 83].

Peer-reviewed publications appear both in the biomedical and healthcare literature, mainly in journals, and also in the engineering and ICT literature, often in conference proceedings [84, 85]. To do a thorough search of the literature thus requires you to consult the major databases in both fields [86], and to use a multiplicity of search terms. The field of health technology assessment offers some parallels [87].

Your digital health research design may not factor in major advances if you rely on peer-reviewed sources entirely. You may benefit from information in sources such as project reports, blogs, policy documents, industry white papers, and so on. This non-peer-reviewed literature—also known as 'grey' literature—is important because the rate of technological change is much faster than the rate of scholarly research reporting. There are a number of ways to ensure that you have done a thorough job of scanning this grey literature. Examples are shown in Table 2.3.

Table 2.3   Example sources of digital health grey literature

Databases		
Canadian Agency for Drugs and Technologies in Health	Search health technology assessment reports	https://www.cadth.ca/reports
National Technical Reports Library USA	Use keywords to search a comprehensive collection of government-sponsored research	https://ntrl.ntis.gov/NTRL/
Primary Health Care Research & Information Service Australia	Outlines a method for manual searching	http://www.phcris.org.au/guides/grey_literature.php
Consulting companies		
Accenture Consulting Health	Blogs and reports	https://www.accenture.com/au-en/health-industry-index
Ernst and Young	"Health reimagined: A new participatory health paradigm" report, 2016	http://www.ey.com/au/en/industries/health/

(continued)

Table 2.3 (continued)

Gartner Group	"Hype Cycles" for Telemedicine, Healthcare Provider Applications, Analytics and Systems, AND Healthcare Provider Technologies and Standards	http://www.gartner.com/ technology/consulting/ healthcare-providers.jsp
Government and non-government agencies		
Austrade	Digital Health. Health IT Industry Capability Report, 2016	http://www.austrade.gov.au/ ArticleDocuments/2814/Digital Health. Industry Capability Report.pdf.aspx
Capital Markets Cooperative Research Centre	"Flying blind" Australian consumers and digital health report, 2016	https://flyingblind.cmcrc.com/
National Broadband Network	Health blog	http://www.nbnco.com.au/blog/ health
News aggregators		
Computer World	Healthcare IT news section	http://www.computerworld.com/ category/healthcare-it/
European EHealth News		http://www.ehealthnews.eu/
Healthcare IT News		http://www.healthcareitnews com/
Information Week	Healthcare IT news section	http://www.informationweek. com/healthcare.asp
Pulse + IT Magazine		http://www.pulseitmagazine. com.au
Industry interest groups		
Australasian TeleHealth Society		http://www.aths.org.au
Health Informatics Society of Australia		www.hisa.org.au
Health Information and Management Systems Society		http://www.himss.org

2.3.2 Developing Tools as Part of Digital Health Research

If, as part of your research project, you need to develop a new tool or portal or platform to support participation by patients and consumers, be aware that international and national standards cover some but not all aspects of this work. If you develop a tool that does not meet the standards that apply in the jurisdiction where you intend it to be used, it may work well enough for your immediate research purposes. However it may not be suitable for scaled-up use.

Compliance and governance are specialised and rapidly evolving aspects of digital health. Be aware that they often are not on the radar of technology developers or sponsors of hackathons and other competitions, even those with experience in the health sector. Ultimately it is your role as the responsible researcher to specify your requirements so that the tools developed through your research funding can support rigorous and transparent research. If you are developing digital health resources for the Australian health sector for example, at the minimum you should be aware of the following kinds of information:

Australian Digital Health Agency (formerly National EHealth Transition Authority) Resources for implementers and developers https://www.digitalhealth. gov.au/implementation-resources; and Standards Australia. EHealth Standards http://www.ehealth.standards.org.au/Home/Publications.aspx.

Anyone can 'have a go' at developing digital health tools, by taking advantage of open source resources for developers (for example, OpenmHealth http://www. openmhealth.org). Technology developers can be found among people in the maker movement or start-up companies. In a university, development may be in the skillset of your research colleagues or their students; or there may be specialised research support services (such as the E-Research Group http://eresearch.unimelb. edu.au/ and the Research Information Technology Unit http://www.grhanite.com at the University of Melbourne). Established businesses which are members of industry organisations (for example the Medical Software Industry Association https://www.msia.com.au) are able to offer a grounded perspective on the viability of what you plan to develop as part of your digital health research.

If you anticipate that the tool you develop will be implementable in routine professional practice, be aware of advisory services from professional and provider organisations, such as:

Pharmacy Guild of Australia Pharmacy innovations in digital health (eHealth). https://www.guild.org.au/docs/default-source/public-documents/issues-and-resources/ Fact-Sheets/factsheet-pharmacy-innovations-in-ehealth.pdf?sfvrsn=4

Royal Australian College of General Practitioners Digital Business Kit. One. Using technology to deliver healthcare. http://www.racgp.org.au/digital-business-kit/one/

Royal College of Physicians. Using Apps in Clinical Practice: Important things that you need to know about apps and CE marking. https://www.rcplondon.ac.uk/ file/175/download?token=5nTJceC1

Victoria. Health Department Health Technology Program. https://www2.health. vic.gov.au/hospitals-and-health-services/patient-care/speciality-diagnostics-therapeutics/ health-technology-program

If you are interested in the broad commercialisation prospects for the tool you develop, you should seek professional advice through an incubator or accelerator program. Some programs are offered through research institutions (for example the University of Melbourne Accelerator Program http://themap.co and Murdoch Childrens Research Institute Bytes4Health https://www.mcri.edu.au/bytes4health). The Australian Government provides links to other reputable resources that it sponsors (at http://www.innovation.gov.au/page/incubator-support-programme); one example is the MTPConnect MedTech and Pharma Growth Centre (http://www.mtpconnect.org.au).

2.3.3 Working with Data Collected from Digital Devices and Online Services

Digital health devices and online services used by patients or citizens who are research participants may generate biometric data automatically or may prompt the user to enter this data manually into a system. A common acronym for this is PGHD, that is, patient- or person-generated health data [88]. In general, research methodologies are still immature in this area, and thought-provoking accounts are worth weighing up before you wade in (for example [89–93]). Further guidance on working with data from specific technologies appears in Part 4 of this paper.

Data analytics methods for working with patient generated health data are a topic of great current interest. By definition the data are likely to be 'patient-centred' but such research may not engage with patients at all (for example [94, 95]). A few analytics research approaches envisage fuller engagement with the patients whose data are under study (for example [96]). There are particular challenges in enabling active patient participation in analytics research (touched on by [97–99] and the subject of a popular work by Tailor [100]).

2.3.4 Collecting Data for Research About Digital Health

Apart from biometric data captured by devices or input prompts, there are other sorts of data that you may wish to collect by other means to find out about the use of digital health technologies as a form of human behaviour. People's attitudes to and experiences with digital health technology are often of particular interest to researchers. It is inadvisable to create data collection metrics and scales from scratch for this purpose, unless you are certain that there is not a validated method or tool already in existence that will meet your needs.

Evaluation is often the aim of digital health research, and evaluation resources abound (examples are [101–107]). Less abundant but equally important are methods for doing implementation research (for example [108, 109]).

There is great scope for original work in digital health research that deploys instruments recognised in health science along with others from information science, and correlates the findings from both. From the health sciences perspective, an important source of patient-reported health outcome measures is PROMIS www.nihpromis.com. The many other ways of eliciting patient health effects include narratives [110] and activation measures [111]. The health science literature also offers several systematic reviews of research on digital health patient engagement and impact, from which accepted methods can be derived (examples are [112, 113]).

From the information science perspective, studying the user experience is a highly sophisticated type of research. A neat summary of user-centred design study methods has been produced by the US government (https://www.usability.gov/what-and-why/user-research.html) and there are many other resources from science and industry (such as [114–116]).

2.3.5 *Research Data Management and Storage Planning*

As early as possible in your research, it is advisable to construct a formal data management plan. This specifies why, where and how you will organise, secure, store and potentially share the data collected in your research. This information is required as part of many ethics and funding application pro formas, and it will also be useful for the core task of organising your research data.

General guidance is offered by research institutes (for example, University of Melbourne Research Data Management http://research.unimelb.edu.au/infrastructure/doing-data-better/how) and funding bodies such as the Australian Research Council (http://www.arc.gov.au/research-data-management) and the National Health & Medical Research Council (through its *Code for the Responsible Conduct of Research* https://www.nhmrc.gov.au/guidelines-publications/r39; note this Code is currently under review and a new version is in preparation).

Digital health projects have the potential to generate terabytes of raw patient health data. This scenario will require that you arrange large-volume storage. You may be able to apply to your research infrastructure support unit for this; or you may need to discuss it with your institution's IT services unit; or you may need to identify and pay for an appropriately secure private storage provider. Be ready with a detailed soutline of the data quantity, format, retention period, and details of collaborators who will require access over the lifetime of the study.

Research that relies on patient-generated health data raises many familiar issues of safeguarding personal health data, plus a few new ones. There are many ethical and legal aspects of this research that are unclear. Laws governing the use of health data differ from country to country even though, especially in commercial services, the data may flow across national borders and be held and owned in a jurisdiction different from where the data are generated.

The Privacy Rule in the USA's Health Insurance Portability and Accountability Act (HIPAA) is often mentioned in the published literature (for example [117]). Be aware that it is not the most applicable law for researchers elsewhere. The Australian Privacy Act, for example, gives special protection to health data (https://www.oaic.gov.au/privacy-law/privacy-act/health-and-medical-research) and the Australian Information Commissioner reports annually on particular digital health data activities (https://www.oaic.gov.au/about-us/corporate-information/annual-reports/ehealth-and-hi-act-annual-reports/).

You should also consider policies and recommendations on data sharing; indeed this may be mandated by some research funders. These are designed to extend the value of the data collected in the study, by encouraging re-use of de-identified or anonymised datasets (e.g. NH&MRC https://www.nhmrc.gov.au/grants-funding/policy/nhmrc-statement-data-sharing). If you are aware of this option at the outset, you will have the opportunity to incorporate re-use provisions in the phrasing of recruitment material, project information sheets, and participant consent forms. The Australian National Data Service offers guides and examples for sharing sensitive data at http://www.ands.org.au/working-with-data/sensitive-data/sharing-sensitive-data.

2.3.6   Writing up and Reporting Digital Health Research

The range of journal, conference and monograph resources cited in this chapter offers many pointers to the presentation and publication forums that accept reports of digital health research. In writing up your research for publication, you may seek to communicate its contribution to theory or its contribution to practice, or both.

Your research can be assured of being readable and replicable if you report it in a structured manner. Specific reporting protocols exist that are applicable to many types of digital health research. There are key papers that you should know about, that can help you to describe formally all of the important elements of your digital research project [118–124].

2.4   Technology-Specific Resources

2.4.1   Looking into Specific Settings of Digital Health Research

So far, this paper has provided broad-brush suggestions and high-level resources for digital health researchers. There are three more specific aspects to consider in pursuing rigorous digital health research in your actual setting.

First, familiarise yourself with the barriers and facilitators in the healthcare setting where your research is conducted. For example: Your hospital may have a bring-your-own-device policy; how does this govern what you plan to do with patient-generated data? Your clinic may have a substantial investment in proprietary record management or knowledge management software; does this offer any of the functionality that you need for your intervention? A patient advocacy organisation in your field of healthcare already may be using online health platforms expertly; how might your project complement these activities?

Second, do a scoping review of the scientific literature and industry resources that are specific to your field of health, to augment the more generic ones mentioned in Parts 1 to 3. In some fields, for example diabetes and mental health, there is already a substantial body of published digital health research and an array of clinically tested digital health products and services. Be clear about what pressing needs you are going to fill and what new knowledge you are going discover.

Third, investigate the existence of guidelines that pertain to the specific form of technology that interests you. To support the way you plan and carry out your digital health research project, the final sections of this paper list a cross-section of recent resources related to health apps, health social media, and healthcare wearables.

2.4.2   Health Apps

Health apps vary widely in quality and functionality, and many fail rapidly. Guidelines after the fact for health app users would be needed less if these apps were developed more consistently and more rigorously in the first place. An enormous number of scientific papers have been published about health app research, from design and development to implementation and evaluation, and there is voluminous report literature from commercial and government sources. However, closer inspection of the literature reveals surprisingly few sources of evidence-based advice that apply generically across the stages of an app's lifecycle. Guidelines and advisory documents may emanate from many different countries, that is, from many different healthcare system jurisdictions. Given the variation in legislation and in technology policies related to health data, for example, this makes it difficult for researchers and developers to easily determine which of these sources of advice are applicable to comply with local health data privacy laws, and to address local clinical guidelines. There is much scope for further systematic research to bring refinement to this area of digital health [125–140].

2.4.3   Health Social Media

Web 2.0, the social web and social media are terms used interchangeably since 2005 to describe how the World Wide Web has become a medium for content, interactions and transactions, using channels such as audio and video podcasting, blogging and microblogging, social networking, virtual worlds and more. This change in the Internet has generated huge innovation in health. Examples include direct-to-consumer diagnostic services online, platforms for personal health data sharing, patient experience video channels, and new approaches to public health promotion and surveillance. Digital health research may consider social media to be a tool for research, a setting for research, or a focus for research. Often researchers fail to clarify their research orientation to health social media, resulting in research designs and methods that hinder best use of participants' data and reflections on data. Examples of strong methodological contributions are captured in the selection of papers provided here [141–163].

2.4.4 Health Self-Tracking with Consumer Wearables

Consumers have direct off-the-shelf access to a range of wearable devices that offer tools for self-quantification of diet, exercise, sleep, medication, mood, blood pressure, body temperature, environmental exposure, and other indicators of health and wellness. Wearables are worn on or sometimes in the body (as distinct from carrying a mobile phone); examples include watches, glasses, contact lenses, e-textiles and smart fabrics, headbands, beanies and caps, jewellery such as pendants, rings, bracelets, earrings etc. They use sensors to completely or partially automate data collection, and they have built-in functionality to ship the collected data to a cloud-based server for analysis and sharing. There are many angles that this area of digital health research may explore: wearables are developing capabilities that rival those inside an intensive care unit; they are not categorised as regulated medical devices nor are the data from them readily able to integrate with clinical information systems; their use is assuming fad proportions and generating volumes of data that may or may not lead to sound decisions about healthcare. The selection of papers provided here illustrates some of the scholarly and scientific approaches that are emerging to advance research in this field [164–178].

2.5 Conclusion

Digital health is a popular and convenient way to convey the changes that new technologies are bringing to healthcare in the twenty-first century. However, digital health can mean many things to many people, and digital health is not a scientific discipline.

Further, many researchers from clinical, IT, engineering and similar domains, are new to conducting investigations in this field, and may be unfamiliar with the sorts of engagement with patients, clients and consumers that new technologies enable.

Therefore, this chapter has provided an overview of digital health research and offered health and biomedical informatics foundations of research in this field. It has outlined theoretical and conceptual frameworks, ethical considerations, research methods, and examples of tools that may be included in studies of digital health interventions. Although many of the examples are from the Australian context, there may be equivalents in other jurisdictions; if not, the Australian examples are edifying.

Digital health research is important, and conducting this research by building upon what is already known is also important, so that investments in digital health in our healthcare systems may be informed by the strongest possible evidence base.

References

1. Taylor, K.: Connected Health: How Digital Technology is Transforming Health and Social Care. Deloitte Centre for Health Solutions, London (2015)
2. IOM Institute of Medicine (US), Grossmann, C., Powers, B., McGinnis, J.M.: Digital Infrastructure for the Learning Health System—The Foundation for Continuous Improvement in Health and Health Care: Workshop Series Summary. National Academy of Sciences, Washington, DC (2011)
3. Frank, S.R., Williams, J.R., Veiel, E.L.: Digital health care: where health care, information technology, and the Internet converge. Manag. Care Q. **8**(3), 37–47 (2000)
4. Iyawa, G.E., Herselman, M., Botha, A.: Digital health innovation ecosystems: from systematic literature review to conceptual framework. In: International Conference on ENTERprise Information Systems/International Conference on Project MANagement/ International Conference on Health and Social Care Information Systems and Technologies. CENTERIS/ProjMAN/HCist, no. (100), pp. 244–252 (2016). doi:10.1016/j.procs.2016.09.149
5. Hagens, S., Zelmer, J., Frazer, C., Gheorghiu, B., Leaver, C.: Valuing national effects of digital health investments: an applied method. Stud. Health Technol. Inf. **208**, 165–169 (2015)
6. World Health Organisation: Atlas of eHealth Country Profiles 2015: The Use of eHealth in Support of Universal Health Coverage; Based on the Findings of the 2015 Global Survey on eHealth. WHO, Geneva (2016)
7. Australia: Public Governance, Performance and Accountability (Establishing the Australian Digital Health Agency) Rule. vol F2016L00070 (2016)
8. Accenture: Digital Health Tech Vision, 2016. (2016)
9. Agarwal, R., Gao, G., DesRoches, C., Jha, A.K.: Research commentary—the digital transformation of healthcare: current status and the road ahead. Inf. Syst. Res. **21**, 796–809 (2010). doi:10.1287/isre.1100.0327
10. Burrill, G.S.: Digital health investment opportunities abound, but standouts deliver disruptive change. J. Commerc. Biotechnol. **18**(1), 495 (2012). doi:10.5912/jcb495
11. Flores, M., Glusman, G., Brogaard, K., Price, N.D., Hood, L.: P4 medicine: how systems medicine will transform the healthcare sector and society. Per. Med. **10**(6), 565–576 (2013). doi:10.2217/PME.13.57
12. Martin-Sanchez, F., Lopez-Campos, G., Gray, K.: Biomedical informatics methods for personalized medicine and participatory health. In: Sarkar, I.N. (ed.) Methods in Biomedical Informatics, pp. 347–394. Academic Press, Oxford (2014). doi:10.1016/B978-0-12-401678-1.00011-7
13. Demiris, G.: Consumer health informatics: past, present, and future of a rapidly evolving domain. Yearb. Med. Inf. **1**, 42–47 (2016). doi:10.15265/IYS-2016-s005
14. Flaherty, D., Hoffman-Goetz, L., Arocha, J.F.: What is consumer health informatics? A systematic review of published definitions. Inf. Health Soc. Care **40**(2), 91–112 (2015). doi:10.3109/17538157.2014.907804
15. Greaves, F., Millett, C., Nuki, P.: England's experience incorporating "anecdotal" reports from consumers into their national reporting system: lessons for the United States of what to do or not to do? Med. Care Res. Rev. MCRR **71**(5 Suppl), 65S–80S (2014). doi:10.1177/1077558714535470

16. Okun, S., Caligtan, C.A.: The engaged ePatient. In: Nelson, R., Staggers, N. (eds.) Health Informatics: An Interprofessional Approach, vol. 2, pp. 204–219. Elsevier, St Louis, MO (2018)

17. Kushniruk, A.W., Turner, P.: Who's users? Participation and empowerment in sociotechnical approaches to health IT developments. Stud. Health Technol. Inf. **164**, 280–285 (2011)

18. Stephanie, F.L., Sharma, R.S.: Health on a cloud: modeling digital flows in an e-health ecosystem. J. Adv. Manag. Sci. Inf. Syst. **2**, 1–20 (2016)

19. Awori, J., Lee, J.M.: A maker movement for health: a new paradigm for health innovation. JAMA Pediatr. **171**(2), 107–108 (2017). doi:10.1001/jamapediatrics.2016.3747

20. Filonik, D., Bednarz, T., Rittenbruch, M., Foth, M.: Collaborative data exploration interfaces —from participatory sensing to participatory sensemaking. In: 2015 Big Data Visual Analytics (BDVA), Hobart, Tas., 22–25 Sep 2015. IEEE, pp 1–2. doi:10.1109/BDVA.2015.7314289

21. Swan, M.: Health 2050: the realization of personalized medicine through crowdsourcing, the quantified self, and the participatory biocitizen. J. Pers. Med. **2**(3), 93–118 (2012). doi:10.3390/jpm2030093

22. Vayena, E., Tasioulas, J.: Adapting standards: ethical oversight of participant-led health research. PLoS Med. **10**(3), e1001402 (2013). doi:10.1371/journal.pmed.1001402

23. Lane, T.S., Armin, J., Gordon, J.S.: Online recruitment methods for web-based and mobile health studies: a review of the literature. J. Med. Internet Res. **17**(7), e183 (2015). doi:10.2196/jmir.4359

24. O'Connor, S., Hanlon, P., O'Donnell, C.A., Garcia, S., Glanville, J., Mair, F.S.: Understanding factors affecting patient and public engagement and recruitment to digital health interventions: a systematic review of qualitative studies. BMC Med. Inf. Decis. Mak. **16** (1):120-016-0359-0353 (2016). doi:10.1186/s12911 016-0359-3

25. Karnoe, A., Kayser, L.: How is eHealth literacy measured and what do the measurements tell us? A systematic review. Knowl. Manag. E-Learn. **7**(4), 576–600 (2015)

26. Mackert, M., Champlin, S.E., Holton, A., Muñoz, I.I., Damásio, M.J.: eHealth and health literacy: a research methodology review. J. Comput.-Mediat. Commun. **19**(3), 516–528 (2014). doi:10.1111/jcc4.12044

27. Agarwal, R., Anderson, C., Crowley, K., Kannan, P.T., Westat: Improving Consumer Health IT Application Development: Lessons From Other Industries, Background Report. vol AHRQ Publication No. 11–0065-EF. Agency for Healthcare Research and Quality, Rockville, MD (2011)

28. Elwyn, G., Kreuwel, I., Durand, M.A., Sivell, S., Joseph-Williams, N., Evans, R., Edwards, A.: How to develop web-based decision support interventions for patients: a process map. Patient Educ. Couns. **82**(2), 260–265 (2011). doi:10.1016/j.pec.2010.04.034

29. Eyles, H., Jull, A., Dobson, R., Firestone, R., Whittaker, R., Te Morenga, L., Goodwin, D., Mhurchu, C.N.: Co-design of mHealth delivered interventions: a systematic review to assess key methods and processes. Curr. Nutr. Rep. **5**(3), 160–167 (2016). doi:10.1007/s13668-016-0165-7

30. Johnson, C.M., Turley, J.P.: A new approach to building web-based interfaces for healthcare consumers. e-J. Health Inf. **2**(2) (2007)

31. Kayser, L., Kushniruk, A., Osborne, R.H., Norgaard, O., Turner, P.: Enhancing the effectiveness of consumer-focused health information technology systems through eHealth literacy: a framework for understanding users' needs. JMIR Human Factors **2**(1), e9 (2015). doi:10.2196/humanfactors.3696

32. Marquard, J.L., Zayas-Caban, T.: Commercial off-the-shelf consumer health informatics interventions: recommendations for their design, evaluation and redesign. J. Am. Med. Inf. Assoc. JAMIA **19**(1), 137–142 (2012). doi:10.1136/amiajnl-2011-000338

33. Mummah, S.A., Robinson, T.N., King, A.C., Gardner, C.D., Sutton, S.: IDEAS (Integrate, Design, Assess, and Share): a framework and toolkit of strategies for the development of

more effective digital interventions to change health behavior. J. Med. Internet Res. **18**(12), e317 (2016). doi:10.2196/jmir.5927

34. Valdez, R.S., Holden, R.J., Novak, L.L., Veinot, T.C.: Transforming consumer health informatics through a patient work framework: connecting patients to context. J. Am. Med. Inf. Assoc. JAMIA **22**(1), 2–10 (2015). doi:10.1136/amiajnl-2014-002826

35. van Gemert-Pijnen, J.E., Nijland, N., van Limburg, M., Ossebaard, H.C., Kelders, S.M., Eysenbach, G., Seydel, E.R.: A holistic framework to improve the uptake and impact of eHealth technologies. J. Med. Internet Res. **13**(4), e111 (2011). doi:10.2196/jmir.1672

36. Andersen, T., Kensing, F., Kjellberg, L., Moll, J.: From research prototypes to a marketable eHealth system. Stud. Health Technol. Inf. **218**, 40589 (2015)

37. Baker, T.B., Gustafson, D.H., Shah, D.: How can research keep up with eHealth? Ten strategies for increasing the timeliness and usefulness of eHealth research. J. Med. Internet Res. **16**(2), e36 (2014). doi:10.2196/jmir.2925

38. Holden, R.J., Bodke, K., Tambe, R., Comer, R.S., Clark, D.O., Boustani, M.: Rapid translational field research approach for eHealth R&D. Proc. Int. Symp. Hum. Factors Ergon. Health C **5**(1), 25–27 (2016). doi:10.1177/2327857916051003

39. Patrick, K., Hekler, E.B., Estrin, D., Mohr, D.C., Riper, H., Crane, D., Godino, J., Riley, W. T.: The pace of technologic change: implications for digital health behavior intervention research. Am. J. Prev. Med. **51**(5), 816–824 (2016). doi:10.1016/j.amepre.2016.05.001

40. Gray, K., Sockolow, P.: Conceptual models in health informatics research: a literature review and suggestions for development. JMIR Med. Inf. **4**(1), e7 (2016). doi:10.2196/medinform.5021

41. Cano, I., Lluch-Ariet, M., Gomez-Cabrero, D., Maier, D., Kalko, S., Cascante, M., Tegner, J., Miralles, F., Herrera, D., Roca, J., Synergy-COPD Consortium.: Biomedical research in a digital health framework. J. Transl. Med. 12(Suppl 2), S10–5876-5812-S5872-S5810. Epub 2014 Nov 5828 (2014). doi:10.1186/1479-5876-12-S2-S10

42. Hu, Y.: Health communication research in the digital age: a systematic review. J. Commun. Healthc. **8**(4), 260–288 (2015). doi:10.1080/17538068.2015.1107308

43. Lupton, D.: Towards critical digital health studies: reflections on two decades of research in health and the way forward. Health **20**(1), 49–61 (2016). doi:10.1177/1363459315611940. (London, England: 1997)

44. Sjostrom, J., von Essen, L., Gronqvist, H.: The origin and impact of ideals in eHealth research: experiences from the U-CARE research environment. JMIR Res. Protoc. **3**(2), e28 (2014). doi:10.2196/resprot.3202

45. Ammenwerth, E.: Evidence-based health informatics: how do we know what we know? Methods Inf. Med. **54**(4), 298–307 (2015). doi:10.3414/ME14-01-0119

46. Mookherji, S., Mehl, G., Kaonga, N., Mechael, P.: Unmet need: improving mHealth evaluation rigor to build the evidence base. J. Health Commun. **20**(10), 1224–1229 (2015). doi:10.1080/10810730.2015.1018624

47. Rigby, M., Ammenwerth, E., Beuscart-Zephir, M.C., Brender, J., Hypponen, H., Melia, S., Nykanen, P., Talmon, J., de Keizer, N.: Evidence based health informatics: 10 years of efforts to promote the principle. Joint contribution of IMIA WG EVAL and EFMI WG EVAL. Yearb. Med. Inf. **8**, 34–46 (2013)

48. Black, A.D., Car, J., Pagliari, C., Anandan, C., Cresswell, K., Bokun, T., McKinstry, B., Procter, R., Majeed, A., Sheikh, A.: The impact of eHealth on the quality and safety of health care: a systematic overview. PLoS Med. **8**(1), e1000387 (2011). doi:10.1371/journal. pmed.1000387

49. Parthasarathy, R., Steinbach, T.: Health Informatics for Healthcare Quality Improvement: A literature review of issues, challenges and findings. In: AMCIS 2015: Twenty-first Americas Conference on Information Systems, Fajardo, Puerto Rico, Citeseer, pp. 1–23 (2015)

50. Ravka, N.: Informatics and health services: the potential benefits and challenges of electronic health records and personal electronic health records in patient care, cost control, and health research—an overview. In: El Morr, C. (ed.) Research Perspectives on the Role of

Informatics in Health Policy and Management, pp. 89–114. IGI Global, Hershey, PA, USA (2014). doi:10.4018/978-1-4666-4321-5.ch007

51. Barratt, H., Campbell, M., Moore, L., Zwarenstein, M., Bower, P.: Randomised controlled trials of complex interventions and large-scale transformation of services. Health Serv. Deliv. Res. **4**(16), 19–36 (2016). doi:10.3310/hsdr04160-19

52. Hubner, U.: What are complex eHealth innovations and how do you measure them? Position paper. Methods Inf. Med. **54**(4), 319–327 (2015). doi:10.3414/ME14-05-0001

53. Liao, P., Klasnja, P., Tewari, A., Murphy, S.A.: Sample size calculations for micro-randomized trials in mHealth. Stat. Med. **35**(12), 1944–1971 (2016). doi:10.1002/sim.6847

54. Washington P, Kumar M, Tibrewal A, Sabharwal A ScaleMed: A methodology for iterative mHealth clinical trials. In: 17th International Conference on E-health Networking, Application & Services, HealthCom 2015, Boston, MA, USA, October 14–17, 2015. IEEE, pp. 139-143 (2015). doi:10.1109/HealthCom.2015.7454487

55. Georgiou, A., Whetton, S.: Broadening the socio-technical horizons of health informatics. Open Med. Inf. J. **4**, 179–180 (2010). doi:10.2174/1874431101004010179

56. Scott, P.J., Briggs, J.S.: STAT-HI: a socio-technical assessment tool for health informatics implementations. Open Med. Inf. J. **4**, 214–220 (2010). doi:10.2174/1874431101004010214

57. Levati, S., Campbell, P., Frost, R., Dougall, N., Wells, M., Donaldson, C., Hagen, S.: Optimisation of complex health interventions prior to a randomised controlled trial: a scoping review of strategies used. Pilot Feasibility Stud. **2**, 17 (2016). doi:10.1186/s40814-016-0058-y

58. Orsmond, G.I., Cohn, E.S.: The distinctive features of a feasibility study: objectives and guiding questions. OTJR Occup. Participation Health **35**(3), 169–177 (2015)

59. Bowling, A.: Research Methods in Health: Investigating Health and Health Services, 4th edn. Open University Press, Maidenhead (2014)

60. Grassel, E., Donath, C., Hollederer, A., Drexler, H., Kornhuber, J., Zobel, A., Kolominsky-Rabas, P.: Evidence-based health services research–a short review and implications. Gesundheitswesen **77**(3), 193–199 (2015). doi:10.1055/s-0034-1382042

61. Scutchfield, F.D., Perez, D.J., Monroe, J.A., Howard, A.F.: New public health services and systems research agenda: directions for the next decade. Am. J. Prev. Med. **42**(5 Suppl 1), S1–S5 (2012). doi:10.1016/j.amepre.2012.01.027

62. Smith, P.C., Anell, A., Busse, R., Crivelli, L., Healy, J., Lindahl, A.K., Westert, G., Kene, T.: Leadership and governance in seven developed health systems. Health Policy **106**(1), 37–49 (2012). doi:10.1016/j.healthpol.2011.12.009

63. Australian Institute of Health and Welfare: National Health Reform: Performance and Accountability Framework. AIHW. http://www.aihw.gov.au/health-performance/performance-and-accountability-framework/ (2017) Accessed 15 July 2017

64. National Health Information Standards and Statistics Committee: Revised National Health Performance Framework, 2nd edn. Australian Institute of Health and Welfare, Canberra (2009)

65. Consumers Health Forum of Australia, George Institute for Global Health Putting the Consumer First: Creating a Consumer Centred Health System for a 21st Century Australia—A Health Policy Report. Sydney (2016)

66. Australian Commission on Safety and Quality in Health Care Practice-Level Indicators of Safety and Quality for Primary Health Care Specification. Version 1.0 edn. ACSQHC, Sydney (2012)

67. ACHS Australian Council on Healthcare Standards (n.d.) EQUiP National Table. http://www.achs.org.au/media/38984/table_equipnational_standards.pdf

68. Australian College of Rural and Remote Medicine: ACRRM Telehealth Advisory Committee Standards Framework. 05/16 edn. Australian College of Rural and Remote Medicine, Brisbane (2016)

69. Mora, M., Steenkamp, A.L., Gelman, O., Raisinghani, : On IT and SwE research methodologies and paradigms: a systemic landscape review. In: Manuel, M., Ovsei, G., Annette, L.S., Mahesh, R. (eds.) Research Methodologies, Innovations and Philosophies in

Software Systems Engineering and Information Systems, pp. 149–164. IGI Global, Hershey, PA (2012). doi:10.4018/978-1-4666-0179-6.ch008

70. Riedl, R., Rueckel, D.: Americas Conference on Information S Historical Development of Research Methods in the Information Systems Discipline. In: AMCIS 2011 Proceedings—All Submissions, Detroit, MI, (2011). AIS Electronic Library, p. 28

71. Venkatesh, V., Brown, S.A., Bala, H.: Bridging the qualitative-quantitative divide: guidelines for conducting mixed methods research in information systems. MIS Q. **37**(1), 21–54 (2013)

72. Coiera, E.: Guide to Health Informatics, 3rd edn. CRC Press, Taylor & Francis Group, Boca Raton, FL (2015)

73. Shortliffe, E.H., Cimino, J.J.: Biomedical Informatics: Computer Applications in Health Care and Biomedicine, 4th edn. Springer, London (2014). doi:10.1007/978-1-4471-4474-8

74. Venot, A., Burgun, A., Quantin, C.: Medical Informatics, e-Health: Fundamentals and Applications. Springer, Paris (2014). doi:10.1007/978-2-8178-0478-1

75. Weaver, C.A., Ball, M.J., Kim, G.R., Kiel, J.M.: Healthcare Information Management Aystems: Cases, Strategies and Solutions, 4th edn. Springer International Publishing, Geneva (2016). doi:10.1007/978-3-319-20765-0

76. Li, F., Li, M., Guan, P., Ma, S., Cui, L.: Mapping publication trends and identifying hot spots of research on Internet health information seeking behavior: a quantitative and co-word biclustering analysis. J. Med. Internet Res. **17**(3), e81 (2015). doi:10.2196/jmir.3326

77. Zuccon, G., Palotti, J., Goeuriot, L., Kelly, L., Lupu, M., Pecina, P., Mueller, H., Budaher, J., Deacon, A.: The IR task at the CLEF eHealth evaluation lab 2016: user-centred health information retrieval. In: CLEF 2016-Conference and Labs of the Evaluation Forum, Evora, Portugal, 5–8 Sep 2016. CEUR Workshop Proceedings. CEUR, pp. 15–27

78. Arora, S., Yttri, J., Nilse, W.: Privacy and security in mobile health (mHealth) research. Alcohol Res. Curr. Rev. **36**(1), 143–151 (2014)

79. Kotz, D., Gunter, C.A., Kumar, S., Weiner, J.P.: Privacy and security in mobile health: a research agenda. Computer **49**(6), 22–30 (2016). doi:10.1109/MC.2016.185

80. Lennon, M., Baillie, L., Hoonhout, J., Robertson, J., Fitzpatrick, G.: Crossing HCI and health: advancing health and wellness technology research in home and community settings. In: CHI EA '15 Proceedings of the 33rd Annual ACM Conference Extended Abstracts on Human Factors in Computing Systems, pp. 2353–2356. ACM, Seoul (2015). doi:10.1145/2702613.2702652

81. Wilson, V., Djamasbi, S.: Human-computer interaction in health and wellness: research and publication opportunities. AIS Trans. Hum.-Comput. Interact **7**(3), 97–108 (2015)

82. Gallivan, M., Tao, Y.: Value of co-Citation analysis for understanding a field's intellectual structure: an application to Healthcare Information Technology (HIT) Research. In: AMCIS 2014: Twentieth Americas Conference on Information Systems, p. 3. Savannah, Georgia, USA (2014)

83. Guise, J.M., Chang, C., Viswanathan, M., Glick, S., Treadwell, J., Umscheid, C.A., Whitlock, E., Fu, R., Berliner, E., Paynter, R., Anderson, J.: Systematic Reviews of Complex Multicomponent Health Care Interventions, vol. 14-EHC003-EF. Agency for Healthcare Research and Quality, Rockville, MD (2014)

84. Deshazo, J.P., Lavallie, D.L., Wolf, F.M.: Publication trends in the medical informatics literature: 20 years of "Medical Informatics" in MeSH. BMC Med. Inform. Decis. Mak. **9**, 7 (2009). doi:10.1186/1472-6947-9-7

85. Weigel, F.K., Rainer, R.K., Hazen, B.T., Cegielski, C.G., Ford, F.N.: Uncovering research opportunities in the medical informatics field: a quantitative content analysis. Commun. Assoc. Inf. Syst. **33**, 15–32 (2013)

86. Urquhart, C., Currell, R.: Systematic reviews and meta-analyses of health IT. In: Ammenwerth, E., Rigby, M. (eds.) Evidence-Based Health Informatics. Studies in Health Technology and Informatics, pp. 262–274. IOS Press, Amsterdam (2016). doi:10.3233/978-1-61499-635-4-262

87. Pfadenhauer, L., Rohwer, A., Burns, J., Booth, A., Lysdahl, K.B., Hofmann, B., Gerhardus, A., Mozygemba, K., Tummers, M., Wahlster, P., Rehfuess, E.: Guidance for the Assessment of Context and Implementation in Health Technology Assessments (HTA) and Systematic Reviews of Complex Interventions: The Context and Implementation of Complex Interventions (CICI) Framework Project report. Integrate-HTA (2016)
88. Rosenbloom, S.T.: Person-generated health and wellness data for health care. J. Am. Med. Inf. Assoc. JAMIA 23(3), 438–439 (2016). doi:10.1093/jamia/ocw059
89. Clark, K., Duckham, M., Guillemin, M., Hunter, A., McVernon, J., O'Keefe, C., Pitkin, C., Prawer, S., Sinnott, R., Warr, D., Waycott, J.: Guidelines for the Ethical use of Digital Data in Human Research. University of Melbourne, Parkville (2015)
90. Gray, K.: Like, comment, share: should you share your genetic data online? Australas. Sci. 37(6), 24 (2016)
91. Lupton, D.: Digital health technologies and digital data: new ways of monitoring, measuring and commodifying human bodies. In: Olleros, F.X., Zhegu, M. (eds) Research Handbook on Digital Transformations, pp. 85–102. Edward Elgar Publishing (2016) doi:10.4337/9781784717766.00011
92. Taylor, P.L., Mandl, K.D.: Leaping the data chasm: structuring donation of clinical data for healthcare innovation and modeling. Harvard Health Policy Rev. Stud. Publ. Harvard Interfaculty Initiative Health Policy 14(2), 18–21 (2015)
93. Winickoff, D.E., Jamal, L., Anderson, N.R.: New modes of engagement for big data research. J. Res. Innov. 3(2), 169–177 (2016). doi:10.1080/23299460.2016.1190443
94. Ajorlou, S., Shams, I., Yang, K.: An analytics approach to designing patient centered medical homes. Health Care Manag. Sci. 18(1), 3–18 (2015). doi:10.1007/s10729-014-9287-x
95. Gachet Páez, D., Morales Botello, M.L., Puertas, E., de Buenaga, M.: Health sensors information processing and analytics using big data approaches. In: Mandler B (ed) Internet of Things. IoT Infrastructures: Second International Summit, IoT 360° 2015, Rome, Italy, October 27–29, 2015. Revised Selected Papers, Part I. pp 481–486. Springer International Publishing (2016) Cham. doi:10.1007/978-3-319-47063-4_52
96. Khan, W.A., Idris, M., Ali, T., Ali, R., Hussain, S., Hussain, M., Amin, M.B., Khattak, A. M., Weiwei, Y., Afzal, M., Lee, S., Kang, B.H.: Correlating health and wellness analytics for personalized decision making. In: 2015 17th International Conference on E-health Networking, Application & Services (HealthCom), pp. 256–261. (2015) doi:10.1109/HealthCom.2015.7454508
97. Cohen, I.G., Amarasingham, R., Shah, A., Xie, B., Lo, B.: The legal and ethical concerns that arise from using complex predictive analytics in health care. Health Aff. (Project Hope) 33(7), 1139–1147 (2014). doi:10.1377/hlthaff.2014.0048
98. Koster, J., Stewart, E., Kolker, E.: Health care transformation: a strategy rooted in data and analytics. Acad. Med. J. Assoc. Am. Med. Coll. 91(2), 165–167 (2016). doi:10.1097/ACM.0000000000001047
99. Weiser, P., Ellis, A.: The Information Revolution Meets Health: The Transformative Power and Implementation Challenges of Health Analytics. Silicon Flatirons Center, Boulder, CO (2015). doi:10.2139/ssrn.2593879
100. Tailor, K.: The Patient Revolution: How Big Data and Analytics are Transforming the Health Care Experience. Wiley, Hoboken, NJ (2015)
101. Bergmo, T.S.: How to measure costs and benefits of eHealth interventions: an overview of methods and frameworks. J. Med. Internet Res. 17(11), e254 (2015). doi:10.2196/jmir.4521
102. Eslami Andargoli, A., Scheepers, H., Rajendran, D., Sohal, A.: Health information systems evaluation frameworks: a systematic review. Int. J. Med. Inf. 97, 195–209 (2017). doi:10.1016/j.ijmedinf.2016.10.008
103. Jacobs, M.A., Graham, A.L.: Iterative development and evaluation methods of mHealth behavior change interventions. Curr. Opin. Psychol. 9, 33–37 (2016). doi:10.1016/j.copsyc.2015.09.001

104. Kumar, S., Nilsen, W.J., Abernethy, A., Atienza, A., Patrick, K., Pavel, M., Riley, W.T., Shar, A., Spring, B., Spruijt-Metz, D., Hedeker, D., Honavar, V., Kravitz, R., Lefebvre, R. C., Mohr, D.C., Murphy, S.A., Quinn, C., Shusterman, V., Swendeman, D.: Mobile health technology evaluation: the mHealth evidence workshop. Am. J. Prev. Med. **45**(2), 228–236 (2013). doi:10.1016/j.amepre.2013.03.017

105. McGee-Lennon, M., Bouamrane, M., Grieve, E., O'Donnell, C.A., O'Connor, S., Agbakoba, R., Devlin, A.A.: Flexible toolkit for evaluating person-centred digital health and eellness at scale. In: Duffy, V.G., Lightner, N. (eds) Advances in Human Factors and Ergonomics in Healthcare: Proceedings of the AHFE 2016 International Conference on Human Factors and Ergonomics in Healthcare. pp. 105–118. Springer International Publishing, Walt Disney World, Florida, USA, Cham (2017) doi:10.1007/978-3-319-41652-6_11

106. Murray, E., Hekler, E.B., Andersson, G., Collins, L.M., Doherty, A., Hollis, C., Rivera, D. E., West, R., Wyatt, J.C.: Evaluating digital health interventions: key questions and approaches. Am. J. Prev. Med. **51**(5), 843–851 (2016). doi:10.1016/j.amepre.2016.06.008

107. World Health Organisation: Monitoring and Evaluating Digital Health Interventions: A Practical Guide to Conducting Research and Assessment. WHO, Geneva (2016)

108. Chaudoir, S.R., Dugan, A.G., Barr, C.H.: Measuring factors affecting implementation of health innovations: a systematic review of structural, organizational, provider, patient, and innovation level measures. Implementation Sci. IS 8:22–5908-5908-5922. (2013) doi:10.1186/1748-5908-8-22

109. Ross, J., Stevenson, F., Lau, R., Murray, E.: Factors that influence the implementation of e-health: a systematic review of systematic reviews (an update). Implementation Sci. IS **11** (1), 146 (2016). doi:10.1186/s13012-016-0510-7

110. Grob, R., Schlesinger, M., Parker, A.M., Shaller, D., Barre, L.R., Martino, S.C., Finucane, M.L., Rybowski, L., Cerully, J.L.: Breaking narrative ground: innovative methods for rigorously eliciting and assessing patient narratives. Health Serv. Res. **51**(Suppl 2), 1248–1272 (2016). doi:10.1111/1475-6773.12503

111. Hibbard, J.H., Stockard, J., Mahoney, E.R., Tusler, M.: Development of the Patient Activation Measure (PAM): conceptualizing and measuring activation in patients and consumers. Health Serv. Res. **39**(4 Pt 1), 1005–1026 (2004). doi:10.1111/j.1475-6773.2004.00269.x

112. Barello, S., Triberti, S., Graffigna, G., Libreri, C., Serino, S., Hibbard, J., Riva, G.: eHealth for patient engagement: a systematic review. Front Psychol. **6**, 2013 (2016). doi:10.3389/fpsyg.2015.02013

113. Sawesi, S., Rashrash, M., Phalakornkule, K., Carpenter, J.S., Jones, J.F.: The impact of information technology on patient engagement and health behavior change: a systematic review of the literature. JMIR Med. Inf. **4**(1), e1 (2016). doi:10.2196/medinform.4514

114. Albert, W., Tullis, T.: Measuring the User Experience: Collecting, Analyzing and Presenting Usability Metrics, 2nd edn. Morgan Kaufmann, Burlington, MA (2013)

115. Klein, L.: UX for Lean Start-ups: Faster, Smarter User Experience Research and Design. O'Reilly Media, Sebastopol, CA (2013)

116. Sauro, J.: The challenges and opportunities of measuring the user experience. J. Usability Stud. **12**(1), 1–7 (2016)

117. Petersen, C., DeMuro, P.: Legal and regulatory considerations associated with use of patient-generated health data from social media and mobile health (mHealth) devices. Appl. Clin. Inf. **6**(1), 16–26 (2015). doi:10.4338/ACI-2014-09-R-0082

118. Agarwal, S., LeFevre, A.E., Lee, J., L'Engle, K., Mehl, G., Sinha, C., Labrique, A., W. H. O. mHealth Technical Evidence Review Group: Guidelines for reporting of health interventions using mobile phones: mobile health (mHealth) evidence reporting and assessment (mERA) checklist. BMJ (Clin. Res. Ed.) **352**, i1174 (2016). doi:10.1136/bmj.i1174

119. Ammenwerth, E., de Keizer, N.F.: Publishing health IT evaluation studies. In: Ammenwerth, E., Rigby, M. (eds.) Evidence-Based Health Informatics: Promoting Safety and Efficiency Through Scientific Methods and Ethical Policy. Studies in Health Technology and

Informatics, pp. 304–311. IOS Press, Amsterdam (2016). doi:10.3233/978-1-61499-635-4-304

120. Brender, J., Talmon, J., de Keizer, N., Nykanen, P., Rigby, M., Ammenwerth, E.: STARE-HI - statement on reporting of evaluation studies in health informatics: explanation and elaboration. Appl. Clin. Inf. **4**(3), 331–358 (2013). doi:10.4338/ACI-2013-04-RA-0024

121. Eysenbach, G.: Improving the quality of Web surveys: the Checklist for Reporting Results of Internet E-Surveys (CHERRIES). J. Med. Internet Res. **6**(3), e34 (2004). doi:10.2196/jmir.6.3.e34

122. Eysenbach G, Group C-E: CONSORT-EHEALTH: improving and standardizing evaluation reports of Web-based and mobile health interventions. J. Med. Internet Res. **13**(4), e126 (2011). doi:10.2196/jmir.1923

123. Khanal, S., Burgon, J., Leonard, S., Griffiths, M., Eddowes, L.A.: Recommendations for the improved effectiveness and reporting of telemedicine programs in developing countries: results of a systematic literature review. Telemed. J. E-health Official J. Am. Telemed. Assoc. **21**(11), 903–915 (2015). doi:10.1089/tmj.2014.0194

124. Niederstadt, C., Droste, S.: Reporting and presenting information retrieval processes: the need for optimizing common practice in health technology assessment. Int. J. Technol. Assess. Health Care **26**(4), 450–457 (2010). doi:10.1017/S0266462310001066

125. SMART—An App platform for healthcare. Boston Children's Hospital Computational Health Informatics Program; Harvard Medical School Department for Biomedical Informatics. http://smarthealthit.org/an-app-platform-for-healthcare/about/. Accessed 15 July 2017

126. Apple Inc. Apple HealthKit. https://developer.apple.com/healthkit/

127. Armstrong, S.: What happens to data gathered by health and wellness apps? BMJ **353**, i3406 (2016). doi:10.1136/bmj.i3406

128. Australia. Therapeutic Goods Administration: Regulation of Medical Software and Mobile Medical 'Apps'. https://www.tga.gov.au/node/4316 (2013)

129. BinDhim, N.F., Trevena, L.: Health-related smartphone apps: regulations, safety, privacy and quality. BMJ Innov. **1**(2), 43–45 (2015). doi:10.1136/bmjinnov-2014-000019

130. Boulos, M.N., Brewer, A.C., Karimkhani, C., Buller, D.B., Dellavalle, R.P.: Mobile medical and health apps: state of the art, concerns, regulatory control and certification. Online J. Public Health Inf. **5**(3), 229 (2014). doi:10.5210/ojphi.v5i3.4814

131. Cerner CODE: Cerner Open Developer Experience. http://www.healthcareitnews.com/news/cerner-launches-open-platform-spur-development-smart-fhir-apps (2016)

132. Chindalo, P., Karim, A., Brahmbhatt, R., Saha, N., Keshavjee, K.: Health apps by design: a reference architecture for mobile engagement. Int. J. Handheld Comput. Res. (IJHCR) **7**(2), 34–43 (2016). doi:10.4018/IJHCR.2016040103

133. Dialogue Consulting: Guidelines for Developing Healthy Living Apps. VicHealth. https://www.vichealth.vic.gov.au/media-and-resources/app-developers (2015)

134. Heffernan, K.J., Chang, S., Maclean, S.T., Callegari, E.T., Garland, S.M., Reavley, N.J., Varigos, G.A., Wark, J.D.: Guidelines and recommendations for developing interactive eHealth apps for complex messaging in health promotion. JMIR Mhealth Uhealth **4**(1), e14 (2016). doi:10.2196/mhealth.4423

135. Hillebrand, U., von Jan, U., Albrecht, U.V.: Concepts for quality assurance of health related apps. Stud. Health Technol. Inf. **226**, 209–212 (2016). doi:10.2196/mhealth.4423

136. Martinez-Perez, B., de la Torre-Diez, I., Lopez-Coronado, M.: Privacy and security in mobile health apps: a review and recommendations. J. Med. Syst. **39**(1):181-014-0181-0183. Epub 2014 Dec 0187 (2015). doi:10.1007/s10916-014-0181-3

137. Research2Guidance (2016) mHealth App Developer Economics: The Current Status and Trends of the mHealth App Market. http://research2guidance.com/r2g/r2g-mHealth-App-Developer-Economics-2016.pdf (2016)

138. Schnall, R., Rojas, M., Bakken, S., Brown, W., Carballo-Dieguez, A., Carry, M., Gelaude, D., Mosley, J.P., Travers, J.: A user-centered model for designing consumer mobile health (mHealth) applications (apps). J. Biomed. Inf. **60**, 243–251 (2016). doi:10.1016/j.jbi.2016.02.002

139. Stoyanov, S.R., Hides, L., Kavanagh, D.J., Zelenko, O., Tjondronegoro, D., Mani, M.: Mobile app rating scale: a new tool for assessing the quality of health mobile apps. JMIR mHealth uHealth **3**(1), e27 (2015). doi:10.2196/mhealth.3422

140. UK. National Health Service App Development: An NHS guide to developing mobile healthcare applications. NHS Innov. South East, [n.p.] (2014)

141. Abbasi, A., Adjeroh, D., Dredze, M., Paul, M.J., Zahedi, F.M., Zhao, H., Walia, N., Jain, H., Sanvanson, P., Shaker, R., Huesch, M.D., Beal, R., Zheng, W., Abate, M., Ross, A.: Social media analytics for smart health. IEEE Intell. Syst. **29**(2), 60 (2014). doi:10.1109/MIS.2014. 29

142. Ben-Harush, O., Carroll, J.-A., Marsh, B.: Using mobile social media and GIS in health and place research. Continuum **26**(5), 715–730 (2012). doi:10.1080/10304312.2012.706460

143. Bond, C.S., Ahmed, O.H., Hind, M.: Implications for research methods when conducting studies with the users of online health communities. Comput. Inf. Nurs. CIN **32**(3), 101–104 (2014). doi:10.1097/CIN.0000000000000049

144. Bradley, M., Braverman, J., Harrington, M., Wicks, P.: Patients' motivations and interest in research: characteristics of volunteers for patient-led projects on PatientsLikeMe. Res. Involvement Engagem. **2**(1), 33 (2016). doi:10.1186/s40900-016-0047-6

145. Capurro, D., Cole, K., Echavarria, M.I., Joe, J., Neogi, T., Turner, A.M.: The use of social networking sites for public health practice and research: a systematic review. J. Med. Internet Res. **16**(3), e79 (2014). doi:10.2196/jmir.2679

146. Cyrus, J.W.: A review of recent research on internet access, use, and online health information seeking. J. Hosp. Librariansh **14**(2), 149–157 (2014). doi:10.1080/15323269. 2014.888630

147. Ekberg, J., Gursky, E.A., Timpka, T.: Pre-launch evaluation checklist for online health-promoting communities. J. Biomed. Inform. **47**, 11–17 (2014). doi:10.1016/j.jbi. 2013.10.004

148. Heidelberger, C.A., El-Gayar, O., Sarnikar, S.: Online health social networks and patient Health decision behavior: a research agenda. In: 2011 44th Hawaii International Conference on System Sciences, New York, 2011. IEEE, pp. 1–7. doi:10.1109/HICSS.2011.328

149. Ho, K., Workshop, Peter Wall: Harnessing the social web for health and wellness: issues for research and knowledge translation. J. Med. Internet Res. **16**(2), e34 (2014). doi:10.2196/ jmir.2969

150. Ji, X., Chun, S.A., Cappellari, P., Geller, J.: Linking and using social media data for enhancing public health analytics. J. Inf. Sci. (Online First):1–25. (2016) doi:10.1177/ 0165551515625029

151. Kim, Y., Huang, J., Emery, S.: Garbage in, garbage out: data collection, quality assessment and reporting standards for social media data use in health research, infodemiology and digital disease detection. J. Med. Internet Res. **18**(2), e41 (2016). doi:10.2196/jmir.4738

152. Martínez, P., Martínez, J.L., Segura-Bedmar, I., Moreno-Schneider, J., Luna, A., Revert, R.: Turning user generated health-related content into actionable knowledge through text analytics services. Nat. Lang. Process. Text Analytics Ind. **78**, 43–56 (2016). doi:10.1016/j. compind.2015.10.006

153. McKee, R.: Ethical issues in using social media for health and health care research. Health Policy (Amsterdam, Netherlands) **110** (2–3):298-301. (2013) doi:10.1016/j.healthpol.2013. 02.006

154. Merolli, M., Martin-Sanchez, F.J., Gray, K.: Social media and online survey: tools for knowledge management in health research. In: HIKM'14 Proceedings of the Seventh Australasian Workshop on Health Informatics and Knowledge Management, Auckland, New Zealand, 2014. Australian Computer Society, Inc, pp. 21–29

155. Merolli, M.A.: Participatory Health Through Social Media in Chronic Disease: a Framework for Research and Practice [PhD thesis]. University of Melbourne, Parkville (2015)

156. Mitra, S., Padman, R., Yang, H., Lee, E.K.: Understanding the role of social media in healthcare via analytics: a health plan perspective. In: Yang, H., Lee, E.K. (eds.) Healthcare

Analytics: From Data to Knowledge to Healthcare Improvement, pp. 555–587. John Wiley & Sons, Inc, Hoboken, NJ (2016). doi:10.1002/9781118919408.ch19

157. Murray, P.J., Wright, G.: Towards a research agenda for Web 2.0 and social media in health and informatics. In: Proceedings of the 4th WSU—International Research Conference, East London, South Africa, 2014. Walter Sisulu University, pp. 89–98

158. Myneni, S., Iyengar, S.: Socially influencing technologies for health promotion: translating social media analytics into consumer-facing health solutions. In: 49th Hawaii International Conference on System Sciences (HICSS), pp. 3084–3093. EEE, New York (2016). doi:10.1109/HICSS.2016.388

159. Neighbors, C., Lewis, M.A.: Editorial overview: Status update: current research on social media and health. Soc. Media Appl. Health Behav. **9**, iv–vi. (2016) doi:10.1016/j.copsyc.2016.04.021

160. Smith, B.G., Smith, S.B.: Engaging Health: Health Research and Policymaking in the Social Media Sphere. AcademyHealth, Washington, DC (2015)

161. Straton, N., Hansen, K., Mukkamala, R., Hussain, A., Grønli, T., Langberg, H., Vatrapu, R.: Big social data analytics for public health: Facebook engagement and performance. In: HealthCom '16: 18th IEEE International Conference on e-Health Networking, Applications and Services, Munich, Germany (2016). doi:10.1109/HealthCom.2016.7749497

162. Taylor, H.A., Kuwana, E., Wilfond, B.S.: Ethical implications of social media in health care research. Am. J. Bioeth. AJOB **14**(10), 58–59 (2014). doi:10.1080/15265161.2014.947820

163. Thirumalai, M., Ramaprasad, A.: Ontological analysis of the research on the use of social media for health behavior change. In: 2015 48th Hawaii International Conference on System Sciences, pp. 814–823.(2015). doi:10.1109/HICSS.2015.103

164. Almalki, M., Gray, K., Martin-Sanchez, F.: Activity theory as a theoretical framework for health self-quantification: a systematic review of empirical studies. J. Med. Internet Res. **18** (5), e131 (2016). doi:10,2196/jmir 5000

165. Almalki, M., Gray, K., Martin-Sanchez, F.J.: Refining the concepts of self-quantification needed for health self-management. A thematic literature review. Methods Inf. Med. **56**(1), 46–54 (2017). doi:10.3414/ME15-02-0007

166. Banaee, H., Ahmed, M.U., Loutfi, A.: Data mining for wearable sensors in health monitoring systems: a review of recent trends and challenges. Sensors (Basel, Switzerland) **13** (12):17472–17500. (2013) doi:10.3390/s131217472

167. Casper, G.R., McDaniel, A.: Introduction to theme issue on technologies for patient-defined and patient-generated data. Pers. Ubiquit. Comput. **19**(1), 1–2 (2015). doi:10.1007/s00779-014-0803-2

168. Chiauzzi, E., Rodarte, C., DasMahapatra, P.: Patient-centered activity monitoring in the self-management of chronic health conditions. BMC Med. **13**, 77-015-0319-0312. (2015) doi:10.1186/s12916-015-0319-2

169. Choe, E.K., Lee, N.B., Lee, B., Pratt, W., Kientz, J.A.: Understanding quantified-selfers' practices in collecting and exploring personal data. In: SIGCHI Conference on Human Factors in Computing Systems, pp. 1143–1152. ACM, Toronto, Ontario, Apr 26–May 1 (2014). doi:10.1145/2556288.2557372

170. Custodio, V., Herrera, F.J., Lopez, G., Moreno, J.I.: A review on architectures and communications technologies for wearable health-monitoring systems. Sensors (Basel, Switzerland) **12**(10),13907–13946. (2012) doi:10.3390/s121013907

171. De Mooy, M., Yuen, S.: Towards privacy-aware research and development in wearable health. In: Proceedings of the 50th Hawaii International Conference on System Sciences (HICSS), pp. 3658–3667 (2017)

172. Deering, M.J.: ONC Issue Brief: Patient-Generated Health Data and Health IT. Office of the National Coordinator for Health Information Technology, Washington, DC (2013)

173. Gray, K., Martin-Sanchez, F.J., Lopez-Campos, G.H., Almalki, M., Merolli, M.: Person-generated data in self-quantification. A health informatics research program. Methods Inf. Med. **56**(1), 40–45 (2017). doi:10.3414/ME15-02-0006

174. Kumara, S., Cui, L., Zhang, J.: Sensors, networks and internet of things: research challenges in health care. In: IIWeb '11: Proceedings of the 8th International Workshop on Information Integration on the Web: in conjunction with WWW 2011, pp. 1–4. ACM, Hyderabad, India, March 28 (2011) doi:10.1145/1982624.1982626
175. Piras, E.M., Ellingsen, G.: International workshop on Infrastructures for health care: patient-centered care and patient generated data. J. Particip. Med. **8**, e4 (2016)
176. Rich, E., Miah, A.: Mobile, wearable and ingestible health technologies: towards a critical research agenda. Health Soc. Rev. **26**(1), 84–97 (2017). doi:10.1080/14461242.2016. 1211486
177. Shapiro, M., Johnston, D., Wald, J., Mon, D.: Patient-Generated Health Data: White Paper. RTI International, Research Triangle Park, NC (2012)
178. Woods, S.S., Evans, N.C., Frisbee, K.L.: Integrating patient voices into health information for self-care and patient-clinician partnerships: veterans affairs design recommendations for patient-generated data applications. J. Am. Med. Inf. Assoc. JAMIA **23**(3), 491–495 (2016). doi:10.1093/jamia/ocv199

Chapter 3
Machine Learning for Structured Clinical Data

Brett Beaulieu-Jones

Abstract Research is a tertiary priority in the EHR, where the priorities are patient care and billing. Because of this, the data is not standardized or formatted in a manner easily adapted to machine learning approaches. Data may be missing for a large variety of reasons ranging from individual input styles to differences in clinical decision making, for example, which lab tests to issue. Few patients are annotated at a research quality, limiting sample size and presenting a moving gold standard. Patient progression over time is key to understanding many diseases but many machine learning algorithms require a snapshot, at a single time point, to create a usable vector form. Furthermore, algorithms that produce black box results do not provide the interpretability required for clinical adoption. This chapter discusses these challenges and others in applying machine learning techniques to the structured EHR (i.e. Patient Demographics, Family History, Medication Information, Vital Signs, Laboratory Tests, Genetic Testing). It does not cover feature extraction from additional sources such as imaging data or free text patient notes but the approaches discussed can include features extracted from these sources.

Keywords Missing data · Semi-supervised machine learning · Longitudinal modeling · Machine learning interpretability

3.1 Introduction

Precision medicine has the potential to substantially change the way patients are treated in many facets of health care. Precision medicine is the idea of delivering personalized treatment and prevention strategies by considering the holistic patient,

B. Beaulieu-Jones (✉)
Institute of Biomedical Informatics, Perelman School of Medicine, University of
Pennsylvania, D200 Richards Hall, 3700 Hamilton Walk, Philadelphia, PA 19104, USA
e-mail: brettbe@med.upenn.edu

© Springer International Publishing AG 2018
D.E. Holmes and L.C. Jain (eds.), *Advances in Biomedical Informatics*,
Intelligent Systems Reference Library 137,
https://doi.org/10.1007/978-3-319-67513-8_3

including their genetics, environment, and lifestyle [1]. Machine learning using structured clinical data will likely play a large role in the success or failure of precision medicine. Specifically, machine learning using structured data can help in finding associations between a patient's genotype and phenotype, identifying similar patients and predicting the efficacy of different clinical treatment strategies on a personalized level.

The amount of digital data collected in the clinic has rapidly expanded, the first EHRs are now more than 20 years old. The United States federal government mandated meaningful use of EHRs by 2014. According to the American Hospital Association, by 2015, 96% of acute care hospitals had implemented a certified EHR. Correspondingly, several top research institutions across the country have recently established or are currently establishing departments or institutes in biomedical informatics using the EHR as a major data.

Smartphones, wearable devices and in-clinic diagnostic tools offer the ability to stream accurate measurements in real time. AliveCor received FDA approval in 2012 for its iPhone-based heart monitor using machine learning to detect Atrial Fibrillation in seconds. Billions of dollars in venture capital are currently being invested in companies, such as Grail, Foundation Medicine, and Guardant health, promising less invasive biopsies, or liquid biopsies, using machine learning to classify patients from circulating tumor cells in the bloodstream. Preventative wellness clinics, such as Forward, are emerging to characterize and track what it means to be healthy.

These are only a few examples of the many opportunities centered on patient data. Data for both evidence-based clinical decision making and computational research is becoming increasingly available and we must now develop new methods to preprocess and analyze this data at a matching rate.

3.2 Uses of Machine Learning for Structured Clinical Data

Each time a patient interacts with a health system, actions, notes, and measurements are recorded in the EHR. This wealth of data has made the EHR the primary source of structure clinical data. Three popular research applications of EHR data are:

1. Patient clustering to identify similar cases.
2. Electronic phenotyping for genetic studies.
3. Advising clinical treatment strategies.

These tasks can be performed using machine learning, but each task requires careful preprocessing of data and appropriate phrasing of the problem to utilize traditional machine learning methods. The nature of EHR data places emphasis on unsupervised clustering and semi-supervised classification. In this section, we discuss these common tasks and show examples where researchers have utilized machine learning effectively to guide discovery. There exist many great resources for understanding machine learning approaches as applied to general problems [2]. We concentrate on how to position relevant clinical questions and the challenges specific to the EHR that need to be solved in order to apply these powerful techniques.

3.2.1   Patient/Disease Stratification

As we learn more about the mechanisms and etiology of a disease, our diagnoses can become more precise, leading to the creation of disease subtypes. Historically, cancers were diagnosed based on their occurrence location and their reaction to different treatments. As the mechanisms of cancer are better understood, they are further categorized by their physiological nature. The progression of subtypes in lung cancer illustrates the increases in resolution over time for a previously poorly defined disease (Kreybe L. [3]. Beginning with a single diagnosis based on occurrence in the lung, it was later differentiated as small cell lung cancer and non-small cell lung cancer [4, 5]. Non-small cell lung cancer was then broken up into squamous cell carcinoma, adenocarcinoma, and large cell carcinoma. Today these subtypes continue to be broken up based on the genetic locations and pathways of associated risk variants.

What happens when physiological differences cannot easily be used to subtype disease? This is true with several metabolic disorders, for example, metabolic syndrome has been redefined numerous times. It is associated with a wide range of comorbidities and presents in a clinically heterogeneous manner. These comorbidities, including coronary heart disease, diabetes, and stroke, represent an oversized risk to public health and increasingly unwieldy burden on the health care system. Despite this, metabolic syndrome's predictive value for cardiovascular events, disease prediction and progression is disputed and may not outperform the individual components it's made up of [6]. While the concept of identifying patients at high risk of developing diseases such as heart disease and diabetes for early intervention is an important one, metabolic syndrome in its current form fails to do this effectively.

Li et al. demonstrated the ability to identify disease subtypes of patients with a metabolic disorder, type 2 diabetes [7]. To do this they performed a topological analysis of 11,210 patients with type 2 diabetes at Mount Sinai Medical Center in New York. This topological analysis constructed a network of patients by connecting those most similar to each other. Using this they found three unique subtypes. Subtype 1 demonstrated the traditional observations of type 2 diabetes, hyperglycemia, obesity, and eye and kidney diseases. Subtype 2's main comorbidity was cancer, and subtype 3's unique comorbidities were neurological diseases. These subtypes are likely enriched for etiological differences; the disease likely operates differently in someone who develops cancer than someone who develops kidney disease. By developing a machine learning classifier to identify which subtype a patient is in as early as possible, clinicians may be able personalize treatment to reduce the odds of developing these more serious comorbidities.

Multiple sclerosis (MS) illustrates an additional area machine learning for disease stratification could be particularly useful. Multiple sclerosis was traditionally subtyped into Relapsing-Remitting MS and Progressive MS. In 2014, it was recommended that these subtypes be further divided into six total subtypes [8]. Unfortunately, the current strategies for determining subtype and thus treatment strategy require looking at the progression of the disease. This is essentially a retrospective diagnosis and means personalized treatment plans cannot be started until progression has been observed. Could unsupervised clustering be used to identify subtypes earlier on?

3.2.2 Electronic phenotypes for Genetic Associations

Genetic associations examine whether a genetic variant is associated with a specific trait (Fig. 3.1). This specific trait, a phenotype, can be a moving target when dealing with the complexity of human disease. The trait is often a human defined disease. Those with the disease are labeled the case and those without the disease are considered controls. Early genetic associations using the electronic health record were performed with raw International Classification of Diseases (ICD) codes. ICD codes are recorded by physicians when diagnosing a patient with a condition, and are used to ensure proper billing and insurance reimbursement. ICD codes are published and updated by the World Health Organization and are primarily used for clinical billing purposes. Despite ICD-10 being initially published in 1994, ICD-9 codes are still commonly used in both clinical and research settings.

While ICD codes provide a clear, discrete endpoint for genetic associations, the use of billing codes can introduce unintentional biases to analyses. An ICD code may be added to an EHR in order to issue and receive insurance reimbursement for a test to screen for the disease the ICD code represents. In this case, not only is the timing of diagnosis difficult to determine, but solely looking at the ICD codes for a patient is likely to introduce false positives. In addition, certain ICD codes are more easily reimbursed than others. When a clinician determines that a patient requires a

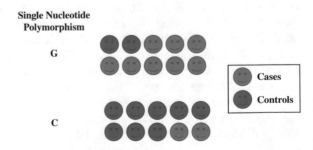

Fig. 3.1 In this example, the guanine (G) nucleotide is more common in the cases than the cytosine nucleotide

treatment or test to increase their odds of a successful outcome, the clinician is incentivized to choose the ICD code most likely to allow them to effectively treat their patient.

Phenotype algorithms can be developed using the structured EHR to leverage both ICD codes and the rest of the of a patient's record. The eMERGE project is a national network which has deployed phenotype algorithms for over 40 diseases, over 500,000 EHRs and 55,000 patients with genetic data. Many of the phenotype algorithms are simple rules-based systems, for example: Type 2 Diabetes (Fig. 3.2).

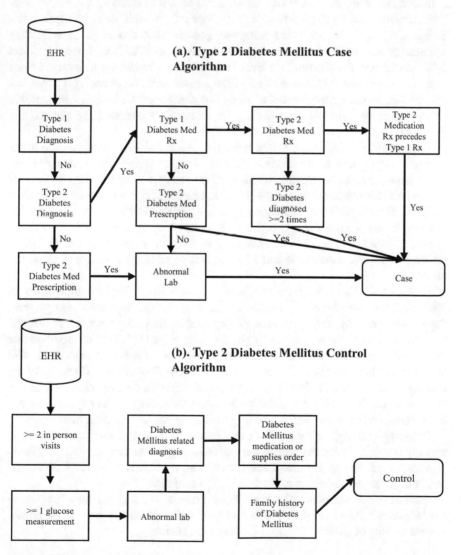

Fig. 3.2 Phenotype Algorithms for Type 2 Diabetes Mellitus. **a** Case selection from the EHR. **b** Control Selection from the EHR. Adapted from: [28]

An approach to study phenotype-genotype associations from the EHR are Phenome-wide association studies or PheWAS [9]. PheWAS use EHRs to define a phenome that can be linked back to individual genetic variants. The approach can discover gene-disease associations while identifying pleiotropic effects of individual SNPs. PheWAS generally uses the ICD9 codes to construct a phenotype. While primarily used for billing, these codes provide a set of discrete variables that can represent many phenotypes for a patient at the same time, providing greater resolution. Besides the repurposing of billing codes, a major challenge of PheWAS is in understanding the functional mechanisms at work behind GWAS SNP matches. Stratification by the 4,841 different codes creates wide data, presenting statistical challenges in achieving adequate power. This challenge of achieving adequate power will be exacerbated by the transition to ICD10, with less historical data built up and the potential for over 16,000 codes. Continuing to increase open data access will allow researchers to utilize a more accurate phenotypic representation while lessening the burden of statistical challenges. Coding systems, unlike patient notes or genomic data should be easier to anonymize, aggregate, and distribute.

Because ICD billing codes can be biased, as evidenced by phenotype algorithms having multiple steps to catch errors for both case and control status, using billing codes alone may cause misclassification of phenotypes. The misclassification of phenotypes substantially reduces the power to detect linkage in case-control studies. With 1% phenotypic misclassification up to 10% of the power is lost, and with 5% phenotypic misclassification, the power is reduced by approximately two-thirds [10–12]. Misclassification can occur for a variety of reasons including misdiagnosis/clinical error, clerical error, or lack of scientific knowledge about the disease in question.

Labbe et al. showed increased linkage by clustering lifetime symptoms in schizophrenia and bipolar disease to form more homogenous phenotypes. Separating cases by the symptoms of psychiatric diseases compensates for the inability to subtype these diseases by physical properties [13]. This is important due to the deficit of physiological understanding for these diseases. Labbe et al. also included familial information to understand the heritability of these diseases. When looking at subtypes that show a strong familial aggregation they observed higher linkage scores. By looking at ancestral histories for subtypes, the expected heritability could be better estimated resulting in a reduction of "missing heritability."

Phenotypic subtyping was also used successfully in the analysis of genetic variants responsible for the severe development regression and stereotypical hand movements of Rett syndrome. Causal mutations were found in the FOXG1 and MECP2 genes and deletions at the 22q11.2 locus [14].

Each of these examples point towards the promise of using machine learning to cluster patients based on their EHRs to identify disease subtypes or more homogenous groups of patients for use in association studies.

3.2.3   Clinical Recommendations

The availability of data and advances in biomedical informatics have helped to make medicine increasingly evidence based and in some cases entirely data driven. Clinicians and researchers now have the ability to leverage millions of data points when designing and determining treatment best practices. The New England Journal of Medicine recently held the SPRINT data analysis to"use the data underlying a recent article to identify a novel clinical finding that advances medical science." The original clinical trial sought to see whether intensive management of systolic blood pressure (<120 mm Hg) was more effective than standard management (<140 mm Hg). The original trial was stopped early due to the success of the intensive management strategy in reducing cardiovascular events. The data from the trial was released as a challenge where teams used machine learning approaches (primarily rules based) to provide personalized recommendations.

More personalized treatment strategies are a popular use of machine learning in the EHR. This can be driven by genomics (pharmacogenomics), or simply by sub setting patients based of attributes (race, BMI, etc.). Wiley et al. demonstrate the importance of training an algorithm on a population similar to the application population [15]. In their case it was necessary to extract the perecent African ancestry from the genome instead of self reported race in order to improve the model fit.

Due to the inherent risk of adjusting clinical treatment strategies, many of the early applications of machine learning in health systems have been seen in academic research (retrospective analysis, drug development, pharmacogenomics) and for things like resource usage. For example, how likely is a patient coming into the ER to need an ICU bed? Increasingly machine learning methods are likely to be applied to clinical decisions including providing prognosis information for shared decision making strategies. Deep learning, in particular, is becoming an increasingly tool for drug discovery and development [29].

3.3   Challenges of using Machine Learning in the Structured EHR

3.3.1   Limited "gold-standards"

Large institutions and health care systems can have EHRs containing millions of patients and billions of measurements. Despite the size of these data, electronic phenotyping requires a gold standard to validate accuracy. This gold standard often requires time consuming, manual clinician review and is thus expensive.

In addition, the selection of cases and controls can unintentionally create biases in downstream algorithms and analyses. It is often easiest to select the most severe cases and the healthiest controls. In these circumstances researchers can have the greatest confidence they are accurately selecting a true case or control. Unfortunately, this creates a biased training set where it is difficult to differentiate between less severe cases and less healthy controls. Fig. 3.3 shows an illustration of a simulated dataset where the first two principal components happen to represent the degree of the case phenotype. If the most severe cases are selected, a classifier trained to distinguish between cases and controls is unlikely to generalize well. If less severe cases are chosen, there may be issues with mislabeled cases. An example of another bias can be seen in the Type 2 Diabetes Mellitus algorithm, controls must have at least 1 glucose measurement (Fig. 3.2B). For a young patient, this means that a clinician must have had reason to suspect that the patient's glucose could be abnormal and thus could bias controls to patients who "look" like they are at a high risk for developing Type 2 Diabetes.

Because patients move between health systems, a patient may not be diagnosed in the system they are treated in and may only have a partial history. Some methods for controlling for incomplete histories can result in smaller sample sizes. It is common to include only patients who have a visit in the system prior to the

Fig. 3.3 Simulated disease severity plot where the 2 principal components stratify patients according to severity

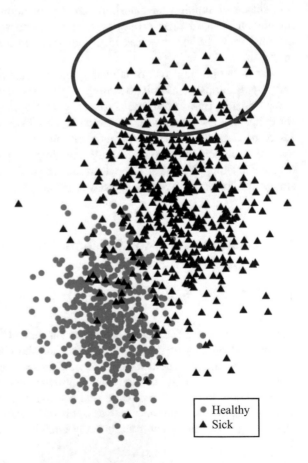

diagnosis of the phenotype of interest. While this can help to determine the diagnosis date for a disease, it excludes anyone who was diagnosed on their first visit to a particular health system.

3.3.2 Missing Data

The average patient is unlikely to have measurements for the clear majority of fields in the EHR. It does not make sense logistically or economically to administer every test to a seemingly healthy patient. There are three primary types of missing data:

1. Missing Completely at Random (MCAR) – when data is missing in a completely unrelated way to the values of both the observed and the unobserved data.
2. Missing at Random (MAR) – when the data is missing based on the observed data, when other fields in the EHR indicate whether the value will be present or absent.
3. Missing Not at Random (MNAR) – when the data is missing based on the values of the unobserved data

Fig. 3.4 shows the ability to use a random forest to predict whether a lab value will be present or absent based on other lab values. Unsurprisingly data that is MCAR cannot be predicted (Fig. 3.4A), and data that is either MAR or MNAR (Fig. 3.4B, C) can be predicted with an accuracy significantly greater than random. In practice, most missing data in the EHR tends to be of the MAR variety. Clinicians must decide which measurements are relevant and fiscally responsible, irrelevant tests are wasteful and it does not make sense to subject patients to unnecessary discomfort. The clinician is making these decisions based on the observations they make, so when data is missing it is related to the observed data.

MCAR data is less likely to present issues to downstream analyses than data that is either MAR or MNAR, but if not handled correctly all three types of missing data

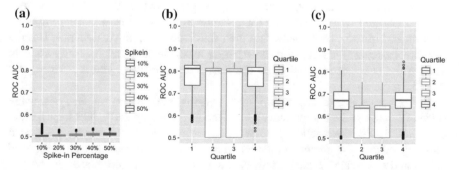

Fig. 3.4 A.) Data that is missing completely at random cannot be predicted. B & C.) Data that is missing at random and missing not at random can be predicted. (To appear)

can introduce unintentional biases to all sorts of downstream analyses including machine learning. Machine learning algorithms often expect a complete matrix as input and are not designed to handle null values. This often leads to researchers performing one of three options:

1. Perform feature selection of relevant features and use only complete cases, or patients that have values for all features.
2. Modify the algorithm to accept null inputs (often by ignoring them) or
3. Perform imputation to predict what the value for a feature would be.

Each of these options have several pros and cons and can have unintended effects on machine learning. When performing complete case analysis after feature selection, the features included can lead to including either more severe cases or cases that were harder to diagnose. Imagine a disease that is diagnosed by a laboratory measurement where values over 10 conclusively indicate you have the disease but values between 8 and 10 require an additional test. If the additional test is included in the features selected, the complete cases are now only the patients that were harder to diagnose. When modifying an algorithm to accept null inputs, the researcher needs to be careful that the algorithm does not disproportionately learn to depend on patients that have all of the measurements or only the most complete measurements. If the algorithm relies on patients with all of the measurements, many of the same issues that arise in complete case analysis repeat. If the algorithm learns to ignore rare measurements it can miss signal. For example, in an analysis of different treatment options, say 40 of 10,000 patients suffered a fairly rare but severe adverse event. Without careful monitoring the algorithm may not place enough importance on this feature despite the fact this outcome is disproportionately important.

Imputation can be effective in the EHR because many missing values can be inferred by omission and just knowing whether a value was present or absent can be useful. If a patient has never had a chest X-ray, it is unlikely that their physician suspects a broken rib cage. This information can be provided to downstream machine learning algorithms by performing imputation. It is, however, very important to carefully analyze the results of imputation. Oftentimes much can be learned simply by looking at which methods are the most accurate. Direct accuracy can be measured by spiking in missing values to replace known values, imputing these spiked-in values and measuring their difference from the real values. Despite this direct accuracy should only be used as a benchmark, and it is important to analyze the effect of imputation on the downstream analyses you are performing (Fig. 3.5).

For example, mean imputation may perform strongly in an analysis using spiked-in missingness but remove all variance from the imputed values for a feature (Fig. 3.5C). Other evaluation criteria, such as, comparing the variance between imputed values of different imputation runs and the difference between imputed values and real values. Ideally these values would be highly correlated in order to maintain the variance structure. Popular imputation methods for EHRs include

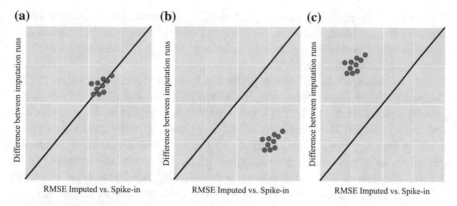

Fig. 3.5 Comparison between spike in accuracy and variation between imputation runs

K-Nearest Neighbors, Singular Value Decomposition and Multiple Imputation by Chained Equations.

All three of these strategies for handling missing data may introduce bias when performing EHR-based analyses. It is important to consider potential effects and ideally to utilize multiple strategies and examine the differences.

3.3.3 Privacy, Reproducibility and Data Sharing

Patient privacy needs to be a focus of any secondary use of EHRs. Because a patients EHR is ‚de-identified‘ does not mean that it is anonymous. Latanya Sweeney demonstrated this emphatically when the Massachusetts Group Insurance Commission released de-identified data on state employees [16]. These records included each hospital visit and Sweeney was able to re-identify several patients including the former Governor of Massachusetts. Sweeney was able to do this from his birth date, zip code and sex alone, and to prove a point mailed the Governor a copy of his personal records. The task of re-identification has been shown possible in several other cases where data holders attempted to share their data, including the Netflix challenge. Narayana and Shmatikov were able to de-anonymize users in the Netflix challenge by linking their viewing histories with popular movie review sites. For users who had rated more than 6 movies, they were able to do this with greater than 90% accuracy [17]. This, in part, led to Netflix canceling the second iteration of its popular recommendation contest following a privacy lawsuit.

Caution needs to be taken even when the actual data is not released. Deep learning models can have many millions of parameters, allowing adversaries to perform membership inference attacks in order to determine whether a user was a member of the study or not. Shokri et al. [18]) demonstrate greater than 90% precision even with 30,000 examples in the training set. They do this by examining the trained parameters of a deep neural network trained on the CIFAR-100 dataset.

Even without the model, enterprising adversaries performed membership inference attacks with only black-box access to the target model through an API. Shokri et al. again demonstrated this on various purchase history datasets made available through Amazon and Google APIs.

One approach to adding privacy protection is called "Differential Privacy" [19]. Differential Privacy is a robust, meaningful and mathematical rigorous definition of privacy which operates under the knowledge that data cannot be fully anonymized and remain useful. If you remove all of the signal in a dataset to anonymize, machine learning methods are fruitless. If you keep any signal at all, there is a chance an adversary will able to discover information about the members of the dataset. The goal of differential privacy is to find a balance between an acceptable risk, the privacy budget, and usefulness of the data. It attempts to minimize the likelihood an adversary can perform a membership inference attack to determine if a subject is in a dataset. It works by adding a plausible deniability of any outcome by inserting random noise into the information made available. If balanced, meaningful answers can be interrogated from the data while greatly reducing the risk that any member of a study is harmed by de-identification. A classic example and simple way to think about differential privacy is to imagine a study where participants are told to answer a question. Before answering the question they flip a coin, if the coin lands heads, they give the real answer, the truth. If the coin lands tails, they answer randomly by flipping an additional coin and responding yes if it lands heads and no if it lands tails.

Simmons et al. used a variant of differential privacy to enable privacy preserving genome wide association studies even when there is significant population stratification. Genomic data has a high dimensionality and relatively low signal to noise ratio making de-identification or other attempts at masking individual records impractical. They demonstrate the ability to allow users to query summary statistics while minimizing privacy risks. This is a particularly interesting application because while genome sequencing prices have rapidly decreased, the combined costs of recruitment and sequencing are a major barrier to this type of research.

Beaulieu-Jones et al. recently showed the ability to train deep neural networks under differential privacy to generate synthetic data that closely resembles the SPRINT clinical trial data [20]. This method allows for increased sharing of valuable, difficult to obtain datasets. Differential privacy is a rapidly growing area, we suggest "The Algorithmic Foundations of Differential Privacy" [21] as a starting point if interested in implementing differential privacy.

Privacy challenges can make sharing data prohibitively difficult. This in turn presents challenges in reproducing work from other researchers. Even if source code is shared, researchers attempting to reproduce original research generally can only compare final results. This means that even if a protocol of a paper is well written and described, if it has 100 steps, a researcher attempting to reproduce cannot be sure where their results diverged. Because of this challenge we strongly advocate publishing intermediate results. This can help narrow down divergences to a few steps, was it the data? The preprocessing? The actual analysis? The plotting into charts? One way to release intermediate results without adding a large amount of additional work is to use continuous integration to run the analysis and export the log file [22].

3.3.4 Longitudinal Data

A key attribute and potential strength of EHRs is the ability to track the way a patient progresses over time. Early moving caregivers such as Geisinger Health System implemented initial EHRs over twenty years ago but fully utilizing this longitudinal presents challenges to researchers.

Longitudinal EHR data are often irregular time series. Measurements are recorded at irregular times, can be mixed type (continuous, ordinal, categorical), require feature extraction (images, free text). It is common for researchers to take a single time point (i.e. current time, set time after diagnosis etc.) and use this as the single end point or label for machine learning analyses. This can be problematic when patients have arrived at that point through very different routes. For example, if using a systolic blood pressure as an end point, one patient may be on an intensive blood pressure management protocol while another with the same blood pressure may have never taken medication. In the SPRINT clinical trial there were patients on as many as seven medications to manage blood pressure (Group [23], if unmedicated these patients would almost definitely have significantly higher measurements. One method researchers use to remediate this issue is to derive statistics to represent the time series, such as taking the median value. This can be insufficient when the way clinicians choose to observe and treat patients based on data either not recorded in the electronic health record, in the unstructured data or in fields not selected for inclusion can also bias the labels. For example, if patient A has a single normal white blood cell count, and patient B has had a monthly count every month for the past 5 years. A clinician could have been checking to see if patient A showed an increased white blood cell count after a surgery suspecting a possible infection. In contrast, the repeated measurements for patient B indicate the clinician may have a reason to believe patient B is immunocompromised or may become immunocompromised due to a virus or adverse reaction to a medication and is using the white blood cell count to monitor this. Despite the patients having relatively equal white blood cell counts, using this single value as a label is clearly inadequate to represent the complete state of the patient. For this specific case deriving a panel of statistics including features such as the count and variance of the measurement could help to better represent the current state of a patient. Recent work takes this further to calculate disease and patient trajectories by generating networks of the way a patient or disease progresses over time. Jensen et al. demonstrated this using 6.2 million patients from Danish National Patient Registry to cluster patients based on time dependent disease diagnoses [24]. These disease diagnoses were extracted from patterns of ICD-9 codes on patient's EHRs. This method creates a visualization of patient trajectories and allows for analyses of co-morbidities observed in health systems in order to identify important patterns that indicate the potential for more severe outcomes. Further work in this field could move beyond billing codes in order to allow for increased resolution of patient trajectories.

3.4 Future and evolving opportunities for Machine Learning in the Structured EHR

3.4.1 Quantitative Electronic Phenotyping

Traditionally, genetic association studies relied on binary outcomes as target phenotypes for the association. Quantitative trait loci studies provide the ability to measure correlation between DNA variation and a phenotype. Quantitative traits occur on a continuum and are driven by multiple genes in conjunction with the environment. By using clustering and other machine learning techniques, researchers can represent disease as quantitative rather than binary values. This has several advantages. Patients with a common disease that present with different symptoms, different levels of effect or different paths of progression can be clustered into homogenous subgroups with similar patients. These clusters are likely enriched etiologically, meaning the reasons the disease is causing each cluster are different. Within each cluster or across the entire spectrum of diseases, a phenotype can be constructed to better represent how severe of a case a patient has. Diseases where using a binary case control status has been effective are likely etiological homogenous, or so disruptive to a particular system that the severity is irrelevant. These represent the low hanging fruit, but many diseases present in heterogeneous manners (Cancers, Amyotrophic Lateral Sclerosis, Multiple Sclerosis, Alzheimers etc.). Fine-tuned quantitative phenotyping could have the ability to resolve homogenous subgroups, greatly increasing statistical power and creating a better target for association.

3.4.2 Deep Learning, Unsupervised and Semi-Supervised Learning Approaches

Deep Learning has already led to state of the art results in a variety of fields including image processing, speech recognition, and gameplay. Many of the early "wins" using deep learning and more generally machine learning in the EHR involve applications of algorithms proven successful in other domains. This has been particularly true in unstructured EHR data such as images (cancer tumor detection etc.) and natural language processing for free text. This is sufficient when EHR learning tasks resemble tasks popular among general machine learning researchers but long term advances will require specialized algorithms customized to the unique challenges presented by EHR-based research. Algorithmic development is only one part of the equation. The proper phrasing of problems and preprocessing of data will likely have as much if not more importance than algorithmic development.

Three of the pioneers of deep learning, Yann Lecun, Yoshua Bengio and Geoffrey Hinton wrote, "Unsupervised learning had a catalytic effect in reviving interest in deep learning, but has since been overshadowed by the successes of purely supervised learning… we expect unsupervised learning to become far more important in the longer term" [25]. The challenge of collecting labeled data for supervised learning in the EHR may be an ideal environment for a reemergence of unsupervised and semi-supervised learning approaches. Early examples show that deep autoencoders are adept at this task [26, 27].

3.4.3 Interpretation and the "Right to explanation"

For some clinical decisions, a black box algorithm with high accuracy is sufficient to improve medical care. An algorithm that can more accurately identify a tumor in imaging than a human has obvious benefits. For other problems, a black box is insufficient, it is unlikely to help researchers understand the physiology or etiology of a disease. In this setting, outside of a clinical decision, a less accurate but interpretable algorithm may be preferred. In addition, clinicians are likely to be skeptical and slower to adopt algorithms whose decisions cannot be rationally explained.

Furthermore, in April 2016, the European Union passed a data protection law entitled the "General Data Protection Regulation (GDPR) which will begin in 2018. The GDPR provides stricter conditions for sensitive data collection and storage, including, for example genetic and biometric data. It also sets regulation on privacy policies and further formalizes the "right to be forgotten." Of particular interest to the machine learning community is the language prohibiting decisions "based solely on automated processing and which produces adverse legal effects concerning, or significantly affects, him or her" and provides the right "to obtain an explanation of the decision reached after such assessment or to challenge the decision." It remains to be seen how this would be applied and if it will have any effect on the usage of artificial intelligence in the clinic but it demonstrates the fact that people want to understand how and why a decision affecting their wellbeing is made. Work to create high performing algorithms that provide interpretable, explainable decisions is increasingly important as clinicians increasingly rely on the aid of artificial intelligence.

References

1. Collins, F.S., Varmus, H.: A new initiative on precision medicine. N. Engl. J. Med. **363**, 1–3 (2010). doi:10.1056/NEJMp1002530
2. Bishop, C.M.: Pattern recognition and machine learning. Springer, Berlin (2006)
3. Kreybe, L.: Histological lung cancer types. A morphological and biological correlation. Acta Pathol Microbiol Scand Suppl **157**, 1–92 (1962)
4. Mountain, C.F.: Revisions in the international system for staging lung cancer. Chest **111**, 1710–1717 (1997). doi:10.1378/chest.111.6.1710
5. West, L., Vidwans, S.J., Campbell, N.P., et al.: A novel classification of lung cancer into molecular subtypes. PLoS ONE **7**, 1–11 (2012). doi:10.1371/journal.pone.0031906
6. Shin, J.-A., Lee, J.-H., Lim, S.-Y., et al.: Metabolic syndrome as a predictor of type 2 diabetes, and its clinical interpretations and usefulness. J Diabetes Investig **4**, 334–343 (2013). doi:10.1111/jdi.12075
7. Li, L., Cheng, W., Glicksberg, B.S., et al.: Identification of type 2 diabetes subgroups through topological analysis of patient similarity. Sci. Transl. Med. **7**, 1–16 (2015). doi:10.1126/scitranslmed.aaa9364
8. Lublin, F.D., Reingold, S.C., Cohen, J.A., et al.: Defining the clinical course of multiple sclerosis: The 2013 revisions. Neurology **83**, 278–286 (2014). doi:10.1212/WNL.0000000000000560
9. Denny, J.C., Ritchie, M.D., Basford, M.A., et al.: PheWAS: Demonstrating the feasibility of a phenome-wide scan to discover gene-disease associations. Bioinformatics **26**, 1205–1210 (2010). doi:10.1093/bioinformatics/btq126
10. Buyske, S., Yang, G., Matise, T.C., Gordon, D.: When a case is not a case: Effects of phenotype misclassification on power and sample size requirements for the transmission disequilibrium test with affected child trios. Hum. Hered. **67**, 287–292 (2009). doi:10.1159/000194981
11. Gordon D, Yang Y, Haynes C, et al: Increasing power for tests of genetic association in the presence of phenotype and/or genotype error by use of double-sampling. Stat Appl Genet Mol Biol. 3: Article 26 (2004). doi: 10.2202/1544-6115.1085
12. Manchia, M., Cullis, J., Turecki, G., et al.: The Impact of phenotypic and genetic heterogeneity on results of genome wide association studies of complex diseases. PLoS ONE **8**, 1–7 (2013). doi:10.1371/journal.pone.0076295
13. Labbe, A., Bureau, A., Moreau, I., et al.: Symptom dimensions as alternative phenotypes to address genetic heterogeneity in schizophrenia and bipolar disorder. Eur. J. Hum. Genet. **20**, 1182–1188 (2012). doi:10.1038/ejhg.2012.67
14. Chaste, P., Klei, L., Sanders, S.J., et al.: A genome-wide association study of autism using the Simons Simplex Collection: Does reducing phenotypic heterogeneity in autism increase genetic homogeneity? Biol. Psychiatry **77**, 775–784 (2015). doi:10.1016/j.biopsych.2014.09.017
15. Wiley, L.K., Vanhouten, J.P., Samuels, D.C., et al.: strategies for equitable pharmacogenomic-guided warfarin dosing among european and african american individuals in a clinical population. Pac Symp Biocomput **22**, 545–556 (2016)
16. Shaw, J.: The erosion of privacy in the internet era (2009)
17. Narayanan, A., Shmatikov, V.: Robust de-anonymization of large sparse datasets (2008)
18. Shokri, R., Stronati, M., Song, C., Shmatikov, V. Membership inference attacks against machine learning models (2016)
19. McSherry, F., Talwar, K.: Mechanism design via differential privacy. 48th Annual IEEE Symposium on Foundations of Computer Science (FOCS'07). IEEE, pp. 94–103 (2007)
20. Beaulieu-Jones, B.K., Wu, Z.S., Williams, C., Greene, C.S.: Privacy-preserving generative deep neural networks support clinical data sharing. bioRxiv (2017). doi:10.1101/159756
21. Dwork, C., Roth, A.: The algorithmic foundations of differential privacy. Found trends®. Theor Comput Sci **9**, 211–407 (2013). doi:10.1561/0400000042

22. Beaulieu-Jones, B.K., Greene, C.S.: Reproducibility of computational workflows is automated using continuous analysis. Nat Biotech **35**, 342–346 (2017)
23. Group TSR: A randomized trial of intensive versus standard blood-pressure control. N. Engl. J. Med. **373**, 2103–2116 (2015). doi:10.1056/NEJMoa1511939
24. Jensen, A.B., Moseley, P.L., Oprea, T.I., et al.: Temporal disease trajectories condensed from population-wide registry data covering 6.2 million patients. Nat Commun **5**, 1769–1775 (2014). doi:10.1038/ncomms5022
25. LeCun, Y., Bengio, Y., Hinton, G., et al.: Deep learning. Nature **521**, 436–444 (2015). doi:10.1038/nature14539
26. Beaulieu-Jones, B.K., Greene, C.S.: Semi-supervised learning of the electronic health record for phenotype stratification. J. Biomed. Inform. **64**, 168–178 (2016). doi:10.1016/j.jbi.2016.10.007
27. Miotto, R., Li, L., Kidd, B.A., et al.: Deep patient: An unsupervised representation to predict the future of patients from the electronic health records. Sci Rep **6**, 26094 (2016). doi:10.1038/srep26094
28. Khardori, R.M. Type 2 Diabetes Mellitus. PhekKB 1–24 (2014)
29. Ching, T. et al. Opportunities And Obstacles For Deep Learning In Biology And Medicine. bioRXiv. 102 (2017). doi: 10.1101/142760

Chapter 4
Defining and Discovering Interactive Causes

Xia Jiang and Richard Neapolitan

Abstract The problem of learning causal influences from passive data has attracted a good deal of attention in the past 30 years, and techniques have been developed and tested. These techniques assume the *composition property*, which entails that they cannot in general learn interactive causes with little marginal effects. However, such interactions are fairly commonplace. One notable example is genetic epistasis, which is the interaction of two or more genetic loci to affect phenotype. Often the genes exhibit little marginal effects. Another important example is the interaction of a treatment with patient features to affect outcomes. Even though efforts have recently been made towards developing new algorithms that discover such interactions from data, to our knowledge no definition of a discrete causal interaction has been forwarded. Using information theory, we develop a fuzzy definition of a discrete causal action, called *Interaction Strength* (*IS*). The *IS* is bounded above by 1 and equals 1 if the causes in the interaction exhibit no marginal effects. Using the *IS* and BN scoring, we develop an exhaustive search algorithm, Exhaustive-IGain, which learns interactions from low-dimension datasets, and a heuristic search algorithm, called MBS-IGain, which learns interactions from high-dimensional datasets. Using simulated high-dimensional datasets, based on models of genetic epistasis, we compare MBS-IGain to 7 algorithms that learn genetic epistasis from high-dimensional datasets, and show that MBS-IGain's discovery performance is notably better than the other methods. We apply MBS-IGain to a real LOAD dataset, and obtain results substantiating previous research and new results. Using low-dimensional simulated datasets, we show Exhaustive-IGain can learn 4-cause interactions with no marginal effects. We apply Exhaustive-Gain to a real clinical breast cancer datasets, and learn interactions that agree with the judgements of a breast cancer oncologist. Our algorithms are only directly applicable to problems where we have a specified target and its candidate causes. However, our algorithms

X. Jiang (✉)
Department of Biomedical Informatics, University of Pittsburgh, Pittsburgh, PA, USA
e-mail: xij6@pitt.edu

R. Neapolitan
Department of Preventive Medicine, Northwestern University Feinberg School of Medicine, Chicago, IL, USA

© Springer International Publishing AG 2018
D.E. Holmes and L.C. Jain (eds.), *Advances in Biomedical Informatics*,
Intelligent Systems Reference Library 137,
https://doi.org/10.1007/978-3-319-67513-8_4

could be used for general causal learning by being a front end to a standard causal learning algorithm.

Keywords Bayesian network · Interaction · Causal learning · Information gain · Entropy · Epistasis · SNP · GWAS

4.1 Introduction

The problem of learning causal influences from passive data has attracted a good deal of attention in the past 30 years, and techniques have been developed and tested. The constraint-based technique for learning Bayesian networks is a well-known method [1], and has been implemented in the Tetrad package [2]. This method orients edges which are compelled to be causal influences. Another method for learning Bayesian networks is the *greedy equivalent search* (*GES*) [3], which does not in itself distinguish which edges are compelled to be causal. However, post-processing of its resultant network can compel edges. Both these (and other) strategies assume the *composition property*, which states that if a variable Z and a set of variables S are not independent conditional on T, then there exists a variable X in S such that X and Z are not independent conditional on T [3]. When T is the empty set, this property simply states if Z and S are not independent then there is an X in S such that Z and X are not independent. So, at least one variable in S much be correlated with Z. However, if two or more variables interact in some way to affect Z, there could be little marginal effect for each variable, and the observed data could easily not satisfy the composition property. Furthermore, if interacting variables have strong marginal effects, the causal learning algorithms do not distinguish them as interactions, but only as individual causes.

So, the standard methods for learning causal influences do not learn that causes are interacting to cause a target, and do not even discover causes that are interacting with little or no marginal effect. An important task then is to learn such interactions from data. A method that does this could be a preliminary step before applying a standard causal learning algorithm.

We make progress towards developing algorithms that learn causal interactions from data by first developing a fuzzy definition of a discrete causal interaction. To our knowledge, such a definition has not been previously forwarded. Using that definition, we then develop algorithms that learn discrete interactions from data. Before proceeding, we provide some examples of situations where discrete variables interact.

An example, which has recently received a lot of attention, is gene-gene interactions, called *epistasis*. Biologically, epistasis describes a situation where a variant at one locus prevents the variant at a second locus from manifesting its effect [4]. Epistasis between n loci is called *pure* epistasis if none of the loci individually are predictive of phenotype and is called *strict* epistasis if no proper multi-locus subset of the loci is predictive of phenotype [5]. Epistasis has been defined statistically as a deviation from additivity in a model summarizing the relationship between multi-locus genotypes and phenotype [6]. It is believed that much of genetic risk for disease is due to epistatic interactions [7–10]. A *Single nucleotide polymorphism* (*SNP*) is a substitution of one base for another. *Genome-wide association studies* (*GWAS*) investigate many SNPs, often numbering in the millions, along with a phenotype such as disease status. By investigating single-locus associations, researchers have identified over 150 risk loci associated with 60 common diseases and traits [11–14]. However, these single-locus investigations would miss epistatic interactions with little marginal effect.

Another important example is the interaction of clinical or genomic variables with treatments to affect patient outcomes. For example, Herceptin is a treatment for breast cancer patients which is effective for HER2 + patients. So, Herceptin and HER2 status interact to affect survival. This is a well-known relationship. However, we now have large scale breast cancer and other datasets [15] from which we can learn treatment-variable interactions that are not yet known. This knowledge will enable us to better provide precision medicine.

As another example, we are now obtaining abundant hospital data concerning workflow. These data can be analysed to determine good personnel combinations and sequencing [16].

4.2   Background

Our discussion assumes a knowledge of Bayesian networks and information gain. So we first provide background on these two.

4.2.1   Bayesian Networks

Bayesian networks [17–20] are an important architecture for reasoning under uncertainty in machine learning. They have been applied to many domains including biomedical informatics [21–26]. A *Bayesian network* (*BN*) represents a joint probability distribution by a *directed acyclic graph* (*DAG*) $G = (V, E)$, where the nodes in V are random variables and the edges in E represent relationships among the variables, and by the conditional probability distribution of every node $X \in V$ given every combination of values of the node's parents. The edges in the DAG often represent causal relationship [17]. A BN modeling causal relationship among variables related to respiratory diseases appears in Fig. 4.1. In a BN each node in $X \in V$ is probabilistically independent of all other nodes in the network conditional on the parents of X.

Using a BN, we can determine probabilities of interest with a BN inference algorithm [17]. For example, using the BN in Fig. 4.1, if a patient has a smoking history (H = yes), a positive chest X-ray (X = pos), and a positive CAT scan

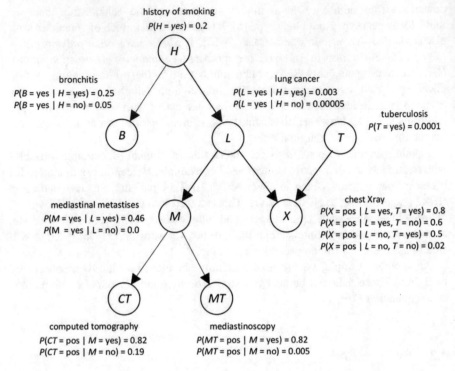

history of smoking
$P(H = yes) = 0.2$

H

bronchitis
$P(B = yes \mid H = yes) = 0.25$
$P(B = yes \mid H = no) = 0.05$

lung cancer
$P(L = yes \mid H = yes) = 0.003$
$P(L = yes \mid H = no) = 0.00005$

tuberculosis
$P(T = yes) = 0.0001$

B   L   T

mediastinal metastises
$P(M = yes \mid L = yes) = 0.46$
$P(M = yes \mid L = no) = 0.0$

M   X

chest Xray
$P(X = pos \mid L = yes, T = yes) = 0.8$
$P(X = pos \mid L = yes, T = no) = 0.6$
$P(X = pos \mid L = no, T = yes) = 0.5$
$P(X = pos \mid L = no, T = no) = 0.02$

CT   MT

computed tomography
$P(CT = pos \mid M = yes) = 0.82$
$P(CT = pos \mid M = no) = 0.19$

mediastinoscopy
$P(MT = pos \mid M = yes) = 0.82$
$P(MT = pos \mid M = no) = 0.005$

Fig. 4.1 A Bayesian network modeling relationships among variables related to respiratory disorders

(CT = pos), we can determine the probability of the patient having lung cancer (L = yes). That is, we can compute $P(L = yes \mid H = Yes, X = pos, CT = pos)$. Inference in BNs is NP-hard [27]. So, approximation algorithms are often employed [17].

Learning a BN from data concerns learning both the parameters and the structure (called a DAG model). In the score-based structure-learning approach, a score is assigned to a DAG based on how well DAG model G fits the. *Data* The Bayesian score [28] is the probability of the *Data* given G This score, which uses a Dirichlet distribution to represent prior belief concerning each conditional probability distribution in the BN, follows:

$$score_{Bayes}(G : Data) = P(Data|G)$$
$$= \prod_{i=1}^{n} \prod_{j=1}^{q_i} \frac{\Gamma(\sum_{k=1}^{r_i} a_{ijk})}{\Gamma(\sum_{k=1}^{r_i} a_{ijk} + \sum_{k=1}^{r_i} s_{ijk})} \prod_{k=1}^{r_i} \frac{\Gamma(a_{ijk} + s_{ijk})}{\Gamma(a_{ijk})}$$

where n is the number of variables in the model, r_i is the number of states of X_i, q_i is the number of different values that the parents of X_i can jointly assume, a_{ijk} is a hyperparameter, and s_{ijk} is the number of times X_i assumed its k th value when the parents of X_i assumed their j th value. When, $a_{ijk} = \alpha/r_i q_i$ where α represents a prior equivalent sample size, we call the Bayesian score the *Bayesian Dirichlet equivalent uniform* (*BDeu*) score [29].

It has been shown that the problem of learning a BN DAG model from data is NP-hard [30]. Resultantly, heuristic search algorithms have been developed [17].

4.2.2   Information Gain

Information theory [31] concerns the quantification and communication of information. Given a discrete random variable Z with m alternatives, the *entropy* $H(Z)$ is defined as follows:

$$H(Z) = -\sum_{i=1}^{m} P(z_i) \log_2 P(z_i).$$

If we repeat n trials of the experiment having outcome Z, then it is possible to show that the entropy $H(Z)$ is the limit as $n \to \infty$ of the expected value of the number of bits needed to report the outcome of every trial. Entropy provides a measure of our uncertainty in the value of Z in the sense that, as entropy increases, it takes more bits on the average to resolve our uncertainty. Entropy achieves its maximum value when $P(z_i) = 1/m$ for all z_i, and its minimum value (0) when $P(z_j) = 1$ for some z_j.

The expected value of the entropy of Z given random variable X is called the conditional entropy of Z given.X We denote conditional entropy as $H(Z \mid X)$. Mathematically, we have

$$H(Z|X) = \sum_{j=1}^{k} H(Z|x_j)P(x_j),$$

where X has k alternatives. Knowledge of the value of X can reduce our uncertainty in Z. The *Information Gain* of Z relative to X is defined to be the expected reduction in the entropy of Z given Z:

$$IG(Z;X) = H(Z) - H(Z|X).$$

The notion of *IG* extends readily to a set of random variables A, and the Information Gain of Z relative to A is denoted $IG(Z;A)$. Although for simplicity we show results using single random variables, they apply to sets of random variables.

The following are some important properties of Information Gain:

IG1: $IG(Z;X) \geq 0$ with equality holding if and only if Z and X are independent.
IG2: $IG(Z;X) = IG(X;Z)$.
IG3: $IG(Z;X,Y) = IG(Z;X|Y) + IG(Z;Y)$. (chain rule for Information Gain).

IG4: $IG(Z; X, Y) \geq IG(Z; Y)$.

The 4th property follows easily from the 1st and 3rd ones.

4.3   Discrete Causal Interactions

Next we develop a definition of a discrete causal interaction based on information gain.

4.3.1   Statistical Interactions

In statistics, the standard definition of an interaction is a relationship where the simultaneous influence of two or more variables on a target variable is not additive. However, when we leave the domain of regression and deal with the type of non-linear discrete interactions discussed in Sect. 4.1, this definition is limited. For example, researchers have developed the Noisy-Or model to combine the effect of binary causes that are independently causing a binary target [17]. We would either not call this relationship an interaction, or at most call it a weak interaction; yet the rule for combining the individual effects is not additive. When variables combine to affect a target with no marginal effect (e.g. pure, strict epistasis), we definitely can say there is an interaction. Fig. 4.2 shows Bayesian networks illustrating these two disparate situations.

Fig. 4.2a shows a causal relationship with no marginal effects. That is,

$$P(z_1|x_1) = 0 \times 0.25 + 0.1 \times 0.5 + 0 \times 0.25 = 0.05$$
$$P(z_1|x_2) = 0.1 \times 0.25 + 0.0 \times 0.5 + 0.1 \times 0.25 = 0.05$$
$$P(z_1|x_3) = 0.0 \times 0.25 + 0.1 \times 0.5 + 0.0 \times 0.25 = 0.05.$$

By the symmetry of the problem, we see the same result holds for Y. Fig. 4.1b shows a causal relationship developed with the Noisy-Or model. That model assumes each cause has a causal strength that independently affects the target. See [17] for the details of the assumptions. In this case the causal strength of X is $p_x = 0.9$ and the causal strength of Y is $p_y = 0.9$. From these causal strengths, the Noisy-Or model computes the conditional probabilities of Z as follows:

$$P(z_1|x_1, y_1) = 1 - (1-0.9)(1-0.9) = 0.99$$
$$P(z_1|x_1, y_2) = 1-(1-0.9) = 0.9$$
$$P(z_1|x_2, y_1) = 1 - (1-0.9) = 0.9$$
$$P(z_1|x_2, y_2) = 1-1 = 0$$

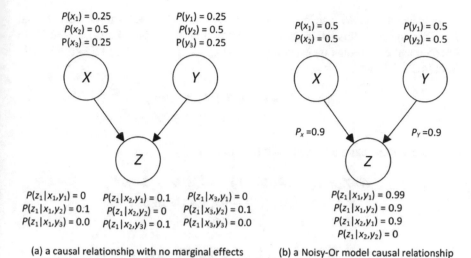

(a) a causal relationship with no marginal effects   (b) a Noisy-Or model causal relationship

Fig. 4.2 Bayesian networks modeling causal relationships

The examples just shown are two extreme cases, providing us with clear examples of a strong interaction and at most a weak interaction. However, in general, there does not appear to be a dichotomous way to classify a discrete causal relationship as an interaction or a non-interaction. So, we propose a fuzzy set membership definition of a discrete causal interaction in the following sections.

4.3.2   Interaction Dividend

Let $IG(Z;X,Y)$ denote the Information Gain of Z relative to the joint probability distribution of X and Y. The *Interaction Dividend* (*ID*) of X and Y relative to Z as then defined as follows:

$$ID(Z; X, Y) = IG(Z; X, Y) - IG(Z; X) - IG(Z; Y).$$

Let $IG(Z;A)$ denote the Information Gain of Z relative to the joint distribution of all variables in set A. The *ID* of a set of variables A and a set of variables B is then defined as follows:

$$ID(Z; A, B) = IG(Z; A \cup B) - IG(Z; A) - IG(Z; B).$$

Interaction Dividend provides a measure of the increase in Information Gain obtained when A and B are known together relative to knowing each of them separately.

In general, the *ID* can be positive or negative. Fig. 4.3 shows three causal BN DAG models illustrating some of the possibilities. In Fig. 4.3 (a), X and Z are independent conditional on Y. We then have by Properties IG1 and IG3 (defined in Sect. 4.2.2) above

$$IG(Z;X,Y) = IG(Z;X|Y) + IG(Z;Y)$$
$$= 0 + IG(Z;Y) = IG(Z;Y).$$

We therefore have by Property IG1 above

$$ID(Z;X,Y) = IG(Z;X,Y) - IG(Z;X) - IG(Z;Y)$$
$$= IG(Z;Y) - IG(Z;X) - IG(Z;Y)$$
$$= -IG(Z;X) \leq 0.$$

On the other hand, in Fig. 4.2(b) X and Y are independent causes of Z. If X and Y are independent, we have the following by Properties IG1 and IG3

$$IG(X;Z|Y) = IG(X;Z,Y) - IG(X;Y)$$
$$= IG(X;Z,Y).$$

So, by Properties IG2, IG3, and IG4

$$ID(Z;X,Y) = IG(Z;X,Y) - IG(Z;X) - IG(Z;Y)$$
$$= IG(Z;X|Y) + IG(Z;Y) - IG(Z;X) - IG(Z;Y)$$
$$= IG(X;Z,Y) - IG(Z;X)$$
$$= IG(X;Z,Y) - IG(X;Z)$$
$$\geq 0.$$

It is a straightforward extension that if we have a set of variables A and B such that A and B are independent, then

$$ID(Z;A,B) \geq 0.$$

We stress that, in general, a high value of the Interaction Dividend does not imply that the investigated variables are causes of the target. Consider the causal DAG model in Fig. 4.2 (c). Since this complete DAG can represent any joint probability distribution of three variables, we could have the same Interaction Dividend with this underlying causal mechanism as we would have when X and Y have a strong interactive causal effect on Z. So, in an agnostic search for interactions, we cannot assume that a discovered interaction is causal. For example, if we are searching for interactions when developing a BN DAG model and make every variable a target, we cannot assume that what appears to be a discovered interaction is a causal interaction.

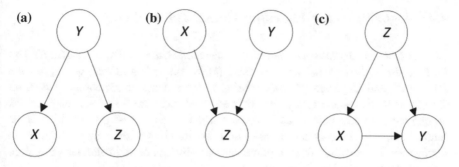

Fig. 4.3 Causal BN DAG models

4.3.3 Interaction Strength: A Discrete Causal Interaction Defined

By causal interaction model M we shall mean a set of possible interacting causes of target Z. When $IG(Z;M) \neq 0$, we define the *interaction strength* (IS) of causal model M for effect Z as follows:

$$IS(Z;M) = \min_{A \subseteq M} \frac{ID(Z;M-A,A)}{IG(Z;M)} = \min_{A \subseteq M} \frac{IG(Z;M) - IG(Z;M-A) - IG(Z;A)}{IG(Z;M)}$$

Since Information Gain (IG) is nonnegative, it is straightforward that $IS(Z; M) \leq 1$. If the causes in M are causing Z with no marginal effects (e.g. pure, strict epistasis), then the IS equals 1. We would consider this a very strong interaction. When the IS is small, the increase in IG obtained by considering the variables together is small compared to considering them separately. We would consider this a weak interaction.

We showed above that if the variables in M are independent causes of Z, then

$$ID(Z;M-A,A) \geq 0.$$

So, in situations where all causes in model M are independent, the IS is between 0 and 1, and therefore satisfies the notion of a fuzzy set [32], where the greater the value of the IS the greater membership the model has in the fuzzy set of interactions.

The ID and IS can be used to discover causal interactions. In this next section we develop algorithms for learning causal interactions that use the ID and the IS.

4.4 Algorithms for Learning Causal Interactions

We present the *Exhaustive Search Information Gain* (*Exhaustive-IGain*) and *Multiple Beam Search Information Gain* (*MBS-IGain*) algorithms, which use the *ID*, the *IS*, and Bayesian network scoring to learn causal interactions of a target *Z* from a set of possible causes *PA*. It is assumed there are no causal relationships among the candidate causes, as is the case when we are investigating which SNPs interact epistatically to affect disease status. So, these algorithms make no effort to distinguish direct causes from indirect causes, which is done in standard Bayesian network causal learning algorithms.

Exhaustive-IGain does an exhaustive search over all possible causal models, and is applicable when the size of *PA* is not large. MBS-IGain does a heuristic search, and is applicable when the size of *PA* is large. We first presented Exhaustive-IGain in [33] and MBS-IGain in [34].

The Exhaustive-IGain algorithm appears in Fig. 4.4. The $score(Z;M)$ in Algorithm MBS-IGain is the BDeu score of the DAG model that has the causes in M being parents of the target Z. The parameter R is the maximum size of the models we are considering. PA is the set of all candidate causes. For each subset $M \subset PA$ of size between 2 and R, Exhaustive-IGain checks every proper subset $A \subset M$ to see if the ratio of $ID(Z;M - A,A)$ to $IG(Z;M)$ exceeds a threshold T. In this way it makes certain that the *IS* exceeds T. It also checks that M yields a higher score than both A and M-A. If M passes these tests for every subset, then M is reported as an interaction. The latter criterion is included because we not only want to discover causes that seem to be interacting, but we also want to discover probable models. On the other hand, the check for a sufficiently large *IS* is performed because a set of causes could score very high as parents of Z when there is no interaction. For example, if X and Y each have strong causal strengths for Z but affect Z independently, the model with them as parents of Z would score high. The Noisy-OR model, discussed in Sect. 4.3.1, is such a model. In this situation the model $X \rightarrow Z \leftarrow Y$ would have a high score without there being a significant interaction.

When the size of the set PA of candidate causes is not small, we cannot exhaustively search all subsets. In this case we need a heuristic search algorithm that searches over promising subsets. The Multiple Beam Search (MBS)-IGain algorithm, which appears in Fig. 4.5, is such an algorithm. Initially, MBS-IGain choses the most promising causes using the scoring criterion. A beam is then started from each of these causes. On each beam, the cause, which has the highest *ID* with the set of causes chosen so far, is greedily chosen. The search ends when either the *ID* is small relative to the *IG* of the model (based on a threshold T), indicating that the *IS* would be small, or when adding the cause decreases the score of the model. We could include a parameter R in MBS-IGain which limits the size of the interactions we consider.

Algorithm *Exhaustive-IGain*

PA = set of all investigated causes of Z;
$Models = \emptyset$;
for every subset M of PA containing between 2 and R predictors
   $flag = 0$;
   for every proper subset $A \subset M$
      if

$$\frac{ID(Z;M - A, A)}{IG(Z;M)} < T \text{ or } score(Z;M) < \max(score(Z;M - A), score(Z;A))$$

         $flag = 1$;
      endif
   endfor
   if $flag = 0$
      add M to $Models$;
   endif
endfor

Fig. 4.4 The Exhaustive-IGain algorithm

Algorithm *MBS-IGain*

PA = set of all investigated causes of Z;
$Models = \emptyset$;
Determine the subset $Best$ of PA containing the n highest scoring causes Y using $score(Z;Y)$;
for each predictor $Y \in Best$
   $M = \{Y\}$;
   $flag = 0$;
   while $flag = 0$
      determine predictor X that maximizes $ID(Z; X, M)$;
      if

$$\frac{ID(Z;X,M)}{IG(Z;X,M)} \leq T \quad \text{or} \quad score(Z;X,M) < score(Z;M)$$

         $flag = 1$;
      else
         add X to M;
      endelse
   endwhile
   while some deletion increases $score(Z;M)$
      delete the cause from M whose deletion increases the score the most;
   endwhile
   if $size(M) \geq 2$
      add M to $Models$;
   endif
endfor

Fig. 4.5 The MBS-IGain algorithm

4.5 Reporting the Noteworthiness of an Interaction

Once we discover an interaction, we need to report its noteworthiness. First, we report its *IS* to indicate its strength as an interaction. However, if the model is unlikely, it is still not very noteworthy even if the *IS* is large. So, we also need to in some way report the significance of the model. Standard *p*-values are not very informative because there is more than one null hypothesis. Consider Fig. 4.6, which shows the DAG model M_{XY} in which X and Y are both parents of Z. The three competing models are on the right. Model M_0 represents that neither variable is a parent of Z, Model M_X has X as a parent of Z and Y not as a parent of Z, and model M_Y has Y as a parent of Z and X not as a parent of Z. Standard statistical techniques do not investigate these multiple competing hypotheses. They only pit the null hypothesis M_0 against M_{XY}. However, if either model M_X or M_Y were the correct model, we would obtain an association of the two variables together with Z (and therefore reject M_0) even though M_{XY} is incorrect. Towards addressing this difficulty, Jiang et al. [35] developed the *Bayesian network posterior probability* (*BNPP*), which provides the posterior probability of a DAG model that has an arbitrary number of parents of a target Z. For the two-parent model, M_{XY} the BNPP is as follows:

$$P(M_{XY}|Data) = \frac{P(Data|M_{XY})P(M_{XY})}{P(Data|M_{XY})P(M_{XY}) + P(Data|M_0)P(M_0) + \sum_k P(Data|M_k)P(M_k)},$$

where k sums over the two 1-cause models. The BNPP extends to larger models, but the number of competing hypotheses grows exponentially with size of the model. However, in general, we usually don't learn an interaction with more than 5 causes. Jiang et al. [35] discuss and provide prior probabilities in the case of interactions learned from GWAS datasets.

4.6 Experiments

Next we discuss experiments evaluating the performance of MBS-IGain and Exhaustive-IGain.

4.6.1 MBS-Gain

As discussed in Sect. 4.1 epistasis is the interaction of two more genes/SNPs to affect phenotype. Biologically, epistasis is believed to occur when the effect of one gene is modified by one or more other genes. SNPs can interact epistatically with little marginal effect. Furthermore, high-dimensional GWAS datasets can concern

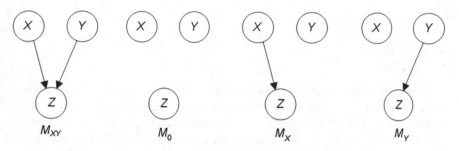

Fig. 4.6 The model that X and Y are both parents of Z is on the left, and its three competing models are on the right

millions of SNPs, and provide researchers unprecedented opportunities to investigate the complex genetic basis of diseases. By looking at single-locus associations using standard statistical analyses, researchers have identified over 150 risk loci associated with 60 common diseases and traits [11–14]. However, such analyses will miss SNPs that are interacting epistatically with little marginal effect. So, researchers endeavored to develop methods for learning epistasis from high-dimensional GWAS datasets. Traditional techniques such as logistic regression (LR) [36], logistic regression with an interaction term (LRIT) [37], penalized logistic regression [38] and Lasso [39] were applied to the task. Other techniques include multifactor dimensionality reduction (MDR) [40], full interaction modeling (FIM) [41], using information gain (IG) alone to investigate only 2-SNP interactions [42], SNP Harvester (SH) [43], the use of ReliefF [44], random forests [45], predictive rule inference [46], Bayesian epistasis association mapping (BEAM) [47], maximum entropy conditional probability modeling (MECPM) [48], and Bayesian network learning [49, 50].

We evaluated MBS-IGain by investigating how well it detects epistatic interactions from simulated datasets and real data.

Simulated Data: Using simulated datasets, we evaluated the performance of MBS-IGain and 7 other methods according to their ability to detect epistatic interactions from GWAS datasets. The 7 methods are as follows: logistic regression (LR) [36], multifactor dimensionality reduction (MDR) [40], full interaction modeling (FIM) [41], simple information gain (IG) [42], SNP Harvester (SH) [44], Bayesian epistasis association mapping (BEAM) [47], and maximum entropy conditional probability modeling (MECPM) [48]. First we describe the simulated dataset; then we present the results of the evaluation.

Chen et al. [51] generated datasets based on two 2-SNP interactions, two 3-SNP interactions, and one 5-SNP interaction, making a total of 15 causative SNPs. The effects of the interactions were combined using the Noisy-OR model [17]. The specific interacting SNPs were as follows:

1. {S1, S2, S3, S4, S5}

2. {S6, S7, S8}
3. {S9, S10, S11}
4. {S12, S13}
5. {S14, S15}

Three parameters were varied to create the interactions: 1) θ, which determined the penetrance; 2) β, which determined the minor allele frequency; and l, which determined the linkage disequilibrium of the true causative SNPs with the observed SNPs. See [51] for details concerning these parameters. For various combinations of these parameters, Chen et al. [51] developed datasets containing 1000 cases and 1000 controls. In our evaluation, we used the 100 1000-SNP datasets they developed with $\theta = 1$, $\beta = 1$, and $l = null$.

We applied the other 7 methods and MBS-IGain to these datasets. We used the BDeu score with $\alpha = 4$ and the threshold $T = 0.1$ in MBS-IGain.

The results were compared using power definition 1, which we describe next. This power definition measures how well the method discovers the causes in the interactions, but does not concern itself with whether the method discovers the actual interactions. First, the learned interactions are ordered by their scores. Then each cause is ordered according to the first interaction in which it appears. Finally, the power is computed as follows:

$$Power_1(K) = \frac{1}{H \times M} \sum_{i=1}^{H} N_K(i) \qquad (4.1)$$

where $N_K(i)$ is the number of true interacting causes appearing in the first K causes learned for the ith dataset, M is the total number of interacting causes in all interactions, and H is the number of datasets. In our comparison experiments using the 100 1000 SNP datasets $M = 15$ and $H = 100$.

Fig. 4.7 shows the results of the comparison. MBS-IGain performed notably better than all other methods.

Real Data: Reiman et al. [52] developed a GWAS *late onset Alzheimer's disease (LOAD)* data set that concerns data on 312,260 SNPs and contained records on 859 cases and 552 controls. We applied MBS-IGain to this dataset to investigate how well it can learn interactions in a real setting. The models investigated were causal models with SNPs as causes and LOAD status as the effect. In order to apply MBS-IGain to this dataset, it was necessary to first filter the SNPs because 312,260 is too many SNPs to handle. One of the 312,260 loci is the APOE gene; however, we still use the terminology "SNP" to refer to the loci. We filtered by choosing the 1000 top-scoring 1-SNP models, the 5000 top-scoring 1-SNP models, and the 10,000 top-scoring 1-SNP models, and ran MBS-IGain with each of these sets of SNPs. We also ran the top-scoring 100 SNP models with all the remaining SNPs. We used the BDeu Score with $\alpha = 4$, and a threshold $T = 0.1$. We limited the size of the models to 5, which means at most 4 SNPs were added on each beam.

MBS-IGain discovered a total of 14,818 unique models. Fig. 4.8a shows a histogram of the BDeu scores and Fig. 4.8b shows the percentile distribution. We

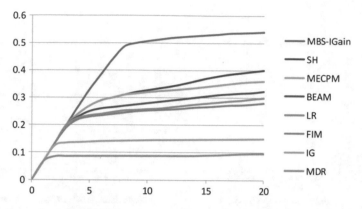

Fig. 4.7 Comparison of MBS-IGain to 7 other methods using the power definition in Eq. 4.1. The number of discovered SNPs K is plotted on the x-axis, and $Power_1(K)$ is plotted on the y-axis. SNP Harvester (SH), maximum entropy conditional probability modeling (MECPM), Bayesian epistasis association mapping (BEAM), logistic regression (LR), full interaction modeling (FIM), information gain (IG), multifactor dimensionality reduction (MDR)

see that the vast majority of the BDeu scores approximately follow a normal distribution; however, there are 14 outliers with much higher scores. The outliers constitute notable findings.

Table 4.1 shows the notable findings that were discovered, along with their BDeu scores. The APOE gene is the strongest genetic predictor of LOAD. We've included the model including only APOE as a parent of LOAD, by way of comparison. The notable findings distribute into two groups. The first group contains higher-scoring interactions involving the APOE gene, and the second group contains lower-scoring interactions involving the APOC1 gene. APOE and APOC1 are in linkage disequilibrium, and APOC1 predicts LOAD almost as well as APOE [53]. Our results indicate that every notable interaction includes one of them.

All the notable findings can provide LOAD researchers with candidate interactions which they can investigate further. We discuss some of the more interesting ones. The highest scoring interaction is a 4-locus interaction containing GAB2, rs10510511, and rs197899; and the second highest scoring interaction is a 3-locus interaction containing GAB2, SPAG16, and APOE. A good deal of previous research has indicated that GAB2 and APOE interact to affect LOAD [52]. Our results support that there might be other loci involved in the GAB2/APOE interaction. We know of no previous research indicating this. Furthermore, there is previous research indicating APOE and the SPAG16 gene interact to affect LOAD, but not with GAB2 [54].

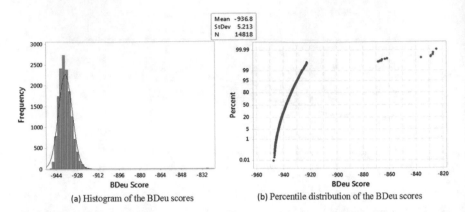

(a) Histogram of the BDeu scores   (b) Percentile distribution of the BDeu scores

Fig. 4.8 Histogram and percentile distribution when determining interactive models from the LOAD dataset

4.6.2   Exhaustive-Gain

If we have three or more causes interacting with no marginal effects (e.g. pure, strict epistasis), MBS-IGain would only discover the interaction by chance, even if we started a beam from every cause. The reason is that any 2-cause models sub-model of the given interaction would exhibit no marginal effect, and therefore would in general not score high in the first iteration of MBS-IGain. Exhaustive-IGain looks at all submodels, and therefore should not have this problem. To test this, we evaluated Exhaustive-IGain using simulated models with no marginal effects. We then applied Exhaustive-IGain to a real breast cancer clinical dataset. We discuss each of these next.

Simulated Data: We evaluated Exhaustive-IGain by comparing it to MBS-IGain using the simulated datasets discussed next. Urbanowicz et al. [55] created GAMETES, which is a software package for generating pure, strict epistatic models with random architectures. We used GAMETES to develop 2-SNP, 3-SNP, and 4-SNP models of pure epistatic interaction. That is, there are no marginal effects. The software allows the user to specify the *heritability* and the *minor allele frequency (MAF)*. We used values of heritability ranging between 0.01 and 0.2, and values of MAF ranging between 0.1 and 0.4. Using these values, we generated 16 datasets based on pure, strict 2-SNP interactions, 16 datasets based on pure, strict 3-SNP interactions, and 16 datasets based on pure, strict 4-SNP interactions. The 2-SNP and 3-SNP based datasets contained 1000 cases and 1000 controls, and the 4-SNP based datasets contained 5000 cases and 5000 controls.

We used both MBS-IGain and Exhaustive-IGain to analyze both sets of datasets. We ran both algorithms with all combination of the following values of the threshold T in the algorithms: $T = 0.1, 0.2$; and the parameter α in the BDeu score: $\alpha = 9, 54, 128$.

Table 4.1 Interactions MBS learned from the LOAD dataset using the BDeu score with $\alpha = 4$ and the threshold $T = 0.1$. The third column shows gene on which the SNP resides if it is located in a gene; otherwise it shows the chromosome. The fourth column shows the BDeu score of the interaction

Rank	Interaction	Genes	BDeu
1	rs10510511, rs197899, rs7115850, APOE	Chrome 3, Chrome 6, GAB2, APOE	- 824.6
2	rs11895074, rs7115850, APOE	SPAG16, GAB2, APOE	- 827.1
3	rs536128, rs7115850, APOE	CALN1, GAB2, APOE	- 827.3
4	rs7101429, rs10510511, rs197899, APOE	GAB2, Chrome 3, Chrome 6, APOE	- 828.7
5	rs11122116, rs16856748, rs734600, rs16992170, APOE	NPHP4, LRP2, EYA2, EYA2, APOE	- 828.9
6	APOE	APOE	- 836.3
7	rs41369150, rs2265264, rs11217838, rs4420638	FNDC3B, Chrome 10, ARHGEF12, APOC1	- 861.8
8	rs7355646, rs41369150, rs4420638	Chrome 2, FNDC3B, APOC1	- 863.4
9	rs7355646, rs41528844, rs4420638	Chrome 2, ADAMTS16, APOC1	- 865.6
10	rs2265264, rs4420638	Chrome 10, APOC1	- 865.9
11	rs41369150, rs4420638, rs6121360	FNDC3B, APOC1, TM9SF4	- 866.1
12	rs10922885, rs7355646, rs4420638	Chrome 1, Chrome 2, APOC1	- 867.3
13	rs41369150, rs4420638	FNDC3B, APOC1	- 867.4
14	rs7355646, rs4420638	Chrome 2, APOC1	- 868.6

The methods were compared using power definition 2, which we describe next. This power definition measures how well a method discovers each of the interactions. It uses the Jaccard index which is as follows:

$$Jaccard(A, B) = \frac{\#(A \cap B)}{\#(A \cup B)}.$$

The Jaccard index equals 1 if the two sets are identical and equals 0 if their intersection is empty. The criterion provides a separate measure for each true interaction. The learned interactions are first ordered by their scores for each dataset i. Denote the jth learned interaction in the ith dataset by $M_j(i)$, and denote the true interaction we are investigating by C. For each i and j we compute $Jaccard(M_j(i), C)$. We then set

$$J_K(i, C) = \max_{1 \leq j \leq K} Jaccard(M_j(i), C),$$

The power for interaction C is then computed as follows:

$$Power_2(K, C) = \frac{1}{H \times M} \sum_{i=1}^{H} J_K(i, C) \qquad (4.2)$$

where H is the number of datasets and M is the total number of interacting causes in interaction C.

Fig. 4.9 shows $Power_2(K,C)$ for Exhaustive-IGain and MBS-IGain for $K \leq 25$ for each of the three pure epistatic interactions. We see from Fig. 4.9a that both methods discover the 2-SNP interaction very well. In fact Exhaustive-IGain ranked the correct interaction first in 15 of the datasets and 3rd in the remaining dataset, while MBS-IGain ranks it first in 15 of the datasets and 4th in the remaining dataset (This information is not in the figure). In the case of a 2-SNP interaction, MBS-IGain effectively does an exhaustive search, explaining why it performs almost as well as Exhaustive-IGain. Its slightly worse performance is due to its different exit criteria concerning the score. It stops adding causes when no cause increases the score. On the other hand, Exhaustive-IGain checks whether any sub-model has a higher score than the model being considered. Exhaustive-IGain achieves this performance with very few false discoveries. The average number of interactions discovered by Exhaustive-IGain is 2.0. On the other hand, the average number of interactions discovered by MBS-IGain is 4.75.

Fig. 4.9b shows that Exhaustive-IGain also discovers the 3-SNP interactions extremely well, while MBS-IGain exhibits poor performance. This poor performance is to be expected. That is, when there are no marginal effects, if {S1,S2,S3} is our interaction, S2 or S3 would be chosen first on the beam initiating from S1 only by chance. In general, Exhaustive-IGain exhibited this good performance with a low false positive rate. The average number of interactions discovered for 15 of the datasets was 2.47. However, for one of the dataset 100 interactions (the maximum reported) were identified.

As Fig. 4.9c shows, Exhaustive-IGain performed well for the 4-SNP interactions, but not as well it did for the smaller models. This result indicates that higher order interactions are more difficult to discover. As expected, MBS-IGain again showed very poor performance. For 14 of the datasets, the average number of interactions discovered by Exhaustive-IGain was 1.85. However, for two of the datasets, 100 interactions were discovered.

Real Data: The METABRIC data set [15] has clinical data and outcomes for 1981 primary breast cancer tumors. Table 4.2 shows the clinical variables and their values used in our analysis. The data in three of these variables were transformed

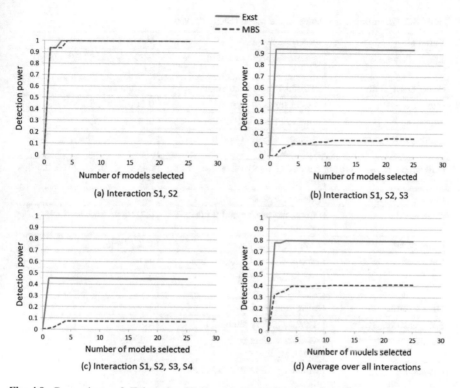

Fig. 4.9 Comparison of Exhaustive-IGain and MBS-IGain, when analysing the simulated datasets based on pure epistatic interactions with no marginal effects using the power definition in Eq. 4.2. The number of models selected is plotted on the x-axis, and $Power_2(K)$ is plotted on the y-axis

from their original METABRIC values using domain knowledge and the equal distribution discretization strategy. The transformations follow:

age_at_diagnosis: This variable was discretized to the five ranges shown using the equal distribution discretization technique and breast cancer expert knowledge.

size: This variable was discretized to the three standard ranges shown.

lymph_nodes_positive: This variable was grouped into the six ranges shown.

The outcome variable is whether the patient died from breast cancer. If the person was known to die from breast cancer, the days after initial consultation that the patient died is recorded. If the person was not known to die from breast cancer, the days after initial consultation that the patient was last seen alive or died from another cause is recorded. If a patient was known to die from breast cancer within *x* years after initial consultation or is known to be alive *x* years after initial consultation, we say their breast cancer survival status is known *x* years after initial consultation. These data provide us with 1698 patients whose breast cancer survival status is known 5 years after initial consultation, 1228 patients whose breast cancer

Table 4.2 The clinical variables in the METABRIC dataset

Variable	Description	Values
age_at_diagnosis	age at diagnosis of the disease	0-39, 39-54, 54-69, 69-84, 84-100
menopausal_status	inferred menopausal status	pre, post
size	size of tumor in cm	0-20, 20-50, 50-180
lymph_nodes_positive	number of positive lymph nodes	0, 1, 2-3, 4-5, 6-9. $\geq$ 10
lymph_nodes_removed	number of lymph nodes removed	0, 1-3, 4-9, 10-20, $\geq$ 21
percent_nodes_positive	percent of removed nodes positive	0-0.2, 0.2-0.4, 0.4-0.6, 0.6-0.8, 0.8-1
grade	grade of disease	1, 2, 3
stage	composite of size and # positive nodes	0,1,2,3,4
histological	tumor histology	IDC, Other
ER_Expr	estrogen receptor expression	+, -
PR_Expr	progesterone receptor expression	+, -
HER2_status	HER2 expression	+, -
P53_mutation_status	whether P53 is mutated	+, -
chemo	whether patient had chemotherapy	yes, no
radiation	whether patient had radiation therapy	yes, no
hormone	whether patient had hormone therapy	yes, no

survival status is known 10 years after initial consultation, and 782 patients whose breast cancer survival status is known 15 years after initial consultation.

We used Exhaustive-IGain to learn interactions that affect 5 year, 10 year, and 15 year breast cancer survival. Table 4.3 shows the correlations of each of the predictors with breast cancer survival according to both the BNPP (see Sect. 4.5), computed by assigning equal prior probabilities to the two competing model, and Pearson's chi-square test. Except for a few exceptions, the two methods are in agreement. Our purpose here is not to discuss these correlations, but rather to provide them as a frame of reference for the learned interactions, which appear in Table 4.4.

Table 4.4 shows the interactions learned from the Metabric dataset that have $IS > 0.4$. The BNPP, computed by assigning equal prior probabilities to all

Table 4.3 The individual variable effects learned from the METABRIC dataset. The p-values were obtained using the chi-square test

Variable	5 year BC death		10 year BC death		15 year BC death	
	BNPP	p-value	BNPP	p-value	BNPP	p-value
P53_mutation_status	1	0	0.97	0.001	0.936	0.0004
HER2_Status	1	0	1	0	0.853	0.0006
chemo	1	0	1	0	0.999	0
PR_category	1	0	1	0	0.971	0.002
hormone	0.880	0.112	0.410	0.120	0.999	0
radiation	0.240	0.320	0.170	1	0.280	0.576
ER_category	1	0	1	0	0.889	0.002
overall_stage	1	0	1	0	1	0
menopausal_status	0.940	0.019	0.190	0.76	0.421	0.554
histological	0.450	0.0250	0.940	0.002	0.913	0.055
lymph_nodes_pos	1	0	1	0	1	0
percent_nodes_positive	1	0	1	0	0.999	0
overall_grade	1	0	1	0	0.999	0.0001
size	1	0	1	0	0.954	0.014
age_at_diagnosis	1	0	1	0	0.950	0.0003
axillary_nodes_removed	0.160	0.113	0.950	0.003	0.147	0.567

Table 4.4 The interactions learned from the METABRIC dataset

Outcome	Interaction	BNPP	IS
5 year BC death	histological, menopausal_status	0.77	0.43
	histological, hormone	0.93	0.47
10 year BC death	hormone, menopausal_status	0.32	0.72
15 year BC death	histological, menopausal status	0.57	0.49

competing models, and the *IS* are shown in the table. The data indicates that *histological* interacts with *menopausal_status* to affect both 5 year and 15 year breast cancer death survival. A consultation with a breast cancer oncologist[1] reveals that invasive ductal carcinoma (IDC) has a worse prognosis in premenopausal women, but other histological types do not. Furthermore, Table 4.3 indicates that neither *histological* nor *menopausal status* is highly correlated with 5 year or 15 year breast cancer death survival by themselves. Table 4.4 also shows that the data indicates *hormone* and *menopausal_status* interact to affect 10 breast cancer death survival. The breast cancer oncologist indicated that hormone therapy is more effective in post-menopausal women. As Table 4.3 shows, neither *hormone* nor *menopausal_status* are highly correlated with 10 year breast cancer death survival

---

[1]Adam Brufsky, MD, PhD, Professor of Medicine at the University of Pittsburgh School of Medicine.

by themselves. Finally, Table 4.4 shows that the data indicates that *histological* and *hormone* interact to affect 5 year breast cancer death survival. The oncologist stated IDC might respond slightly worse to hormone therapy than other types, but that this difference is not well-established.

The BNPP and the *IS* are new concepts. So, we do not have the same intuition for their values as we have for a p-value. That is, we have come to consider a p-value of 0.05 meaningful partly due to Fisher's [56] statement in 1921 that "it is convenient to draw the line at about the level at which we can say: Either there is something in the treatment, or a coincidence has occurred such as does not occur more than once in twenty trials," and also due to years of experience. To provide a context for the results in Table 4.4, Table 4.5 shows the average BNPPs and ISs of all 2, 3, 4, and 5 cause models obtained from the Metabric dataset. As we would expect, the value of the BNPP decreases as the size of the models increases. However, the IS small for models of all sizes. The models we learned (Table 4.4) are all 2-cause models. So we compare those results to the averages for 2-cause models. Our IS results of 0.43, 0.47, 0.72 and 0.49 are all substantially larger than the 2-cause IS average of 0.042. Three of our BNPP results, namely 0.77, 0.93, and 0.57 are much higher than the average 2-cause BNPP of 0.266. However, the value of 0.32, which is obtained for {*hormone, menopausal_status*}, is not much higher than the average. Yet, this model has the largest IS (0.72).

4.7 Discussion

The notion of a discrete causal interaction has recently received substantial attention, largely due to the problem of learning genetic epistasis from GWAS datasets. Even though a great deal of effort was made towards developing algorithms that learn genetic epistasis, no definition of a discrete interaction was forwarded. We developed a fuzzy definition, called Interaction Strength (*IS*), of a discrete causal action, based on information theory. The *IS* bounded above by 1 and equals 1 if the interaction exhibits no marginal effects, and is bounded below by 0 if all causes in the interaction are independent.

Using the *IS* and BN scoring, we developed an exhaustive search algorithm, Exhaustive-IGain, which learns interactions from low-dimension datasets, and a heuristic search algorithm, called MBS-IGain, which learns interactions from high-dimensional datasets. Using simulated high-dimensional GWAS datasets based on genetic interactions with some marginal effects, we compared MBS-IGain to 7 algorithms that learn genetic epistasis from high-dimensional datasets, and showed that MBS-IGain's discovery performance was notably better than the other methods. We applied MBS-IGain to a real LOAD dataset, and obtained results substantiating previous research and new results. Using low-dimensional simulated datasets, we showed Exhaustive-IGain can learn 4-cause interactions with no marginal effects. We applied Exhaustive-Gain to a real clinical breast cancer

Table 4.5 The average BNPPs and ISs of all 2, 3, 4, and 5 cause models obtained from the Metabric dataset

Model	Avg. BNPP	Avg. IS
2-cause models	0.266	0.042
3-cause models	0.005	-0.005
4-cause models	6.13×10^{-7}	0.013
5-cause models	7.04×10^{-16}	0.040

datasets, and learned interactions that agreed with the judgements of a breast cancer oncologist.

Current algorithms for learning causes from data assume the composition property, when entails that they cannot in general learn interactive causes with little marginal effects. Our algorithms assume the candidate causes have no causal relationships among them, and are only directly applicable to problems where we have a specified target and its candidate causes. However, our algorithms could also be applicable to general causal learning by being a front end to a constraint-based or score-based causal learning algorithm.

Our algorithms only learn interactive causes. They do not identify whether the interactive causes are direct causes of the target relative to the set of candidate causes being investigated. To identify whether they are direct causes, they could be combined with an algorithm for learning direct causal influences of a target [57].

Acknowledgements Funding

This work was supported by National Library of Medicine grants number R00LM010822, R01LM011663, and R01LM011962.

References

1. Spirtes, P., Glymour, C., Scheines, R.: Causation, prediction, and search. MIT Press, Boston, MA (2000)
2. http://www.phil.cmu.edu/tetrad/
3. Chickering, D., Meek, C.,: Finding optimal Bayesian networks. In: Darwiche, A., Friedman, N. (eds.) Uncertainty in Artificial Intelligence, Proceedings of the Eighteenth Conference. Morgan Kaufmann, San Mateo, CA (2002)
4. Cheverud, J., Routman, E.: Epistasis and its contribution to genetic variance components. Genetics 139(3), 1455 (1995)
5. Urbanowicz, R., Granizo-Mackenzie, A., Kiralis, J., Moore, J.H.: A classification and characterization of two-locus, pure, strict, epistatic models for simulation and detection. BioData Min. 7, 8 (2014)
6. Fisher, R.: The correlation between relatives on the supposition of mendelian inheritance. Trans R Soc Edinburgh 52, 399–433 (1918)
7. Galvin, A., Ioannidis, J.P.A., Dragani, T.A.: Beyond genome-wide association studies: Genetic heterogeneity and individual predisposition to cancer. Trends Genet. 26(3), 132–141 (2010)
8. Manolio, T.A., Collins, F.S., Cox, N.J., et al.: Finding the missing heritability of complex diseases and complex traits. Nature 461, 747–753 (2009)
9. Mahr, B.: Personal genomics: The case of missing heritability. Nature 456, 18–21 (2008)

10. Moore, J.H., Asselbergs, F.W., Williams, S.M.: Bioinformatics challenges for genome-wide association studies. Bioinformatics **26**, 445–455 (2010)
11. Manolio, T.A., Collins, F.S.: The HapMap and genome-wide association studies in diagnosis and therapy. Annu. Rev. Med. **60**, 443–456 (2009)
12. Herbert, A., Gerry, N.P., McQueen, M.B.: A common genetic variant is associated with adult and childhood obesity. J. Comput. Biol. **312**, 279–384 (2006)
13. Spinola, M., Meyer, P., Kammerer, S., et al.: Association of the PDCD5 locus with long cancer risk and prognosis in smokers. Am. J. Hum. Genet. **55**, 27–46 (2001)
14. Lambert, J.C., Heath, S., Even, G., et al.: Genome-wide association study identifies variants at CLU and CR1 associated with Alzheimer's disease. Nat. Genet. **41**, 1094–1099 (2009)
15. Curtis, C., Shah, S.P., Chin, S.F., et al.: The genomic and transcriptomic architecture of 2,000 breast tumours reveals novel subgroup. Nature **486**, 346–352 (2012)
16. Soulakis, N.D., Carson, M.B., Lee, Y.J., Schneider, D.H., Skeehan, C.T., Scholtens, D.M.: Visualizing collaborative electronic health record usage for hospitalized patients with heart failure. JAMIA **22**(2), 299–311 (2015)
17. Neapolitan, R.E.: Learning Bayesian Networks. Prentice Hall, Upper Saddle River, NJ (2004)
18. Jensen, F.V., Neilsen, T.D.: Bayesian Networks and Decision Graphs. Springer-Verlag, New York (2007)
19. Neapolitan, R.E.: Probabilistic reasoning in expert systems. Wiley, NY, NY (1989)
20. Pearl, J.: Probabilistic reasoning in intelligent systems. Morgan Kaufmann, Burlington, MA (1988)
21. Segal, E., Pe'er, D., Regev, A., Koller, D., Friedman, N.: Learning module networks. Journal of Machine Learning Research **6**, 557–588 (2005)
22. Friedman, N., Linial, M., Nachman, I., Pe'er, D. Using Bayesian networks to analyze expression data. In: Proceedings of the fourth annual international conference on computational molecular biology, Tokyo, Japan (2005)
23. Fishelson, M., Geiger, D.: Optimizing exact genetic linkage computation. J. Comput. Biol. **11**, 263–275 (2004)
24. Neapolitan, R.E.: Probabilistic Reasoning in Bioinformatics. Morgan Kaufmann, Burlington, MA (2009)
25. Jiang, X., Cooper, G.F.: A real-time temporal Bayesian architecture for event surveillance and its application to patient-specific multiple disease outbreak detection. Data Min. Knowl. Disc. **20**(3), 328–360 (2010)
26. Jiang, X., Wallstrom, G., Cooper, G.F., Wagner, M.M.: Bayesian prediction of an epidemic curve. J. Biomed. Inform. **42**(1), 90–99 (2009)
27. Cooper, G.F.: The computational complexity of probabilistic inference using Bayesian belief networks. J. Artif. Intell. Res **42**(2–3), 393–405 (1990)
28. Cooper, G.F., Herskovits, E.: A Bayesian method for the induction of probabilistic networks from data. Mach. Learn. **9**, 309–347 (1992)
29. Heckerman D, Geiger D, Chickering D. Learning Bayesian networks: The combination of knowledge and statistical data. Technical report MSR-TR-94–09. Microsoft Research, 1995
30. Chickering, M.: Learning Bayesian networks is NP-complete. In: Fisher, D., Lenz, H., (eds.) Learning from Data: Artificial Intelligence and Statistics V. Springer-Verlag, NY (1996)
31. Shannon, C.E.: A mathematical theory of communication. The Bell System Technical Journal **27**(3), 379–423 (1948)
32. Zadeh, L.A.: Fuzzy sets. Inf. Control **8**, 338–353 (1965)
33. Zang, Z., Jiang, X., Neapolitan, R.E.: Discovering causal interactions using Bayesian network scoring and information gain. BMC Bioinformatics **17**, 221 (2016)
34. Jiang, X., Jao, J., Neapolitan, R.E. Learning predictive interactions using Information Gain and Bayesian network scoring. PLOS ONE (2015) http://dx.doi.org/10.1371/journal.pone.0143247
35. Jiang, X., Barmada, M.M., Cooper, G.F., Becich, M.J.: A Bayesian method for evaluating and discovering disease loci associations. PLoS ONE **6**(8), e22075 (2011)

36. Kooperberg, C., Ruczinski, I.: Identifying interacting SNPs using Monte Carlo logic regression. Genet. Epidemiol. **28**, 157–170 (2005)
37. Agresti, A.: Categorical data analysis, 2nd edn. Wiley, New York (2007)
38. Park, M.Y., Hastie, T.: Penalized logistic regression for detecting gene interactions. Biostatistics **9**, 30–50 (2008)
39. Wu, T.T., Chen, Y.F., Hastie, T., Sobel, E., Lange, K.: Genome-wide association analysis by lasso penalized logistic regression. Genome Analysis **25**, 714–721 (2009)
40. Hahn, L.W., Ritchie, M.D., Moore, J.H.: Multifactor dimensionality reduction software for detecting gene-gene and gene-environment interactions. Bioinformatics **19**, 376–382 (2003)
41. Marchini, J., Donnelly, P., Cardon, L.R.: Genome-wide strategies for detecting multiple loci that influence complex diseases. Nat. Genet. **37**, 413–417 (2005)
42. Moore, J.H., Gilbert, J.C., Tsai, C.T., Chiang, F.T., Holden, T., Barney, N., et al.: A flexible computational framework for detecting characterizing and interpreting statistical patterns of epistasis in genetic studies of human disease susceptibility. J. Theor. Biol. **241**, 252–261 (2006)
43. Yang, C., He, Z., Wan, X., Yang, Q., Xue, H., Yu, W.: SNPHarvester: a filtering-based approach for detecting epistatic interactions in genome-wide association studies. Bioinformatics **25**, 504–511 (2009)
44. Moore, J.H., White, B.C. Tuning ReliefF for genome-wide genetic analysis. In: Marchiori, E., Moore JH, Rajapakee JC (eds.) Proceedings of EvoBIO 2007. Berlin: Springer-Verlag (2007)
45. Meng Y, Yang Q, Cuenco KT, Cupples LA, Destefano AL, Lunetta KL 2007. Two-stage approach for identifying single-nucleotide polymorphisms associated with rheumatoid arthritis using random forests and Bayesian networks. BMC Proc 2007: 1 Suppl 1:S56
46. Wan, X., Yang, C., Yang, Q., Xue, H., Tang, N.L., Yu, W.: Predictive rule inference for epistatic interaction detection in genome-wide association studies. Bioinformatics **26**(1), 30–37 (2007)
47. Zhang, Y., Liu, J.S.: Bayesian inference of epistatic interactions in case control studies. Nat. Genet. **39**, 1167–1173 (2007)
48. Miller, D.J., Zhang, Y., Yu, G., Liu, Y., Chen, L., Langefeld, C.D., et al.: An algorithm for learning maximum entropy probability models of disease risk that efficiently searches and sparingly encodes multilocus genomic interactions. Bioinformatics **25**(19), 2478–2485 (2009)
49. Jiang X, Barmada MM, Neapolitan RE, Visweswaran S, Cooper GF. A fast algorithm for learning epistatic genomic relationships. AMIA Symposium Proceedings 2010: 341–345
50. Jiang, X., Neapolitan, R.E.: LEAP: biomarker inference through learning and evaluating association patterns. Genet. Epidemiol. **39**(3), 173–184 (2015)
51. Chen, L., Yu, G., Langefeld, C.D., et al.: Comparative analysis of methods for detecting interacting loci. BMC Genom. **12**, 344 (2011)
52. Rieman, E.M., Webster, J.A., Myers, A.J., Hardy, J., Dunckley, T., Zismann, V.L., et al.: GAB2 alleles modify Alzheimer's risk in APOE carriers. Neuron **54**, 713–720 (2007)
53. Tycko, B., Lee, J.H., Ciappa, A., Saxena, A., Li, C.M., Feng, L.: APOE and APOC1 promoter polymorphisms and the risk of Alzheimer disease in African American and Caribbean Hispanic individuals. Arch. Neurol. **61**(9), 1434–1439 (2004)
54. Turner SD, Martin ER, Beecham GW, Gilbert JR, Haines JL, Pericak-Vance MA, et al. Genome-wide Analysis of Gene-Gene Interaction in Alzheimer Disease. Abstract in ASHG 2008 Annual Meeting (2008)
55. Urbanowicz R, Kiralis J, Sinnott-Armstrong NA, et al. GAMETES: a fast, direct algorithm for generating pure, strict, epistatic models with random architectures. BioData Min. 2012; 5 (1):16. doi:10.1186/1756-0381-5-16

56. Fisher, R.A.: On the 'probable error' of a coefficient of correlation deduced from a small sample. Metron **1**, 3–32 (1921)
57. Rathnam, C., Lee, S., Jiang, X.: An algorithm for direct causal learning of influences on patient outcomes. Artif. Intell. Med. **75**, 1–15 (2017)

Author Biography

Xia Jiang Xia Jiang is associate professor of biomedical informatics at the University of Pittsburgh. She has 13 years research experience in Bayesian network modeling, machine learning, and algorithm design, with the focus on solving problems in the clinical and biomedical domains. Richard Neapolitan is professor of biomedical informatics at Northwestern University. He is one of the leading researchers in uncertain reasoning in artificial intelligence, having written the seminal 1989 Bayesian network text *Probabilistic Reasoning in Expert Systems*, and more recently the 2004 text *Learning Bayesian Networks*. Drs. Jiang and Neapolitan collaborated on the 2012 text *Contemporary Artificial Intelligence*

Chapter 5
Bayesian Network Modeling for Specific Health Checkups on Metabolic Syndrome

Yoshiaki Miyauchi and Haruhiko Nishimura

Abstract Metabolic syndrome has become a significant public health problem worldwide, and Specific Health Checkup and Guidance on this syndrome began for people aged 40 to 74 in Japan in 2008. Through this guidance, people considered at high risk of developing metabolic syndrome are expected to be made aware of their own problems in terms of their daily lifestyle choices and to improve their daily life behaviors by themselves. To support this large undertaking with information technology, we have introduced ideas based on the Bayesian estimation in data mining technology and proposed a Bayesian network (BN) scheme connecting the information from physical examinations and daily lifestyle questionnaires. By applying this network model to the field data on 11,947 anonymized individuals, the proposed method was found to provide better performance and show its potentiality for the system of specific health checkup. We introduced a novel 4-bit representation with 16 states, treating body shape, blood lipids, blood glucose, and blood pressure as equal binary factors, and analyzed relationships among the support level, physical examination, and daily lifestyle questionnaire. In addition, we applied this BN to individual cases and showed its utility in allowing an examinee to improve his/her lifestyle by demonstrating individual predictions. Through the efforts described above, we confirmed that the Bayesian network for Specific Health Checkup and Guidance has the potential to be an effective support tool for health promotion regarding metabolic syndrome.

Keywords Metabolic syndrome · Bayesian networks · Specific health checkup · Health guidance · Health promotion

Y. Miyauchi
School of Nursing, Nagoya City University, 1-Kawasumi, Mizuho-Cho, Mizuho-Ku, Nagoya 467-8601, Japan

H. Nishimura (✉)
School of Applied Informatics, University of Hyogo, 7-1-28 Minatojima-Minami, Chuo-Ku, Kobe 650-0047, Japan
e-mail: haru@ai.u-hyogo.ac.jp

© Springer International Publishing AG 2018
D.E. Holmes and L.C. Jain (eds.), *Advances in Biomedical Informatics*,
Intelligent Systems Reference Library 137,
https://doi.org/10.1007/978-3-319-67513-8_5

5.1 Introduction

Healthcare activities called "specific medical checkup" and "specific health guidance" have been carried out in Japan since April 2008 [1, 2]. They focus on metabolic syndromes by paying attention to visceral fat-type obesity, target 50 million people with medical insurance from 40 to 74 years of age, and aim to reduce the number of people suffering from such lifestyle diseases as diabetes and its preliminary stages.

The specific health checkup is a framework different from conventional health examinations and has the following features:

1) It classifies specific health guidance subjects using medical examination data.
2) It examines the lifestyle of the target person by assigning it a weight in questionnaire data in a short time.
3) It provides suitable health guidance to subjects.
4) It accumulates health examination data and questionnaire data for each person, and maintains an annual history of the health guidance provided.

We think that practical studies of health maintenance and enhancement should be closely related to the specific health checkup framework, and should not follow the traditional method of researching only health examination data. Knowledge gained from the lifestyle can be utilized for specific health guidance.

We chose the Bayesian network as a research method to improve accuracy by capturing data according to the accumulation of specific health checkups [3–6]. The Bayesian network is a data mining technology that can respond flexibly to such accumulated data and improve accuracy [7–9]. The application of machine learning methods to the healthcare field is recently beginning to increase [10–14]. In this chapter, we introduce a concrete example of a Bayesian network that conforms to the structure of specific health checkup data and provides health management solutions customized to subjects.

5.2 Specific Health Checkups and the Stratification of Individuals

By introducing the concept of the metabolic syndrome, whereby the accumulation of visceral fat and weight gain causes an increase in blood glucose, neutral fat, and blood pressure, we can show detailed data relating to lifestyle-related diseases of the subjects of the medical checkups. This can help us understand the relationship between people's lifestyles and the results of health examinations, and can motivate them to improve their habits. Therefore, in the specific health checkup and related health guidance, the targets of the guidance are rendered hierarchically according to the number of risk factors due to lifestyle diseases resulting from their medical examination and answers to the questionnaire. For subjects those with low risk factors, motivational support for improving lifestyle habits is provided. At the same time, for people with many risk factors, doctors, public health nurses, nutrition managers, etc., actively intervene and urge them to actively change their lifestyle. This framework helps the recipients of health guidance recognize their condition based on the results of health examinations, and understand the relationship between the mechanisms of the body, such as metabolism, and lifestyle (eating habits and exercise habits, etc.), as well as its link to behavioral changes.

In specific health checkups, examinees are arranged hierarchically into four levels—intensive support level, motivational support level, provision of information level, and non-health guidance—according to the medical examination reference values (RV) and their smoking histories based on the questionnaire. Following this, they receive health guidance (referred to as specific health guidance) according to support level.

The concrete hierarchical procedure is as follows:

Step 1) Risk assessment of visceral fat accumulation

- The risk of visceral fat accumulation is assessed according to the abdominal circumference and BMI.

  Abdominal circumference: Men, $\geq$ 85 cm; Women, $\geq$ 90 cm - > (1)
  Abdominal circumference: Individuals who are not included in the above category (1), and whose BMI exceeds 25 kg/m2 (BMI $\geq$ 25 kg/m2) - > (2)

Step 2 Assessment of the number of additional risks

- Additional risks are enumerated on the basis of laboratory results and descriptions in responses on the questionnaire.
- The risk of metabolic syndrome can be evaluated by determining whether the conditions listed in each of the three Sects. (1–3) have been met. Other related risks are included in the fourth Sect. (4). Smoking history (4) is counted only if at least one of the conditions in each of the previous sections has been met.

1) Increased blood glucose

  a. Fasting blood glucose level $\geq$ 100 mg/dl,
  b. HbA1c level $\geq$ 5.6% (NGSP), or
  c. Currently receiving drug therapy (according to the questionnaire)

2) Lipid abnormalities

  a. Triglyceride level $\geq$ 150 mg/dl or
  b. HDL cholesterol level < 40 mg/dl or
  c. Currently receiving drug therapy (according to the questionnaire)

3) Increased blood pressure

  a. Systolic blood pressure $\geq$ 130 mm Hg or
  b. Diastolic blood pressure $\geq$ 85 mm Hg or

   c. Currently receiving drug therapy (according to the questionnaire)

4) Questionnaire Smoking history

Step 3 Classification of health guidance levels

On the basis of the assessment results obtained in Steps 1 and 2, the health guidance levels are grouped as listed below. As mentioned previously, smoking history (4) is counted only if one or more conditions from each of the previous Sects. (1–3) have been met.

Case (1)

The number of additional risk factors listed in the four Sects. (1–4)

≥ 2 (Individuals with ≥ 2 risk factors are classified as requiring intensive support)
1 (Individuals with one risk factor are classified as requiring motivational support)
0 (Individuals with zero risk factors are classified as requiring the provision of information)

Case (2)

The number of additional risk factors listed in the four Sects. (1–4)

≥ 3 (Individuals with ≥ 3 risk factors are classified as requiring intensive support)
1–2 (Individuals with one-two risk factors are classified as requiring motivational support)
0 (Individuals with zero risk factors are classified as requiring the provision of information)

Step 4 Exceptional applications of specific health guidance

- Individuals aged 65–75 years should improve their lifestyle to prevent the deterioration of the quality of life. Therefore, lifestyle improvement programs for these individuals should be based on their activities of daily living and motor functions. Even if people are classified into the intensive support group, they should be treated as those classified in the motivational support group.

- Individuals who receive antihypertensive therapy and regularly visit a medical institution should receive lifestyle improvement support as part of their continuous medical management at their medical institution. These individuals are not required to receive specific health guidance from the health insurer. However, in collaboration with the treating physicians, the health insurer can provide health guidance to these individuals to ensure steady lifestyle improvements, and effectively prevent treatment discontinuation. If the medical checkup shows an abnormal laboratory result that exceeds the threshold of health guidance, even if the test is unrelated to the disease controlled at the medical institution, the data should be provided to the physicians treating the individual.Constructing a Bayesian network corresponding to specific health checkup using examination data and questionnaire data

5.2.1   Study Subjects

We analyzed examination data and questionnaire data subjected to non-connectable anonymization processing at specific health checkups conducted for two years at a certain establishment. The target data consisted of 11,947 medical checkup data items (7655 males, 4292 females), excluding data taken during oral treatment and inadequate data (missing examination data, no answers to six or more questions on the questionnaire), from 12,230 medical checkup data items (7881 males, 4349 females). Moreover, the establishment, to which the employees of the data to be analyzed belonged, had been carrying out projects ranging in location from urban to mountainous areas with diversified work content, such as administrative systems, business systems, and skills. Therefore, there seemed no major bias in the working conditions of the examinees. Moreover, since employment renewal ranged from 61 to 65 years of age, the majority of the consulted individuals were younger than 65.

As concrete items of medical examination data, abdominal circumference [cm], body mass index; BMI [kg/m2], fasting blood glucose level [mg/dl], HbA1c [%], neutral fat [mg/dl], HDL cholesterol [mg/dl], systolic blood pressure [mmHg], and diastolic blood pressure [mmHg] were used. On the contrary, the questionnaire data handled collections of responses to questions related to lifestyle habits.

The questionnaire consisted of 36 question questions, 12 each related to exercise, nutrition, and living. The choices for each question consisted of four or five alternatives, but the question related to smoking had only two choices. The types of questions were as follows:

1. Regarding exercise: basic exercise habits (4), lifestyle disease prevention (4), musculoskeletal disorder prevention (4).
2. Regarding nutrition: dietary habits (3), nutritional balance (3), vegetable intake (1), salinity (2), lipids (1), calcium (2).
3. Regarding living in general: stress (3), alcohol (2), tobacco (2), lifestyle rhythm (1), health consciousness (2), chewing (1), community activity (1).

We allocated a numerical value of 1 to 5 to each answer based on the appropriateness of the choice. The sum of all 36 questions ranged from 36 to 180 points.

5.2.2   Results of Applying the Hierarchical Procedure

Tables 5.1 and 5.2 show the results of the hierarchical use of target data according to the specific health examination stratification procedure described in the previous section. According to Table 5.1, 51.5% of males were classified as beyond the health guidance target, 31.5% as requiring positive support, 10.1% as requiring information provision, and 6.9% as requiring motivational support. On the other hand, from Table 5.2, 85.8%, the majority of women in single-year data, were classified as beyond the health guidance targets, and the rest were divided into three levels. This result yielded very different stratification from those for men. The fact that the majority of women were classified as beyond health guidance leads to the problem of having to secure a larger number of samples in handling inferences based on probability.

5.2.3 Binary Representation of Physical Examination Data

As is clear from previous reports [15, 16] on risk factors for metabolic syndrome, the following pairs of factors in our examination data are known to contribute to the risk of metabolic syndrome: (1) waist circumference and BMI, (2) fasting blood glucose and HbA1c, (3) neutral fat and HDL cholesterol, and (4) systolic and diastolic blood pressure. Therefore, as an expression of health according to the inspection data, these data (body shape, blood sugar, lipids, and blood pressure) can be rendered in a four-bit representation of the 16 potential outcomes, ranging from the (0000) state to the (1111) state, where 1 is defined as one or both of the two items of each factor outside the reference value/range, and 0 as both items being inside the reference (Fig. 5.1). In this study, we used these 16 states to determine the health conditions of the examinees and classify the learning data in the Bayesian network.

5.2.4 Constructing Bayesian Network Containing Examination and Questionnaire Data

Since metabolic syndrome depends on lifestyle, understanding the interrelationship between questionnaire data capturing lifestyle habits and examination data showing health status is important for health guidance following health examination. Therefore, in the Bayesian network, it is important to link nodes based on questionnaire data and those based on medical examination data. Based on the viewpoint that emphasizes the framework of the specific health checkup, we construct a network to evaluate the nodes whose states are support levels as a result of the hierarchy, which is an essential part of examinee classification.

Table 5.1 Results of applying the hierarchical method on the 7655 males

Case (1) in Step 3		
Intensive support level	Motivational support level	Provision of information
2377 people (31.3%)	466 people (6.1%)	731 people (9.5%)
Case (2) in Step 3		
Intensive support level	Motivational support level	Provision of information
30 people (0.4%)	63 people (0.8%)	46 people (0.6%)

Table 5.2 Results of applying the hierarchical method on 4292 females

Case (1) in Step 3		
Intensive support level	Motivational support level	Provision of information
154 people (3.6%)	112 people (2.6%)	120 people (2.8%)
Case (2) in Step 3		
Intensive support level	Motivational support level	Provision of information
11 people (0.3%)	123 people (2.9%)	91 people (2.1%)

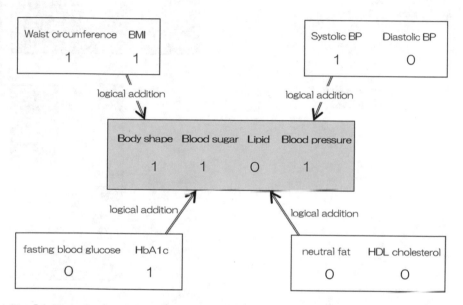

Fig. 5.1 Example of the binary representation of examination data (BMI, body mass index; BP, blood pressure; HbA1c, glycated hemoglobin; HDL, high-density lipoprotein)

Fig. 5.2 Physical examination node: waist circumference (RV, reference value)

Waist_circumference	
in RV	53.3
out of RV	46.7

Thus, we constructed Bayesian networks with each node on a single network, where the linked structures contained an information node from the lifestyle questionnaire, a node for the physical examination results indicating health condition, a node representing the results of the hierarchical data described above, and one representing health status according to the 16 states (Figs. 5.2, 5.3, 5.4, 5.5, 5.6) [17].

Figure 5.2 shows an example of a node for "waist circumference," represented by a two-state medical examination value: > 85 cm (shown as "out of RV")

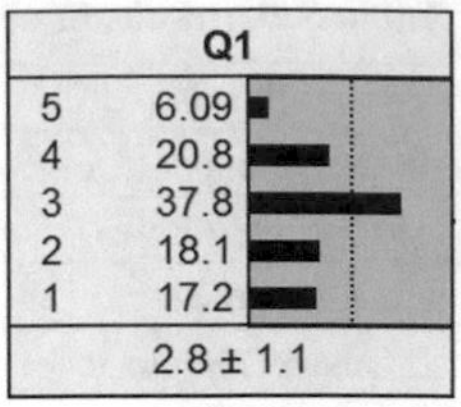

Fig. 5.3 Lifestyle
questionnaire node: Q1

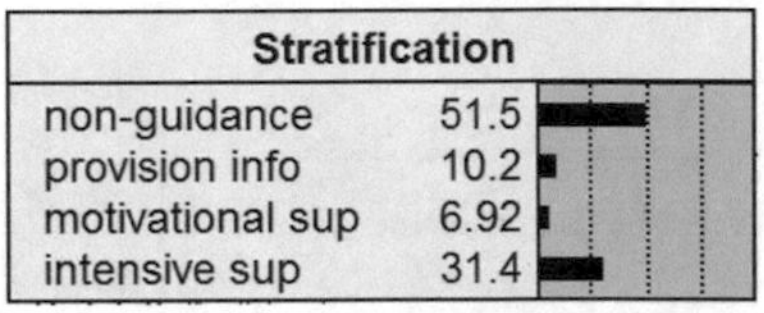

Q1	
5	6.09
4	20.8
3	37.8
2	18.1
1	17.2
2.8 ± 1.1	

Fig. 5.4 Support-level
hierarchical node:
stratification

Stratification	
non-guidance	51.5
provision info	10.2
motivational sup	6.92
intensive sup	31.4

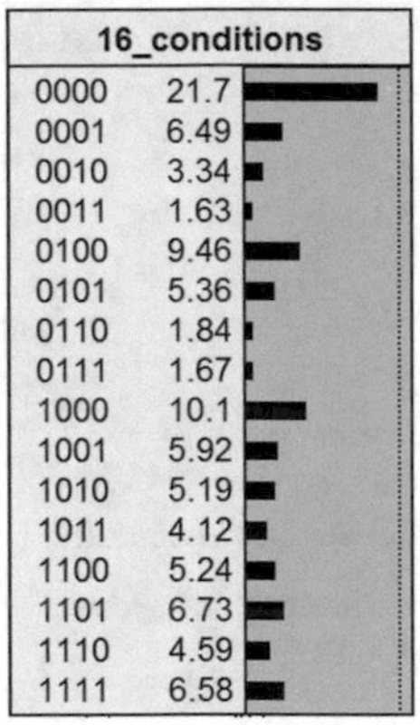

Fig. 5.5 The representation
node for the 16 health states:
16 conditions

16_conditions	
0000	21.7
0001	6.49
0010	3.34
0011	1.63
0100	9.46
0101	5.36
0110	1.84
0111	1.67
1000	10.1
1001	5.92
1010	5.19
1011	4.12
1100	5.24
1101	6.73
1110	4.59
1111	6.58

and < 85 cm ("in RV"). In this case, the probability of the state in RV (expressed as a percentage) indicated that 59.4% of the examinees belonged to the state within the RV.

Figure 5.3 shows an example of the interview node "Q1," and the five states, ranging from 1 to 5 points (Sect. 5.2.1), are represented. For example, the probability of obtaining 3 points for question Q1 was found to be 37.9%. The average value (2.75 ± 1.1 points) is shown at the bottom.

Figure 5.4 shows a hierarchical node ("stratification"); the four support levels in the hierarchy of Sect. 5.2.2 are shown: intensive support level ("intensive sup"), motivational support level ("motivational sup"), provision of information level ("provision info"), and the level of excluded health guidance ("non-guidance"). For example, the probability of the intensive support level was 24.9%, indicating that 24.9% of the 5423 men had been layered into the intensive support level.

Figure 5.5 shows the representation node for the 16 health states ("16 conditions"); it is represented by 16 states ranging from (0000) to (1111), according to the four-bit representation described in Sect. 5.2.3. In the figure, the probability of

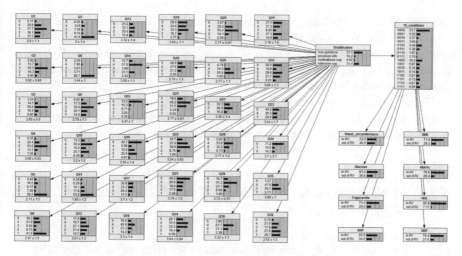

Fig. 5.6 Bayesian network for specific health checkups of metabolic syndrome: (BMI, body mass index; HbA1c, glycated hemoglobin; HDL, high-density lipoprotein; SBP, systolic blood pressure; DBP, diastolic blood pressure)

the (0000) state, which represents the healthiest state, was 28.2%, and the probability of the (1111) state, the least healthy state, was 3.53%.

Figure 5.6 shows the overall picture of the Bayesian network constructed. We calculated the conditional probability using health data of the 5423 examinees. All 36 question nodes (Q1-Q36) are shown on the left side. The lower-right side shows the check node group with eight nodes, namely waist circumference, BMI, fasting blood glucose ("Glucose"), glycated hemoglobin ("HbA1c"), neutral fat ("Triglyceride"), HDL cholesterol ("HDL"), systolic blood pressure ("SBP"), and diastolic blood pressure ("DBP"). The upper right indicates the 16 health states' node ("16 conditions"), and the hierarchical node ("stratification") is shown in the center of the figure.

5.3 Evaluation

5.3.1 Setting Nodes for Risk Assessment

By changing the state of physical examination nodes (setting them inside or outside the RV) or the state of the lifestyle questionnaire nodes (setting them in a good or bad state) for the Bayesian network described in the previous section, it is possible to investigate the overall change in risk or the risk to each examinee. We can set the state of the node by specifying (clicking on) the state in the node.

Figure 5.7 shows the point set to four points for the Q1 node of the lifestyle questionnaire. The state probability of the four points was 20.8% before the setting changed to 100% (darkening of node color indicates that the state was set). In response to this, the probability distribution of all nodes other than Q1 was recalculated and changed. With the same click operation, we can set multiple states at the same time. The probability distribution of the set multiple states was calculated based on the ratio between states prior to the setting.

Figure 5.8 shows the four points and five points simultaneously set in the Q1 node. The graphs of the nodes have varying lengths because the probabilities were distributed based on the ratio between states before the setting.

5.3.2   Risk Evaluation by Setting Lifestyle Questionnaire Nodes

Figure 5.9 shows the Bayesian network where all answers to the questionnaire indicated good lifestyle habits (we set points for all lifestyle questionnaire nodes to 4 or more, implying good lifestyle habits). Figure 5.10 shows the change in the support level of the hierarchical node in this instance. In this case, it can be seen that there was a significant increase (51.5% → 94.7%) in the level of excluded health guidance. On the other hand, other support levels decreased across the board. Particularly, the decrease in the intensive support level (31.4% → 0.66%) was remarkable. These showed that if the answers to the questionnaire concerning the lifestyle of the examinees were all good, the probability of being beyond the health guidance targets, which was the best condition for the specific health checkup, was very high. This was consistent with the idea of lifestyle-related diseases such that the health condition improves if lifestyle does.

Figure 5.11 shows the change in the 16 conditions' nodes in this case. The probability of the state with the first bit as 1 decreased, and the probability of states expressing a break from the metabolic syndrome, such as (0000) and (0001), increased. According to this figure, for each combination of metabolic syndrome factors, it was possible to see the effect of improving lifestyle habits.

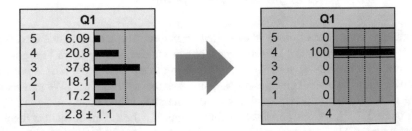

Fig. 5.7 Setting to 4 points for Q1 node

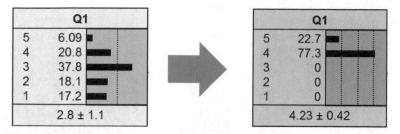

Fig. 5.8 Setting to 4 or more points for Q1 node

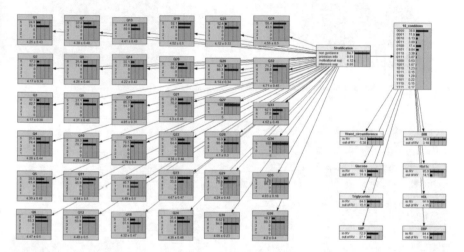

Fig. 5.9 Bayesian network where all answers to the questionnaire were good (all points of the questionnaire were set to 4 or more)

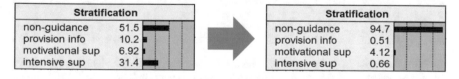

Fig. 5.10 Change in the support level hierarchical node in this instance

5.3.3 Comparison by Setting Changes in Lifestyle Questionnaire Nodes on Exercise Aspect

Figure 5.12 shows the male Bayesian network where all answers to the 12 questions on the exercise aspect were bad (the shaded nodes were set to 2 points or less). Figure 5.13 shows the change in the support level of the hierarchical node in this instance. In this case, the level of excluded health guidance significantly decreased

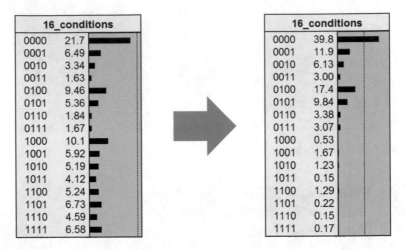

Fig. 5.11 Changes in the 16-condition node in this instance

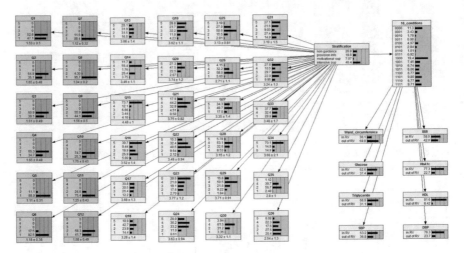

Fig. 5.12 The male Bayesian network, where all answers to the 12 questions on the exercise aspect represented bad lifestyle habits (the shaded nodes were set at 2 points or less)

(51.5% → 26.8%), whereas the intensive support level increased (31.4% → 46.4%). By lowering the points scored in answers to questions on the exercise aspect, the percentage of male examinees beyond the health guidance decreased and those in the intensive support level increased.

On the other hand, Fig. 5.14 shows the change in the support level of the hierarchical node in case of setting similar values for the female Bayesian network. According to this, the number of females excluded from the health guidance decreased (85.7% → 76.0%) and those qualifying for the provision of information increased (4.93% → 14.7%); however, unlike in the case of males, the number of

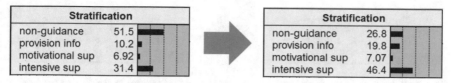

Fig. 5.13 Changes in the support-level hierarchical node in this instance

Stratification		
non-guidance	85.7	
provision info	4.93	
motivational sup	5.49	
intensive sup	3.86	

➡

Stratification		
non-guidance	76.0	
provision info	14.7	
motivational sup	5.44	
intensive sup	3.86	

Fig. 5.14 Changes in the support-level hierarchical node, where all answers to the 12 questions on the exercise aspect were bad (the shaded nodes were set at 2 points or less) for the female Bayesian network

females in the intensive support level did not change (3.86% → 3.86%). From this, it seemed that the influence of changes in lifestyle was much higher for men than for women on the exercise aspect.

Figure 5.15 shows changes in the waist circumference node (changes in male Bayesian network) when all answers to the 12 questions on the exercise plane, as in the previous example, represented bad lifestyle habits. According to this, the increment in the proportion of "out of RV" was 69.9% - 46.7% = 23.2 percentage points; for the male Bayesian network, this indicated that the probability of out-of-health check reference value of waist circumference nodes increased.

The effect on other physical examination nodes in the same case is shown in Fig. 5.16. The difference in "out of RV" (the probability of the state in Fig. 5.12-the probability of the state in Fig. 5.6) is shown for each node of the physical examination. According to Fig. 5.16, in the male Bayesian network, the probability that the physical examination nodes other than the waist circumference node fell outside the medical examination reference values showed an increasing trend. In this way, lifestyle questionnaire nodes and physical examination nodes were not directly connected on the network, but the state of the physical examination nodes was influenced through the support level of the hierarchical node according to the setting of the state of the lifestyle questionnaire nodes. This was due to the basic characteristics of the Bayesian network.

On the other hand, Fig. 5.17 shows the results of the female Bayesian network, where the waist circumference node and the BMI node increased, as in the case of men. However, in the other physical examination nodes, the difference was negligibly small.

This indicates that even if worsening lifestyle habits related to the exercise aspect, in the case of female, it did not significantly affect the physical examination nodes other than body shape. In this way, the Bayesian networks for males and

Fig. 5.15 Changes in the waist circumference node when all answers to the 12 questions on the exercise aspect were bad (changes in male Bayesian network)

Fig. 5.16 Effect on physical examination nodes when all answers to the 12 questions on the exercise aspect were bad (changes in male Bayesian network)

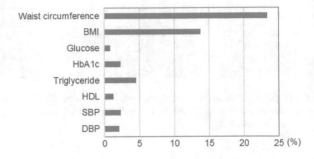

Fig. 5.17 Effect on physical examination nodes of the female Bayesian network where all answers to the 12 questions on the exercise aspect were bad (changes in female Bayesian network)

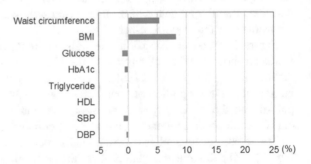

females showed different risk changes. From the above, it can be concluded that a gender-specific Bayesian network should be used for health guidance.

5.3.4 Evaluation of Actual Case

Based on the results of the specific health checkup of examinee A, we evaluated our Bayesian network. Examinee A was a 60-year-old man (height, 174.3 cm; weight, 72.6 kg) who had not received oral treatment for lifestyle-related diseases. All his inspection items, except for BMI and HbA1c and HDL cholesterol, were outside the reference values, indicating a typical metabolic syndrome (1111) state according to the 16 health state representations. He was typical of an individual belonging to the positive support level. Figure 5.18 shows a Bayesian network based on questionnaire data for examinee A.

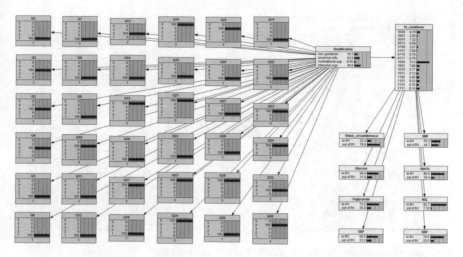

Fig. 5.18 Setting lifestyle questionnaire nodes based on the case of examinee A (BMI, body mass index; HbA1c, glycated hemoglobin; HDL, high-density lipoprotein; SBP, systolic blood pressure; DBP, diastolic blood pressure)

Stratification	
non-guidance	19.3
provision info	32.0
motivational sup	9.95
intensive sup	38.8

Stratification	
non-guidance	40.3
provision info	37.4
motivational sup	8.59
intensive sup	13.7

Fig. 5.19 Outcomes on the hierarchical node when the answers to the 3 questions relating to exercise were changed

To evaluate our Bayesian network, we questioned changes that would occur in the health status of examinee A if he changed his lifestyle habits. In his questionnaire, he provided unsuitable answers to the question group relating to exercise and nutrition. When the scores of the three questions relating to exercise were set as the highest scores, the probability that he would remain in the (1111) state decreased from 8.14% to 2.93%, and the possibility of improvement to the (1000) state increased from 31.4% to 36.7% (Fig. 5.20). That is, improving the lifestyle factors relating to exercise was expected to improve the health of the examinee.

On the other hand, when we set the three questions relating to nutrition to the highest scores, the probability of the examinee staying in the (1111) state changed little, decreasing from 8.14% to 5.56% (Fig. 5.22). This indicated that in case of health guidance to examinee A, encouragement to exercise, rather than improving diet, was expected to yield more positive effects on his health. (Figures 5.19, 5.21)

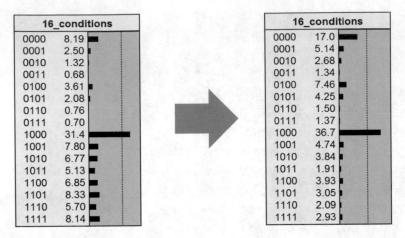

Fig. 5.20 Outcomes on the 16 conditions' node in this instance

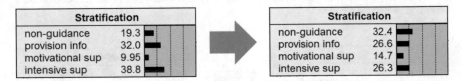

Fig. 5.21 Outcomes on the hierarchical node when the answers to the 3 questions relating to nutrition were changed

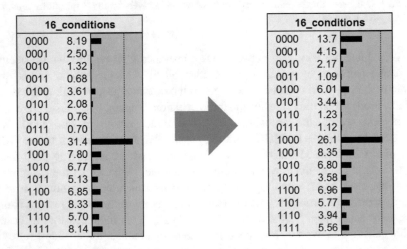

Fig. 5.22 Outcomes on the 16 conditions' node in this instance

5.4 Conclusion

The Data Health Plan was added to health businesses based on the national health insurance law in March 2014. This policy uses health and medical information for all health insurance unions, and enforces effective and efficient health procedures along the lines of the PDCA cycle to determine and execute. This health policy obligates all health insurance associations to develop and implement effective and efficient health business implementation plans in accordance with the PDCA cycle by utilizing health and medical information. It also requires the following: enable examinees to discover problems in their lifestyle and make efforts to encourage improvement, provide health and medical information in an easy-to-understand manner by utilizing information communications technology, and introduce a mechanism that recommends voluntary activities for the promotion of health for examinees. The Data Health Plan Preparation Guide also states that it is desirable to formulate an implementation plan for the specific health checkup and the data health plan through cooperation. In other words, in the future, the healthcare project in Japan should advance according to the framework of the Data Health Plan, and the specific health checkup will be a concrete pillar of this edifice.

References

1. Specific Health Checkups and Specific Health Guidance, The Health Service Bureau of the Ministry of Health, Labour and Welfare, 2007
2. Tamura, T., Kimura, Y.: Specific Health Checkups in Japan: The Present Situation Analyzed Using 5-year Statistics and the Future. Biomedical Engineering Letters 5(1), 22–28 (2015)
3. Y. Miyauchi and H. Nishimura, Bayesian Network for Healthcare of Metabolic Syndrome, IEEE EMBC2013, Osaka, Short paper No. 3164, 2013
4. Y. Miyauchi and H. Nishimura, Construction and Evaluation of Bayesian Networks Related to the Specific Health Checkup and Guidance on Metabolic Syndrome, Innovation in Medicine and Healthcare 2015 (Smart Innovation, Systems and Technologies, Vol.45) Y.W. Chen et al. (Eds.), pp. 183–193, Springer International Publishing, 2015
5. Miyauchi, Y., Nishimura, H., Inada, H.: Analysis of interannual data for the specific health checkup to develop its Bayesian network application. Health Evaluation and Promotion 42(5), 479–491 (2015). (in Japanese)
6. Miyauchi, Y., Nishimura, H., Nakano, Y.: A Study of Bayesian Network Model Related to the Specific Health Checkup based on Lifestyle Factor Analysis. Transactions of Japan Society of Kansei Engineering 15(7), 693–701 (2016). (in Japanese)
7. H. S. Park and S. B. Cho,An Efficient Attribute Ordering Optimization in Bayesian Net-works for Prognostic Modeling of the Metabolic Syndrome, ICIC2006, LNBI4115, pp. 381–391, Springer-Verlag, 2006
8. Maglogiannis, I., Zafiropoulos, E.: A. platis and C. Lambrinoudakis, Risk analysis of a patient monitoring system using Bayesian Network modeling. J. Biomed. Inform. 39(6), 637–647 (2006)
9. S. M. Lee and P. A. Abbott, Bayesian Network for Knowledge Discovery in Large Datasets, Journal of Biomedical informatics, pp. 389–399, 2003

10. Fuster-Parra, P., Tauler, P., Bennasar-Veny, M., Ligęza, A., López-González, A.A., Aguiló, A.: Bayesian Network Modeling: A Case Study of an Epidemiologic System Analysis of Cardiovascular Risk. Comput. Methods Programs Biomed. **126**, 128–142 (2016)
11. F. Sambo, A. Facchinetti, L. Hakaste, J. Kravic, B. Di Camillo, G. Fico, and C. Cobelli, A Bayesian Network for Probabilistic Reasoning and Imputation of Missing Risk Factors in Type 2 Diabetes, In Conference on Artificial Intelligence in Medicine in Europe, pp. 172–176, Springer, Cham, 2015
12. N. Barakat, Diagnosis of Metabolic Syndrome: A Diversity Based Hybrid Model, Machine Learning and Data Mining in Pattern Recognition, Springer International Publishing, pp. 185–198, 2016
13. Zhao, C., Jiang, J., Xu, Z., Guan, Y.: A Study of EMR-based Medical Knowledge Network and Its Applications. Comput. Methods Programs Biomed. **143**, 13–23 (2017)
14. F. Babič, L. Majnarić, A. Lukáčová, J. Paralič, and A. Holzinger, On Patient's Characteristics Extraction for Metabolic Syndrome Diagnosis: Predictive Modelling Based on Machine Learning, In International Conference on Information Technology in Bio-and Medical Informatics, pp. 118–132, Springer, Cham, 2014
15. Shen, B., Todaro, J.F., Niaura, R., McCaffery, J.M., Zhang, J., Spiro III, A., Ward, K.D.: Are Metabolic Risk Factors One Unified Syndrome? Modeling the Structure of the Metabolic Syndrome X, American Journal of Epidemiology **157**, 701–711 (2003)
16. Shah, S., Novak, S., Stapleton, L.M.: Evaluation and Comparison of Models of Metabolic Syndrome Using Confirmatory Factor Analysis. Eur. J. Epidemiol. **21**, 343–349 (2006)
17. Netica User's Guide, http://www.norsys.com/, Application for Belief Network and Influence Diagrams

Chapter 6
Unsupervised Detection and Analysis of Changes in Everyday Physical Activity Data

Gina Sprint, Diane J. Cook and Maureen Schmitter-Edgecombe

Abstract Sensor-based time series data can be utilized to monitor changes in human behavior as a person makes a significant lifestyle change, such as progress toward a fitness goal. Recently, wearable sensors have increased in popularity as people aspire to be more conscientious of their physical health. Automatically detecting and tracking behavior changes from wearable sensor-collected physical activity data can provide a valuable monitoring and motivating tool. In this paper, we formalize the problem of unsupervised physical activity change detection and address the problem with our Physical Activity Change Detection (PACD) approach. PACD is a framework that detects changes between time periods, determines significance of the detected changes, and analyzes the nature of the changes. We compare the abilities of three change detection algorithms from the literature and one proposed algorithm to capture different types of changes as part of PACD. We illustrate and evaluate PACD on synthetic data and using Fitbit data collected from older adults who participated in a health intervention study. Results indicate PACD detects several changes in both datasets. The proposed change algorithms and analysis methods are useful data mining techniques for unsupervised, window-based change detection with potential to track users' physical activity and motivate progress toward their health goals.

Keywords Physical activity monitoring · Wearable sensors · Unsupervised learning · Change point detection · Data mining

G. Sprint (✉)
Gonzaga University, Spokane, WA, USA
e-mail: sprint@gonzaga.edu

D.J. Cook · M. Schmitter-Edgecombe
Washington State University, Pullman, WA, USA
e-mail: cook@eecs.wsu.edu

M. Schmitter-Edgecombe
e-mail: schmitter-e@wsu.edu

© Springer International Publishing AG 2018
D.E. Holmes and L.C. Jain (eds.), *Advances in Biomedical Informatics*,
Intelligent Systems Reference Library 137,
https://doi.org/10.1007/978-3-319-67513-8_6

6.1 Introduction

In recent years, sensors have become ubiquitous in our everyday lives. Sensors are ambient in the environment, embedded in smartphones, and worn on the body. Data collected from sensors form a time series, where each sample of data is paired with an associated timestamp. This sensor-based time series data is valuable when monitoring human behavior to detect and analyze changes. Such analysis can be used to detect seasonal variations, new family or job situations, or health events. Analyzing sensor-based time series data can also be used to monitor changes in human behavior as a person makes progress toward a fitness goal. Making a significant lifestyle change often takes weeks or months of establishing new behavior patterns [13], which can be challenging to sustain. Automatically detecting and tracking behavior changes from sensor data can provide a valuable motivating and monitoring tool.

Recently, wearable sensors have increased in popularity as people aspire to be more conscientious of their physical health. Many consumers purchase a pedometer or wearable fitness device in order to track their physical activity (PA), often in pursuit of a goal such as increasing cardiovascular strength, losing weight, or improving overall health. Physical activity is estimated by pedometers and fitness trackers in terms of the steps taken by the wearer [6]. To track different types of changes in physical activity data, two or more time periods, or windows, of PA data can be quantitatively and objectively compared. If the two time windows contain significantly different sensor data then this may indicate a significant behavior change. Existing off-the-shelf change point detection methods are available to detect change in time series data, but the methods do not provide context or *explanation* regarding the detected change. For PA data, algorithmic approaches to change detection require additional information about what type of change is detected and its magnitude to potentially report progress to users for motivation and encouragement purposes. Furthermore, existing approaches often do not provide a method for determining if a detected change is *significant*, meaning the magnitude of change is high enough to suspect it likely resulted from a lifestyle alteration. A personalized, data-driven approach to significance testing for fitness tracker users is a necessary feature of physical activity change detection.

Currently, there is no clear consensus regarding which change detection approaches are best for detecting and analyzing changes in PA data. Consequently, we formalize the problem of unsupervised physical activity change detection and address the problem with our Physical Activity Change Detection (PACD) approach. PACD is a framework that (1) segments time series data into time periods, (2) detects changes between time periods, (3) determines significance of the detected changes, and (4) analyzes the nature of the significant changes. We review recently proposed change detection methods and we evaluate the ability of four different change detection approaches to capture pattern changes in synthetic PA data. Next, we illustrate how the change approaches are used to monitor, quantify, and explain behavior differences in Fitbit data collected from older adults who participated in a health behavior intervention. Finally, we conclude with discussions about the limitations of current approaches and suggestions for continued research on unsupervised sensor-based change detection.

6.2 Related Work

In the literature, a few studies have aimed to detect change specifically in human behavior patterns. These approaches have quantified change statistically [17, 19], graphically [19, 22, 26], and algorithmically [5, 9, 14, 26]. Recently, Merilahti et al. [17] extracted features derived from actigraphy data collected for at least one year. Each feature was individually correlated with a component of the Resident Assessment Instrument for insights into how longitudinal changes in actigraphy and functioning are associated. While this approach provides insight into the relationship between wearable sensor data and clinical assessment scores, this study does not directly quantify sensor-based change.

Wang et al. [26] introduced another activity-based change detection approach in which passive infrared motion sensors were installed in apartments and utilized to estimate physical activity in the home and time away from home. The data were converted into co-occurrence matrices for computation of image-based texture features. Their case studies suggest the proposed texture method can detect lifestyle changes, such as knee replacement surgery and recovery. Though the approach does not provide explanation of the detected changes over time, visual inspection of the data is suggested with activity density maps. More recently, Tan et al. [22] applied the texture method to data from Fitbit Flex sensors for tracking changes in daily activity patterns for elderly participants. Another approach for activity monitoring is the Permutation-based Change Detection in Activity Routine (PCAR) algorithm [5]. PCAR researchers modeled activity distributions for time windows of size three months. Changes between windows were quantified with probabilities of change acquired via hypothesis testing.

The change detection algorithms described previously are intended for monitoring human activity behavior. There are several additional approaches that are not specific to activity data, but instead represent generic statistical approaches to detecting changes in time series data. Change point detection, the problem of identifying abrupt changes in time series data [28], constitutes an extensive body of research as there are many applications requiring efficient, effective algorithms for reliably detecting variation. There are many families of change detection algorithms that are suitable for different applications [23]. Algorithms appropriately handling two sample, unlabeled data are most relevant to the current study due to their data-driven change score computation and no need for ground truth information. Unsupervised change detection approaches include subspace models and likelihood ratio methods [14]. One particular subgroup of likelihood ratio methods, direct density ratio estimator methods, is used in various applications [7, 12]. Relative Unconstrained Least-Squares Importance Fitting (RuLSIF) [14] is one such

approach used to measure the difference between two samples of data surrounding a
candidate change point. Other recent change point detection research includes work
on multidimensional [11, 27] and streaming time series data [23].

The above approaches are effective methods for detecting change between two
samples of data; however, they are not explanatory methods as they only identify if
two samples are different and do not provide information on how the samples are
different. Once a change is detected and determined significant, additional analyses
are required to explain the change that occurred. Hido et al. [9] formalized this
problem as *change analysis*, a method of examination beyond change detection to
explain the nature of discrepancy. Hido's solution to change analysis utilizes
supervised machine learning algorithms, specifically virtual binary classifiers
(VCs), to identify and describe changes in unsupervised data. Research by Ng and
Dash [18] and Yamada et al. [28] have also explored methods for detecting and
explaining change in time series data.

The aforementioned methods provide several options for change detection and
analysis, each with their own suitability for various applications. In this paper, we
evaluate the following methods for use in our PACD method: (1) RuLSIF [14],
(2) texture-based dissimilarity [22, 26], (3) our proposed adaptation of PCAR [5] to
handle small window sizes (sw-PCAR), and (4) VC-based change analysis [9].

6.3 Methods

Physical activity is often defined as any bodily movement by skeletal muscles that
results in caloric energy expenditure [3]. Physical activity consists of bouts of
movement that are separated by periods of rest. Physical activity bouts are composed of four dimensions [3]:

1. Frequency: the number of bouts of physical activity within a time period, such
   as a day.
2. Duration: the length of time an individual participates in a single bout.
3. Intensity: the physiological effort associated with a particular type of physical
   activity bout.
4. Activity type: the kind of exercise performed during the bout.

To add exercise throughout the day, individuals can increase their number of
bouts (frequency), increase the length of bouts (duration), increase the intensity of
bouts, and vary the type of physical activity performed during the bouts. These four
components of PA represent four distinct types of changes that can reflect progress
toward many different health goals, such as increasing physical activity or consistency in one's daily routine.

We study the problem of detecting and analyzing change in physical activity
patterns. More specifically, we introduce methods to determine if a significant
change exists between two windows of time series step data sampled from a

physical activity sensor. Algorithm 1, PACD, outlines this process. Let X denote a sample of time series step data segmented into days, $D = \{x_1, x_2, \ldots, x_t, \ldots, x_m\}$, where x_t is a scalar number of steps taken at time interval $t = 1, 2, \ldots, m$ and m is the number of equal-sized time intervals in a day. Let t_{mins} denote the number of minutes per time interval, t. For example, if the sampling rate of the wearable sensor device is one reading per minute, $t_{mins} = 1$ minute and $m = 1440$ minutes $/t_{mins}$ = 1440 intervals. Now, let W be a window of n days such that $W \subseteq X$. Furthermore, an aggregate window, $\hat{W}$, represents the average of all days within the window W:

$$\hat{W} = \frac{1}{n}\sum_{i=1}^{n} D_i, D_i \in W \tag{1}$$

We can compare windows of data within time series data X. These windows may represent consecutive times (e.g., days, weeks, months), a baseline window (e.g., the first week) with each subsequent time window, or overlapping windows. Let W_i denote a window starting at day number i of X $(i \geq 1)$ such that $W_i = X[i : i+n-1] = \{D_i, D_{i+1}, \ldots, D_{i+n-1}\}$. Suppose we have two windows of data, W_i and $W_j (i \leq j)$. Windows W_i and W_j can be formed as subsets of X based on the initial value of i and a parameter *offset* that determines the initial value of j $(j = i + offset)$. For change detection and analysis, a function F computes a change score, $CS = F(W_i, W_j)$ between W_i and W_j. Iteration advancements adv_i and adv_j move windows W_i and W_j respectively for the next comparison. Two windows can be compared in either baseline or sliding window mode. For a baseline window comparison, the first window is a reference window that occurs at the beginning of the time series (i is initialized to 1) and is used in each comparison, so $adj_i = 0$. All subsequent windows are compared to the baseline window. Thus j is initialized to $1 + offset$ and is subsequently advanced by adv_j. In the case of a sliding window comparison, both windows used for comparison are advanced through the time series data. Typically $adv_i = adv_j$ for consistently spaced comparisons. In Algorithm 1, PACD, i is initialized to 1 and j is initialized to $1 + offset$. In steps 17 and 18, i is advanced to $i + adv_i$ and j is advanced to $j + adv_j$.

The choice of window size, n, limits the algorithms that can be applied to the data. For example, the PCAR algorithm [5] is designed for longitudinal data comprising several months; consequently sensitivity decreases with small window sizes. For PACD, we categorize choices for window size n into the following descriptors:

1. Small window ($n = 1$ day). Suitable for performing day-to-day comparisons (e.g. $D_{i(Monday)}$ compared to $D_{j(Monday)}$, $D_{i(Tuesday)}$ compared to $D_{j(Tuesday)}$, ...) or aggregate day comparisons (e.g. $\hat{W}_i$ compared to $\hat{W}_j$, $\hat{W}_{i+adv_i}$ compared to $\hat{W}_{j+adv_j}$, ...).
2. Medium window (2 days $\leq n \leq$ 5 days). Suitable for performing weekday-to-weekday comparisons (e.g. W_i compared to W_j where $W_i = \{D_{i(Monday)}, D_{i(Tuesday)}, D_{i(Wednesday)}, D_{i(Thursday)}, D_{i(Friday)}\}$ and $W_j = \{D_{j(Monday)}, D_{j(Tuesday)}, D_{j(Wednesday)}, D_{j(Thursday)}, D_{j(Friday)}\}$) or weekend-to-weekend comparisons.

3. Large window ($n > 5$ days). Suitable for performing week-to-week or month-to-month comparisons.

Algorithm 1 PACD(X, n, *offset*, adv_i, adv_j)

1: Input: X = time series data
2: Input: n = window length in days
3: Input: *offset* = number of days separating windows
4: Input: adv_i = number of days to advance the first window
5: Input: adv_j = number of days to advance the second window
6: Output: V = vector of change scores
7: Initialize: $i = 1$ and $j = 1 + offset$
8: **for each** pair of windows to compare, W_i and W_j of time series X:
9:      $W_i = X[i : i + n - 1]$
10:     $W_j = X[j : j + n - 1]$
11:     Compute $CS = F(W_i, W_j)$
12:     Determine if CS is significant
13:     Identify the type of change that is exhibited
14:          Manual inspection of change
15:          Unsupervised inspection (change analysis)
16:     Append CS to change score vector V
17:     $i = i + adv_i$
18:     $j = j + adv_j$
    end for
19: **return** Change score vector V

6.3.1 Change Detection Algorithms

In the following sections, we describe algorithmic options for the window-based change score function, F. A summary and comparison of the algorithms is listed in Table 6.1.

6.3.1.1 RuLSIF

Non-parametric approaches to change point detection include a family of methods comparing the probability distributions of two time series samples to determine the corresponding dissimilarity. A greater difference between the two distributions implies a higher likelihood that a change occurred between the two samples. Instead of estimating the probability distributions, their ratio can be estimated and used to detect changes in the underlying probability distributions. Direct density ratio estimation between two windows of time series data is substantially simpler to solve than computing the windows' probability densities independently and then

Table 6.1 Window-based change detection algorithms

Approach	Window size	Window preprocessing	Change score	Change significance test
RuLSIF [14]	Any	Hankel matrix	Probability density ratio estimation with Pearson divergence	Threshold learning in supervised applications. N/A for unsupervised applications
Texture-based [22, 26]	Any	Grey-level co-occurrence matrix, texture features	Weighted normalized Euclidean distance	N/A
PCAR [5] Large	Large	$m \times N$ KL distance permutation matrix	Count of time intervals with significant changes (proportion of permuted KL distances greater than observed window)	N/A
sw-PCAR	Any	N KL distance permutation vector	KL distance	Non-parametric outlier detection based on Boxplot analysis
Virtual classifier [9]	Large	Physical activity features (intra-day and inter-day if window size > 1)	Cross validation prediction accuracy of binary classifier	Hypothesis testing based on prediction accuracy exceeding a threshold

KL = Kullback-Leibler, m = number of time intervals, N = number of permutations

using these to compute the ratio. Unconstrained Least-Squares Importance Fitting (uLSIF) [14] is one such ratio estimation approach that measures the difference between two samples of data surrounding a candidate change point. For this approach, the density ratio between two probability distributions is estimated directly with the Pearson divergence dissimilarity measure. Depending upon the data, the Pearson divergence can be unbounded. Consequently, a modification to uLSIF, relative uLSIF (RuLSIF), utilizes an alpha-relative Pearson divergence to bound the change score above by $1/\alpha$ for $\alpha > 0$ [14].

6.3.1.2 Texture-Based Dissimilarity

For the texture-based approach, two windows of PA data, W_i and W_j, are considered 2-dimensional matrices with rows corresponding to time intervals, columns corresponding to days, and cells containing step values measured from a PA device (see Sect. 6.4.1 for visualizations of PA matrices in Figs. 6.2, 6.3, 6.4, and 6.5). In order to extract texture features from the data, each matrix is converted into a grey-level co-occurrence matrix, a histogram of co-occurring grey scale values of an image [1]. Next, texture features are computed from each co-occurrence matrix, including contrast, homogeneity, angular second moment, energy, density, and correlation features [1, 22, 26]. The features from each window produce feature vectors T_i and T_j. Finally, to compare two windows W_i and W_j for changes, a weighted normalized Euclidean distance measure is used as a change score to quantify the differences between the corresponding feature vectors T_i and T_j. The smaller the Euclidean distance between these two vectors, the more similar the two windows of data are. The texture-based approach can operate on small or large window sizes; however, the method lends itself more appropriately to large window sizes (Wang et al. [26] used window size of one month).

6.3.1.3   sw-PCAR

We propose an enhancement of PCAR to allow permutation-based change detection for any window size. Before introducing sw-PCAR, we will provide an overview of the PCAR approach. PCAR utilizes smart home sensor data to detect changes in behavioral routines with an *activity curve* model [5]. The PCAR approach assumes that an activity recognition algorithm [4] is available to label the sensor data with corresponding activity names. Using PCAR, each day within a window is broken into m time intervals. The activities occurring within each time interval are modeled by a probability distribution, which form an activity curve for the corresponding window. To compute a change score CS between two windows W_i and W_j, the two corresponding activity curves are first maximally aligned with dynamic time warping (DTW). Next, the symmetric Kullback-Leibler (KL) divergence is used to compute the distance between each pair of DTW-aligned activity distributions [5]. To test significance of the distance values, W_i and W_j are concatenated to form a window W of length $2n$ days. Next, all days within W are shuffled. The first half of the shuffled days form a new first window, W_i^*, while the second half form a new second window, W_j^*. KL distances for each time interval pair in W_i^* and W_j^* form a vector that is inserted into a matrix. This shuffling procedure is repeated N times, producing a $N \times m$ permutation matrix, M. If N is large enough, M forms an empirical distribution of the possible permutations of activity data within the two windows of time. Next, for each time interval, the number of permuted KL distances that exceed the original DTW-aligned distance is divided by N to form a p-value. After computing a p-value for each time interval, the Benjamini-Hochberg correction [2] is applied for a given α ($\alpha < 0.05$). Finally, the remaining significant p-values are counted to produce the change score, CS.

While the PCAR algorithm is intended for activity distribution data available from activity recognition algorithms, in this paper we adapt PCAR to analyze physical activity data as part of our PACD method. Instead of activity distribution vectors, we use scalar step counts. Additionally, PCAR is suitable for only large window sizes due to the requirement of permuting daily time series data. We propose a version of PCAR that is more suitable for small to medium-sized windows (sw-PCAR) as required by PACD. Finally, PCAR was originally proposed for correlating change scores with standardized clinical assessments to determine if ambient smart home sensor-based algorithms can detect cognitive decline [5]. Consequently, there is not a test for significance of PCAR change scores. In Sect. 6.3.2 we propose an accompanying significance test for sw-PCAR.

Algorithm 2 outlines the sw-PCAR approach. For sw-PCAR, two windows W_i and W_j are averaged to yield aggregate windows $\hat{W}_i$ and $\hat{W}_j$ (see Eq. 1). A change score CS is derived by computing the KL distance between the average number of steps taken in $\hat{W}_i$ and the average number of steps taken in $\hat{W}_j$. Next, $\hat{W}_i$ and $\hat{W}_j$ are concatenated to form a window W of length two days. All time intervals within W are shuffled. The first half of the shuffled intervals form a new first window, W_i^*, while the second half form a new second window, W_j^*. W_i^* and W_j^* are each averaged to produce two step values. The KL distance between the two values is computed and inserted into a vector. This is repeated N times to produce a N-length vector V of KL distances. Vector V is later used for change score significance testing (see Sect. 6.3.2).

Algorithm 2 sw-PCAR(W_i, W_j, N)

1: Input: W_i, W_j = two windows of time series data
2: Input: N – number of permutations
3: Output: CS = change score
4: Output: sig = (Boolean) significance of CS
5: Initialize: $k = 0$
6: Initialize: V as a vector of length N
7: Compute $\hat{W}_i$, $\hat{W}_j$ aggregate windows
8: Compute CS, the KL distance between $\hat{W}_i$ and $\hat{W}_j$
9: **while** $k < N$:
10:    Shuffle the time intervals of $\hat{W}_i$ and $\hat{W}_j$
11:    Generate new aggregate windows W_i^* and W_j^*
12:    Compute the KL distance between W_i^* and W_j^*
13:    Store resulting distance in V
14:    $k = k + 1$
   end while
15: sig = BoxplotOutlierDetection(CS, V) (see Algorithm 3)
16: **return** CS, sig

6.3.1.4   Virtual Classifier

Change analysis, as proposed by Hido et al. [9], utilizes a virtual binary classifier to detect and investigate change. We apply the VC approach as part of PACD for large window sizes. First, a feature extraction step reduces two windows W_i and W_j into two $n \times z$ feature matrices, M_i and M_j, where n is the window size (in days) and z is the number of features that are extracted (see Sect. 6.3.3 for feature descriptions). Next, each daily feature vector of M_i is labeled with a positive class and each daily feature vector of M_j is labeled with a negative class. VC trains a decision tree to learn the decision boundary between the virtual positive and negative classes. The resulting average prediction accuracy based on k-fold cross validation is represented as p_{VC}. If a significant change exists between W_i and W_j, the average classification accuracy p_{VC} of the learner should be significantly higher than the accuracy expected from random noise, $p_{rand} = 0.5$, the binomial maximum likelihood of two equal length windows [9].

6.3.2 Change Significance Testing

Significance testing of change score CS is necessary to interpret change score values. For the VC approach, Hido et al. [9] proposed a test of significance to determine if p_{VC} is significantly greater than p_{rand}. For this test, the inverse survival function of a binomial distribution is used to determine a critical value, $p_{critical}$, at which n Bernoulli trials are expected to exceed p_{rand} at α significance. If $p_{VC} \geq p_{critical}$, a significant change exists between the two windows, W_i and W_j.

The PCAR approach does not have an accompanying test of significance. We address this with our proposed sw-PCAR technique. sw-PCAR computes change significance by comparing CS to the permutation vector V with boxplot-based outlier detection (see Algorithm 3). An outlier can be defined as an observation which appears to be inconsistent with other observations in the dataset [15]. For this method, the interquartile range (75th percentile – 25th percentile) of V is computed. Values outside of the 75th percentile + 1.5· interquartile range are considered outliers [25]. If CS is determined to be an outlier of V, then the change score is considered significant. There are alternative approaches to test membership of an observation (i.e. CS) to a sample distribution (i.e. V) other than boxplot outlier detection. If the sample is normal, statistical tests such as Grubb's test for outliers [8] can be applied. However, the assumption of normality does not hold for all samples of human behavior data. More advanced alternatives include data mining techniques relevant to outlier detection [10, 15]. Exploration and testing of such data mining techniques are outside the scope of this paper.

Algorithm 3 BoxplotOutlierDetection(*CS*, *V*)

1: Input: CS = change score between two windows
2: Input: V = sample distribution vector
3: Output: sig = (Boolean) significance of CS
4: Arrange V in ascending order
5: Compute Q_1, the 25th percentile of V
6: Compute Q_3, the 75th percentile of V
7: Compute the interquartile range of V, $IQR = Q_3 - Q_1$
8: **if** $CS > Q_3 + 1.5 \cdot IQR$:
9:     $sig = True$
10: **else**:
11:     $sig = False$
12: **return** sig

RuLSIF does not explicitly provide a method to determine a cutoff threshold for values of the Pearson divergence function that are considered significant change scores. In supervised applications where ground truth change labels are available, a threshold parameter is typically learned by repeated training and testing with different parameter values. For unsupervised applications, domain knowledge and/or alternative data-driven approaches are necessary. Like RuLSIF, the texture-based method also does not provide a test of change significance. For the RuISIF and texture-based approaches, we propose a large window change significance test based on intra-window variability and outlier detection.

Our proposed change significance test utilizes the existence of day-to-day variability in human behavior patterns [21]. In order to consider a change between two windows as significant, the magnitude of change should exceed the day-to-day variability within each window. To illustrate, consider two adjacent, non-overlapping windows W_1 and W_7, each of length $n = 6$ days. Now run a pairwise sliding window change algorithm over W_1 concatenated with W_7. If there is a significant change between the windows, the magnitude of change should be higher for the inter-window comparison (between days 6 and 7) than any other intra-window comparison. Figure 6.1 shows an example plot of sw-PCAR change scores for real Fitbit data illustrating this phenomenon. There are small, noisy day-to-day changes for all comparisons except the largest maximum occurring for the inter-window comparison (6th change score).

Fig. 6.1 Pairwise sliding window sw-PCAR change scores

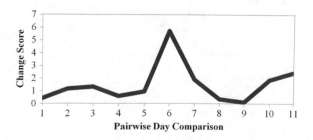

Based on the assumption that a significant inter-window change should exceed intra-window change, we propose an intra-window change significance test (see Algorithm 4). Given a change score CS between two windows, the task is to determine if CS is significant. To do this, first compute a list of all possible daily change scores, DCS, within each window. DCS contains $2 \cdot$ Combination $(n, 2)$ change scores (see Algorithm 5). For example, a week-to-week comparison ($n = 7$) would generate an intra-window daily change score sample of 42 day-to-day variations. Next, apply the outlier detection method (see Algorithm 3) from sw-PCAR to test if CS is an outlier score when compared to the distribution of intra-window daily change scores DCS. Advantages of the proposed test include it is non-parametric and can be coupled with any small window change detection function, F. Furthermore, the candidate change score, CS, can be computed based on any window size (e.g. Monday-to-Monday, aggregate-to-aggregate, week-to-week, etc.).

Algorithm 4 IntraWindowSignificance(W_i, W_j, n, CS, F)

1: Input: W_i, W_j = two windows of time series data
2: Input: n = window size
3: Input: CS = change score between W_i and W_j
4: Input: F = change score function
5: Output: sig = (Boolean) significance of CS
6: Initialize: DCS = vector of daily change scores
7: Append IntraWindowChange(W_i, n, F) to DCS (see Algorithm 5)
8: Append IntraWindowChange(W_j, n, F) to DCS (see Algorithm 5)
9: Compute sig = BoxplotOutlierDetection(CS, DCS) (see Algorithm 3)
10: **return** sig

Algorithm 5 IntraWindowChange(W, n, F)

1: Input: W = window of time series data
2: Input: n = window size
3: Input: F = change score function
4: Output: DCS = vector of daily change scores
5: Initialize: $i = 1$, $j = 1$
6: **while** $i \leq n - 1$:
7:     $j = i + 1$
8:     **while** $j \leq n$:
9:         $CS = F(W[i], W[j])$
10:        Append CS to DCS
11:        $j = j + 1$
12:     **end while**
13:    $i = i + 1$
    end while
14: **return** DCS

6.3.3 Change Analysis

If a change significance test concludes a change score is significant, the next step is to determine the source of change (see Algorithm 1 for an overview of the PACD process). Often this step requires the computation of features that summarize the data and provide a meaningful context for change. For example, the number of daily steps taken is an example of a simple PA feature. The change between daily steps from one window of time to the next can be quantified and used for an explanation of change. Several approaches exist to capture change across time in individual metrics. A straightforward method is to compute the percent change for a feature f from a previous window W_i to a current window W_j: $\Delta\% = (f_{W_j} - f_{W_i})/f_{W_i}$. Statistical approaches such as two sample tests or effect size analyses can also be applied to quantify change; however, in applying repeated statistical tests, the multiple testing problem should be accounted for with a method such as the Bonferroni or Benjamini-Hochberg correction [2].

One of the advantages of the VC approach over other change point detection algorithms is it includes an explanation of the source of change without reliance on statistical tests. Upon significant change detection, retraining a decision tree on the entire dataset and inspecting the tree reveals which features are most discriminatory in learning the differences between two windows. Naturally, this approach requires a pre-processing step to extract relevant features from the windowed PA time series data.

Features extracted from the PA data (see Table 6.2) serve two purposes: (1) as features for the VC approach (RulSIF, texture-based dissimilarity, and sw-PCAR do not make use of features for change detection), and (2) for explanation of changes

Table 6.2 Physical activity features used for training virtual classifiers and for explanation of changes discovered by change detection algorithms. Features are categorized by time period (window size)

Period	Metric	Description
1 Day	Bout steps	Mean and SD of number of steps per bout
	Period steps [19]	Total, mean, and SD per period: (1) 24 h period (full days), (2) Day (9am–9 pm), (3) Night (12am–6am). Normalized by 24 h mean
	Night/day ratio	See period steps definition
	Number of bouts	Count of detected PA bouts
	Bout minutes	Mean and SD of duration of bouts
	Physical activity intensity percentage	Mean percentage of (1) sedentary (< 5 steps/min), (2) low ($5 \leq$ steps/min < 40), (3) moderate ($40 \leq$ steps/min < 100), (4) high (≥ 100 steps/min) activity levels
	Rest minutes	Mean and SD of duration of rest periods
≥ 1 Day	Relative amplitude	Normalized ratio between the most active 8 h and the least active 4 h

(continued)

Table 6.2 (continued)

Period	Metric	Description
	Texture features [1]	See Sect. 6.3.1.2
≥ 2 Days	Inter-daily stability (IS) [17]	Quantifies stability between days
	Intra-daily variability [17]	Quantifies the fragmentation of rhythm and activity. Ratio of variance of consecutive time intervals and overall variance
	Circadian rhythm strength [17]	Ratio of average night-time activity (11 pm–5am) by the average activity of the previous day (8am–8 pm)
	Cosinor mesor [17, 19]	Time series mean from fitting a cosinor functional model with a 24 h period to time series data via least squares method
	Cosinor amplitude	Difference between the mesor and peak (or trough)
	Cosinor acrophase	Time of day at which the peak of a rhythm occurs
	Poincare SD1 [19]	Standard deviation of Poincare data against the axis $x = y$
	Poincare SD2 [19]	Standard deviation of Poincare data against the axis orthogonal to $x = y$ and crosses this axis at the center of mass
	Poincare circadian rhythm preservation (CRP) [19]	Day-to-day circadian rhythm preservation based on dispersion values from SD1 and SD2 with delays of 24 h and 12 h $CRP = SD2_{24h} + SD1_{12h} - SD1_{24h} - SD2_{12h}$

SD = standard deviation

discovered by change detection algorithms (see Sect. 6.3.1). Features are grouped together based on the number of days required for computation: (1) one day, (2) at least one day, or (3) two or more days. Daily features include PA summaries based on intensity, frequency, duration, and variability of PA bouts. Sequences of time series data with steps greater than a threshold, S, are considered a bout of PA. If ground truth activity labels, such as walking, biking, chores, etc., are available from the device user and/or an activity recognition algorithm [4], PA type can be inferred and S can be updated dynamically for different activities. For this study, we assume such labeled information is not available and set $S = t_{mins}$, assuming physical activity is characterized by at least one step per minute. Features requiring at least two days of data summarize activity across or between days or quantify the user's circadian rhythm (the periodicity from day-to-day [21]). Poincare-plot analysis [19] provides an additional set of useful PA features.

6.4   Results

To demonstrate the PACD approach, two datasets are presented, Hybrid-synthetic (HS) and B-Fit (BF). The HS dataset comprises synthetic data and the BF dataset comprises real-world Fitbit data collected from a health intervention study. HS and BF data are subject to pre-processing prior to serving as input to PACD. Pre-processing includes down sampling the data for a given time interval length, t_{mins}, by summing the steps every t_{mins} minutes. For the case of sw-PCAR, add one smoothing is applied to avoid division by zero during KL divergence computations. Furthermore, missing data are identified and handled for BF data. Days with zero steps taken during the day (9am–9 pm) are considered missing data. First, to fill a missing day, $D_{missing}$, the day in the opposite window, D_{other}, with the same day of the week as $D_{missing}$ is identified. Euclidean distance-based clustering is applied to find the k nearest neighbor days, NN_{other}, of $D_{other}(k = 3)$. The days of the week for each day in NN_{other} are then identified. These are used to select days, $NN_{missing}$ in the original window containing $D_{missing}$. The k days of $NN_{missing}$ are aggregated (see Eq. 1) and used to fill $D_{missing}$.

For PACD computations, the following algorithm parameter values are used: window size n: 6 days; window *offset*: 6 days; RuLSIF α: 0.1; RuLSIF cross validation folds: 5; number of sw-PCAR permutations N: 1000; VC cross validation folds: 4; VC prediction threshold $p_{critical}$: 0.75; minimum steps in a bout S: t_{mins}. The time interval aggregation size t_{mins} is tested with values of t_{mins} = {1, 5, 10, 15,..., 60 min}. We hypothesize that PACD will find PA changes (bout frequency, intensity, duration, and variability), using each of the change detection methods. However, we anticipate that the significance of the change will vary depending on the algorithm used, the parameter value choices, and the level of change that is inherent in each dataset.

6.4.1   Hybrid-Synthetic Dataset

To generate the HS dataset, step data collected from a volunteer wearing a Fitbit Charge Heart Rate fitness tracker was re-sampled and modified the data to produce five different synthetic physical activity profiles, each exhibiting a different type of change. The length of HS profiles was set to 12 days, resulting in two equal size windows of 6 days for comparison (days 1–6 compared to days 7–12). Twelve days was chosen for similarity to the BF dataset. The HS profiles with their profile identification (HS0-4) and a description are as follows:

1. HS0: No significant day-to-day or window-to-window change. Data is subject to small daily variation. HS0 represents a baseline for "no change".

2. HS1: Medium day-to-day change and consequently significant window-to-window change. Increased bout duration and intensity from day-to-day.
3. HS2: No significant day-to-day change but significant window-to-window change. Increased activity for days 7–12.
4. HS3: Medium day-to-day change and consequently significant window-to-window change. Increased activity variability from day-to-day.
5. HS4: No significant day-to-day change for days 1–6. Significant day-to-day activity variability for days 7–12. Consequently significant window-to-window change.

Figures 6.2, 6.3, 6.4, and 6.5 show the associated activity density maps for HS profiles HS1-4. An activity density map is a heat map proposed by Wang et al. [26]

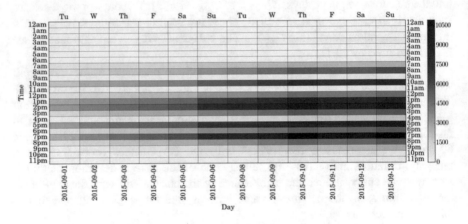

Fig. 6.2 Hybrid-synthetic (HS) step density map for HS1

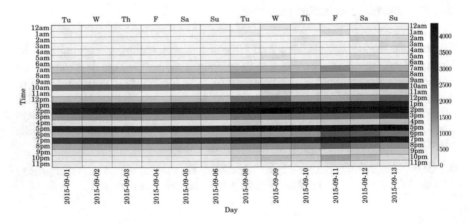

Fig. 6.3 Hybrid-synthetic (HS) step density map for HS2

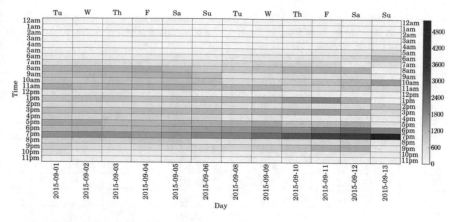

Fig. 6.4 Hybrid-synthetic (HS) step density map for HS3

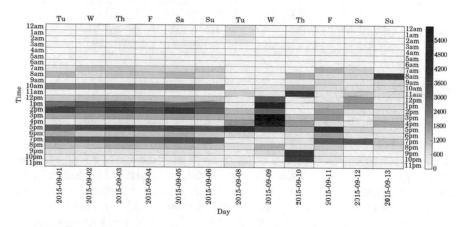

Fig. 6.5 Hybrid-synthetic (HS) step density map for HS4

to visualize daily activity (steps for this study) as a function of 24 h time (Y-axis) and window time (X-axis). Table 6.3 shows RuLSIF, Texture-based, sw-PCAR, and VC significant change results for each HS profile for each time interval length t_{mins}. Window one (days 1–6) and window two (days 7–12) values (mean ± standard deviation) for the contextual features of number of bouts, bout minutes, daily steps, and sedentary minutes percent are listed in Table 6.4. Results in Table 6.4 have time interval length $t_{mins} = 1$ minute in order to report the most detailed feature values. For further change analysis, decision trees are shown in Fig. 6.6 for HS profiles HS1-4.

Table 6.3 Hybrid-synthetic (HS) significant change detection as a function of time interval size t_{mins} for each HS profile. Results are in the form count: Boolean (significant change 0: false, 1: true) {HS0, HS1, HS2, HS3, HS4}

t_{mins}	RulSIF	Texture-based	sw-PCAR	Virtual classifier	Total
1	2:0,0,1,0,1	1:0,0,1,0,0	3:0,1,1,0,1	4:0,1,1,1,1	10
5	3:0,1,1,0,1	2:0,0,1,0,1	3:0,1,1,0,1	4:0,1,1,1,1	12
10	2:0,0,1,0,1	2:0,0,1,1,0	2:0,1,1,0,0	3:0,0,1,1,1	9
15	3:0,1,1,0,1	2:0,0,1,1,0	1:0,1,0,0,0	3:0,0,1,1,1	9
20	3:0,1,1,0,1	3:0,0,1,1,1	1:0,1,0,0,0	3:0,0,1,1,1	10
25	1:0,0,1,0,0	1:0,0,1,0,0	1:0,1,0,0,0	3:0,0,1,1,1	6
30	3:0,1,1,0,1	1:0,0,0,1,0	1:0,1,0,0,0	4:0,1,1,1,1	9
35	2:0,1,1,0,0	1:0,0,0,1,0	1:0,1,0,0,0	3:0,1,1,0,1	7
40	3:0,1,1,1,0	0:0,0,0,0,0	1:0,1,0,0,0	3:0,0,1,1,1	7
45	2:0,0,1,0,1	1:0,0,1,0,0	1:0,1,0,0,0	2:0,0,1,0,1	6
50	2:0,0,1,0,1	0:0,0,0,0,0	1:0,1,0,0,0	4:0,1,1,1,1	7
55	3:0,1,1,1,0	1:0,1,0,0,0	0:0,0,0,0,0	3:0,1,1,0,1	7
60	4:0,1,1,1,1	0:0,0,0,0,0	1:0,1,0,0,0	4:0,1,1,1,1	9
Total	33:0,8,13,3,9	15:0,1,7,5,2	17:0,12,3,0,2	43:0,7,13,10,13	108

6.4.2 B-Fit Dataset

The BF dataset consists of data collected from 11 older adults (Male = 3, Female = 8; age 57.09 ± 8.79 years) participating in a 10-week health intervention. Study inclusion criteria consisted of older adults over the age of 55 who had risk factors for developing dementia. At risk individuals were defined as those who had at least one first degree relative with Alzheimers disease or dementia, or who had cardiovascular disease risk factors (e.g., diabetes, midlife obesity, smoking, hypertension). Participants had to be able to provide their own informed consent. As part of this study, participants' PA profiles were assessed with wrist-worn Fitbit Flex fitness trackers for one week (six full 24 h days) before and after the intervention. During weeks two through nine, the participants were educated in eight different subjects related to health (e.g., exercise, nutrition, sleep) and set personal goals for each subject. To track goal achievement each week, individuals rated themselves on each personalized goal that they set using a 0 to 3 rating scale (0: did not meet goal, 1: partly met goal, 2: completely met goal, 3: exceeded goal). For the BF dataset, each participant's change significance testing results are presented in Table 6.5. Four contextual features (number of bouts, bout minutes, daily steps, and sedentary percent) pre and post-intervention values are listed in Table 6.6. Finally, decision trees are shown in Fig. 6.7 for select BF participants with a significant VC change score (P2, P7, and P10).

Table 6.4 Hybrid-synthetic (HS) feature results (mean ± standard deviation) with t_{mins} = 1 min. First and second window values are separated by a comma

ID	Number of bouts	Bout minutes	Daily steps	Sedentary %
HS0	70.33, 70.00	$5.10 \pm 9.91, 5.13 \pm 9.92$	20601.65, 21274.32	75.65, 75.56
HS1	34.50, 14.17	$19.39 \pm 23.93, 46.82 \pm 56.78$	36409.49, 72769.11	64.57, 54.59
HS2	71.50, 62.50	$5.07 \pm 9.83, 7.63 \pm 11.13$	20755.53, 30037.48	75.62, 67.44
HS3	54.83, 102.83	$18.71 \pm 51.74, 6.49 \pm 14.49$	14395.85, 14746.43	45.22, 63.72
HS4	53.50, 81.33	$8.14 \pm 12.61, 4.56 \pm 8.48$	22048.02, 17327.00	70.66, 77.86

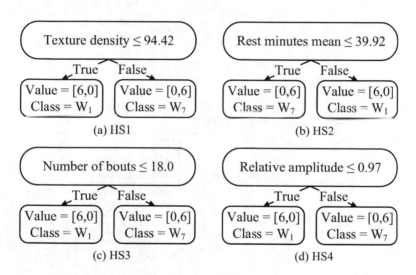

Fig. 6.6 Decision trees for Hybrid-Synthetic (HS) profiles with significant virtual classifier change scores for t_{mins} = 5 min

6.5 Discussion

In this paper, we investigate the PACD approach for unsupervised change detection and analysis of PA time series. The abilities of four presented methods to detect change are evaluated on two original datasets: 1) the HS dataset, comprised of 5 synthetic profiles and 2) the BF dataset, comprised of 11 participants' Fitbit data from an intervention study.

Table 6.5 Health intervention study significant change detection as a function of time interval size t_{mins}. Results are in the sparse form count: {IDs}: Boolean (significant change: 0 false, 1 true)

t_{mins}	RulSIF	Texture-based	sw-PCAR	Virtual classifier	Total
1	1:P2:1	0	10:P10:0	5:P2,6,7,10,11:1	16
5	2:P2,7:1	0	5:P1,2,3,4,8:1	6:P2,6,7,8,10,11:1	13
10	1:P2:1	0	4:P1,2,4,8:1	6:P2,3,6,7,8,10:1	11
15	1:P2:1	0	4:P1,2,4,8:1	4:P2,7,10,11:1	9
20	2:P2,7:1	0	3:P2,4,8:1	4:P2,7,10,11:1	9
25	1:P2:1	0	3:P2,4,8:1	2:P7,10:1	6
30	2:P2,8:1	0	3:P2,4,8:1	2:P6,10:1	7
35	3:P2,4,8:1	0	3:P2,4,8:1	5:P2,6,7,10,11:1	11
40	1:P2:1	0	3:P2,4,8:1	3:P2,7,10:1	7
45	4:P2,3,4,6	0	1:P2:1	2:P2,10:1	7
50	1:P2:1	0	1:P2:1	5:P2,6,7,9,10:1	7
55	1:P10:1	0	1:P2:1	3:P2,4,10:1	5
60	2:P4,9:1	0	1:P2:1	4:P2,6,10,11	7
Total	22	0	42	51	115

Table 6.6 B-Fit (BF) feature results (mean ± standard deviation) with t_{mins} = 1 min. Pre and post-intervention values are separated by a comma

ID	Number of bouts	Bout minutes	Daily steps	Sedentary %
P1	73.50, 89.67	2.35 ± 1.93, 2.63 ± 2.61	3479.00, 4658.33	88.37%, 84.43%
P2	81.00, 15.83	2.57 ± 2.54, 2.75 ± 1.85	4279.50, 1161.44	86.30%, 97.44%
P3	88.50, 27.50	2.72 ± 2.86, 2.36 ± 2.08	5886.67, 4558.50	84.06%, 86.32%
P4	81.17, 60.33	3.90 ± 5.27, 3.71 ± 4.85	11177.00, 7399.67	79.11%, 85.71%
P5	73.67, 76.50	2.96 ± 4.06, 2.60 ± 3.22	6994.17, 5470.50	85.71%, 86.22%
P6	105.33, 88.00	2.63 ± 2.46, 2.66 ± 2.39	7127.00, 6207.67	81.82%, 84.85%
P7	64.33, 63.50	3.30 ± 4.20, 3.45 ± 3.61	7354.67, 6181.17	85.78%, 85.90%
P8	104.17, 102.50	5.52 ± 7.51, 3.35 ± 3.79	17680.78, 11440.00	66.66%, 77.18%
P9	99.50, 116.67	2.40 ± 2.49, 2.40 ± 2.57	5844.50, 6731.83	84.11%, 81.46%
P10	85.00, 80.50	2.99 ± 3.24, 3.58 ± 4.29	1136.51, 1210.85	82.94%, 81.62%
P11	83.00, 89.50	2.51 ± 2.73, 2.31 ± 2.23	5753.50, 4868.83	86.16%, 86.44%

6.5.1 Hybrid-Synthetic Dataset

The HS dataset reveals several insights into the change detection algorithms. First, the time interval length yielding the highest number of significant changes is t_{mins} = 5 min with 12 changes, closely followed by t_{mins} = 1 and t_{mins} = 20 min with 10 changes (see Table 6.3). Since HS profiles are sampled from a volunteer's real Fitbit data, these intervals suggest movement patterns occur in 1, 5, and 20 min

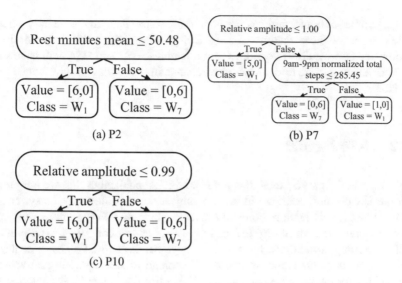

Fig. 6.7 Decision trees for B-Fit (BF) participants with significant virtual classifier change scores for t_{mins} = 5 min

chunks for this individual. For all time interval lengths, the algorithms correctly do not detect a significant change between first and second window data for the HS0 profile. HS0 is generated to exhibit small day-to-day variation in step intensity and is not characterized by large changes between windows.

For HS1-4 profiles, significant changes between windows are detected. For all time interval lengths, the VC approach picks up the most changes (43 changes), followed by RuLSIF (33), sw-PCAR (17), and texture-based (15). As a group, the algorithms' are able to sense changes in value (HS1, HS2) and changes in variability (HS3, HS4), with 64 and 44 changes respectively. Changes for HS2 (36) are the most frequently detected, followed by HS1 (28), HS4 (26), and HS3 (18). The lower number of changes detected for HS3 is possibly due to high intra-window daily change scores for days 7–12 (see Fig. 6.4) used for Boxplot significance testing (see Algorithm 3). Window-based change (HS2, HS4) is perfectly detected for all time intervals by VC (HS2, HS4: 13) and RuLSIF (HS2: 13). Investigating the t_{mins} = 5 min results reveal all four algorithms determine significant changes for HS2 and HS4 (see Table 6.3). For HS1, near perfect detections are made by sw-PCAR (12).

Upon inspection of the associated decision trees for HS1-4 (see Fig. 6.6), the features of texture density, average daily rest minutes, number of bouts, and relative amplitude are discriminatory features. The explanatory power of the features is potentially useful for reporting to the wearable sensor user the dimensions of change in their physical activity. Features useful for such purposes are simple, common features that do not require explanation to the user. For example, texture density or relative amplitude are useful features for detecting changes in PA patterns, but are relatively unimportant to a user. More meaningful features to a user include number of bouts, minutes per bout, daily steps taken, and sedentary percent. Table 6.4 shows these features for the HS profiles. HS0 exhibits quite similar window one and window two values for all features. HS2 and HS4 both have small standard deviations due to window-based change in lieu of day-to-day change (HS1 and HS3).

6.5.2 B-Fit Dataset

Analyzing the BF participants' data poses additional challenges that are not present with the HS profiles. Real-world human subject data is inherently noisy, characterized by seemingly random bouts of PA and rest periods. Furthermore, self-report and direct measurement of physical activity are often not congruent, with previous studies reporting correlations in as wide a range of -0.71 to 0.96 [20]. For the BF group, the participants demonstrated a wide spread of self-reported goal achievement ratings for the exercise category, 1.59 ± 1.05. For example, P6, P9, and P10 rated their exercise goal achievements as low (exercise rating: 1, 0, 0.5 respectively). Due to heart problems, P9's doctor instructed him not participate in exercise-related activities. On the other hand, P2 rated her exercise goal achievement the highest (exercise rating: 3). Upon inspection of P2's data, it is evident there is a discrepancy between the participant's perception of her PA and the steps recorded by the Fitbit. It is not uncommon for self-reported measures of physical activity to be inconsistent with direct measures [20]; therefore, the participants' self-ratings are used for insights into individual goal achievements, not as ground truth information for changes exhibited. The issues with self-reported PA measures exacerbate the need for unsupervised change detection and analysis methods.

Depending on the algorithm, significant changes are commonly detected for 5 out of the 11 BF participants: P2: 35; P10: 14; P4: 13; P8: 13; P7: 12 (see Table 6.3). Virtual classifier and sw-PCAR detect the highest number of changes (51 and 42 changes each). The distribution of detected changes by sw-PCAR is highly influenced by time interval length (sw-PCAR: 3.23 ± 2.42 number of changes detected compared to VC: 3.92 ± 1.44). sw-PCAR is not sensitive for small time intervals ($t_{mins} = 1$ min) or large time intervals ($t_{mins} = \{45, 50, 55, 60$ min$\}$), and the number of changes detected decreases as time interval length increases. Virtual classifier does not appear to be as heavily influenced by the time interval length. The texture-based approach is the least sensitive algorithm and did not detect any changes in the BF data.

Performing change analysis and investigating the detected changes yields insights for several of the participants. P2 rated herself as completely meeting her exercise goal of walking more; however, the Fitbit data tells a different story. Several features in Table 6.4 show decreased PA for P2: average number of bouts (pre: 81.00, post: 15.83), daily steps (pre: 4279.50, post: 1161.44 steps), and percentage of time sedentary (pre: 86.30%, post: 97.44%). Additionally, P2's decision tree (see Fig. 6.7a) provides evidence that she rested more during post-intervention testing. In summary, the features suggest the changes detected by the algorithms are actually changes in the opposite direction of her goal. Contrary to P2, P10 exhibited a significant change (as detected consistently by VC) in the direction toward her goal of walking more. Inspection of P10's features shows an increase in bout minutes and average steps per day. Average daily steps increased from 1136.51 steps pre-intervention to 1210.85 steps post-intervention testing, a 6.54% increase. The remaining participants with significant changes (P4, P7, and P8) demonstrated a decrease in average daily steps taken from pre to post intervention. It should be noted that during the week of post-test data collection the weather conditions were adverse and this may have partially contributed to the decrease observed in average daily steps. Research has shown PA levels can be influenced by adverse weather conditions [24]. It is also worth noting the participants exhibited improvements in other PA features. For example, relative amplitude has been reported to decrease with worsening health [16], thus P7 and P10's increased relative amplitude post-intervention is healthy (see Figs. 6.7b, c). Also, P9 was not planning on increasing exercise; however, P9 increased his daily steps post-intervention by 15.18%.

One of the limitations of this study includes having only one week of pre-intervention Fitbit data for BF participants. With at least two weeks of pre-intervention data, change scores can be computed between week one and two of pre-intervention data to provide an estimate of inter-week variability. With a quantification of inter-week variability, we can determine if the measured change between pre and post-intervention weeks is due to the intervention or natural variability. An additional limitation includes not having a full 7 days of BF data during pre and post-intervention weeks. Finally, more sophisticated methods to fill missing data could be utilized with fitness trackers that include heart rate monitors, due to more reliable detection of sensor donned/doffed. Consequently, future work includes performing change analysis on real-world datasets from different fitness trackers, multidimensional data (e.g. heart rate, elevation, etc.), labeled activity data, and longer windows of time. With time series data longer than two years, several additional analyses could be performed including: daily/weekly/monthly/ yearly period analysis and slicing along different dimensions (e.g. Mondays, weekends, holidays, or activities if labeled information is available).

6.6   Conclusions

We address the problem of unsupervised physical activity change detection and analysis with our proposed Physical Activity Change Detection approach. PACD is a framework we designed to detect and analyze changes in physical activity data. We compare the abilities of three change detection algorithms from the literature and one proposed algorithm, sw-PCAR, to capture different types of changes in synthetic and real-world datasets. Results indicate the approaches detect several changes in both datasets; particularly for physical activity profiles exhibiting large changes between windows instead of incremental day-to-day changes. Contextual features such as average number of daily steps, minutes per bout, and sedentary percent provide an explanation of the changes that are detected. The algorithms and analysis methods are useful data mining techniques for unsupervised, window-based change detection. Future work involves quantifying the change in accuracy (ability to find true positives and not false positives in synthetic data) as parameters such as time window length is incremented or decremented. Additional future work includes implementing our PACD method in an online, smartphone application to track users' physical activity and motivate progress toward their health goals.

Acknowledgements We wish to thank the Department of Psychology at Washington State University for their insights and help with data collection. This material is based upon work supported by the National Science Foundation under Grant No. 0900781.

References

1. Albregtsen, F., et al.: Statistical texture measures computed from gray level coocurrence matrices. Image processing laboratory, Department of informatics, university of oslo pp. 1–14. http://www.uio.no/studier/emner/matnat/ifi/INF4300/h08/undervisningsmateriale/glcm.pdf (2008)
2. Benjamini, Y., Hochberg, Y.: Controlling the false discovery rate: A practical and powerful approach to multiple testing. J. R. Stat. Soc. Series B (Methodological) 57(1):289–300. http://www.jstor.org/stable/2346101 (1995)
3. Caspersen, C.J., Powell, K.E., Christenson, G.M.: Physical activity, exercise, and physical fitness: Definitions and distinctions for health-related research. Public Health Rep 100 (2):126–131. http://www.ncbi.nlm.nih.gov/pmc/articles/PMC1424733/ (1985)
4. Chen, L., Hoey, J., Nugent, C., Cook, D., Yu, Z.: Sensor-based activity recognition. IEEE Trans. Syst. Man Cybern. Part C Appl. Rev. **42**(6), 790–808 (2012). doi:10.1109/TSMCC.2012.2198883
5. Dawadi, P.N., Cook, D.J., Schmitter-Edgecombe M.: Modeling patterns of activities using activity curves. Pervasive and Mobile Computing. doi: 10.1016/j.pmcj.2015.09.007. http://www.sciencedirect.com/science/article/pii/S157411921500173X (2015)
6. Dobkin, B.H.: Wearable motion sensors to continuously measure real-world physical activities. Curr Opin Neurol 26(6):602–608, doi: 10.1097/WCO.0000000000000026. http://www.ncbi.nlm.nih.gov/pmc/articles/PMC4035103/ (2013)
7. Feuz, K., Cook, D., Rosasco, C., Robertson, K., Schmitter-Edgecombe, M.: Automated Detection of Activity Transitions for Prompting. IEEE Transactions on Human-Machine Systems 45(5), 575–585 (2015). doi:10.1109/THMS.2014.2362529
8. Grubbs, F.E.: Procedures for detecting outlying observations in samples. Technometrics 11 (1):1, doi: 10.2307/1266761. http://www.jstor.org/stable/1266761?origin=crossref (1969)
9. Hido, S., Id, T., Kashima, H., Kubo, H., Matsuzawa, H.: Unsupervised change analysis using supervised learning. In: Washio, T., Suzuki, E., Ting, K.M., Inokuchi, A. (eds.) Advances in Knowledge Discovery and Data Mining, no. 5012 in Lecture Notes in Computer Science, Springer Berlin Heidelberg, pp. 148–159. http://link.springer.com/chapter/10.1007/978-3-540-68125-0_15, doi: 10.1007/978-3-540-68125-0_15 (2008)

10. Hodge, V.J., Austin, J.: A survey of outlier detection methodologies. Artif. Intell. Rev. 22 (2):85–126, doi: 10.1007/s10462-004-4304-y. http://link.springer.com/article/10.1007/s10462-004-4304-y (2004)
11. Hu, M., Zhou, S., Wei, J., Deng, Y., Qu, W.: Change-point detection in multivariate time-series data by recurrence plot. WSEAS Transactions on Computers 13:592–599. http://www.wseas.org/multimedia/journals/computers/2014/a305705-716.pdf (2014)
12. Javed, F., Farrugia, S., Colefax, M., Schindhelm, K.: Early warning of acute decompensation in heart failure patients using a noncontact measure of stability index. IEEE Trans. Biomed. Eng. 63(2), 438–448 (2016). doi:10.1109/TBME.2015.2463283
13. Lally, P., Jaarsveld, C. van, Potts, H., Wardle, J.: How are habits formed: Modelling habit formation in the real world. Eur. J. Soc. Psychol. 40(6):998–1009, doi: 10.1002/ejsp.674. http://dx.doi.org/10.1002/ejsp.674 (2010)
14. Liu, S., Yamada, M., Collier, N., Sugiyama, M.: Change-point detection in time-series data by relative density-ratio estimation. Neural Networks 43:72–83, doi: 10.1016/j.neunet.2013.01.012. http://www.sciencedirect.com/science/article/pii/S0893608013000270 (2013)
15. Maimon, O., Rokach, L.: Data mining and knowledge discovery handbook. Springer, New York. http://www.books24x7.com/marc.asp?bookid=16218 (2005)
16. Merilahti, J., Petkoski-Hult, T., Ermes, M., Gils, M.v., Lahti, H., Ylinen, A., Autio, L., Hyvrinen, E., Hyttinen, J.: Evaluation of new concept for balance and gait analysis: Patients with neurological disease, elderly people and young people. Gerontechnology 7(2):164. http://www.gerontechnology.info/index.php/journal/article/viewFile/gt.2008.07.02.101.00/832 (2008)
17. Merilahti, J., Viramo, P., Korhonen, I.: Wearable monitoring of physical functioning and disability changes, circadian rhythms and sleep patterns in nursing home residents. IEEE J. Biomed. Health Inform. PP(99):1–1, doi: 10.1109/JBHI.2015.2420680 (2015)
18. Ng, W., Dash, M.: A change detector for mining frequent patterns over evolving data streams. IEEE International Conference on Systems, Man and Cybernetics, 2008. SMC 2008, pp. 2407–2412, doi: 10.1109/ICSMC.2008.4811655 (2008)
19. Paavilainen, P., Korhonen, I., Ltjnen, J., Cluitmans, L., Jylh, M., Srel, A., Partinen, M.: Circadian activity rhythm in demented and non-demented nursing-home residents measured by telemetric actigraphy. J. Sleep Res. 14(1), 61–68 (2005). doi:10.1111/j.1365-2869.2004.00433.x
20. Prince, S.A., Adamo, K.B., Hamel, M.E., Hardt, J., Gorber, S.C., Tremblay, M.: A comparison of direct versus self-report measures for assessing physical activity in adults: a systematic review. Int. J. Behav. Nutr. Phys. Act. 5:56, doi: 10.1186/1479-5868-5-56. http://dx.doi.org/10.1186/1479-5868-5-56 (2008)
21. Refinetti, R., Lissen, G.C., Halberg, F.: Procedures for numerical analysis of circadian rhythms. Biol. Rhythm. Res. 38(4):275–325, doi: 10.1080/09291010600903692. http://www.ncbi.nlm.nih.gov/pmc/articles/PMC3663600/ (2007)
22. Tan, T.H., Gochoo, M., Chen, K.H., Jean, F.R., Chen, Y.F., Shih, F.J., Ho, C.F.: Indoor activity monitoring system for elderly using RFID and Fitbit Flex wristband. 2014 IEEE-EMBS International Conference on Biomedical and Health Informatics (BHI), pp. 41–44, doi: 10.1109/BHI.2014.6864299 (2014)
23. Tran, D.H., Gaber, M.M., Sattler, K.U.: Change detection in streaming data in the era of big data: Models and issues. SIGKDD Explor Newsl 16(1):30–38, DOI 10.1145/2674026.2674031. http://doi.acm.org/10.1145/2674026.2674031 (2014)
24. Tucker, P., Gilliland, J.: The effect of season and weather on physical activity: A systematic review. Public Health 121(12):909–922, doi: 10.1016/j.puhe.2007.04.009. http://www.sciencedirect.com/science/article/pii/S0033350607001400 (2007)
25. Tukey, J.W.: Exploratory data analysis. http://xa.yimg.com/kq/groups/16412409/1159714453/name/exploratorydataanalysis.pdf (1977)
26. Wang, S., Skubic, M., Zhu, Y.: Activity density map visualization and dissimilarity comparison for eldercare monitoring. IEEE Trans. Inf. Technol. Biomed. 16(4), 607–614 (2012). doi:10.1109/TITB.2012.2196439

27. Xu, Y., Zhang, Z., Yu, P., Long, B.: Pattern change discovery between high dimensional data sets. In: Proceedings of the 20th ACM International Conference on Information and Knowledge Management, ACM, New York, NY, USA, CIKM '11, pp. 1097–1106, DOI 10. 1145/2063576.2063735. http://doi.acm.org/10.1145/2063576.2063735 (2011)
28. Yamada, M., Kimura, A., Naya, F., Sawada, H.: Change-point detection with feature selection in high-dimensional time-series data. In: Proceedings of the Twenty-Third International Joint Conference on Artificial Intelligence, AAAI Press, Beijing, China, IJCAI '13, pp. 1827–1833. http://dl.acm.org/citation.cfm?id=2540128.2540390 (2013)

Chapter 7
Machine Learning Applied to Optometry Data

Beatriz Remeseiro, Noelia Barreira, Luisa Sánchez-Brea, Lucía Ramos and Antonio Mosquera

Abstract Optometry is the primary health care of the eye and visual system. It involves detecting defects in vision, signs of injury, ocular diseases as well as problems with general health that produce side effects in the eyes. Myopia, presbyopia, glaucoma or diabetic retinopathy are some examples of conditions that optometrists usually diagnose and treat. Moreover, there is another condition that we have all experienced once in a while, especially if we work with computers or have been exposed to smoke or wind. Dry eye syndrome (DES) is a hidden multifactorial disease related with the quality and quantity of tears. It causes discomfort and could lead to severe visual problems. In this chapter, we explain how machine learning techniques can be applied in some DES medical tests in order to produce an objective, repeatable and automatic diagnosis. The results of our experiments show that the proposed methodologies behave like the experts so that they can be applied in the daily practice.

Keywords Optometry data · Dry eye syndrome · Image analysis · Feature selection · Classification · Regression

B. Remeseiro (✉)
Departament de Matemàtiques i Informàtica, Universitat de Barcelona,
Gran Via de les Corts Catalanes 585, 08007 Barcelona, Spain
e-mail: remeseirob@gmail.com

N. Barreira · L. Sánchez-Brea · L. Ramos
Departamento de Computación, Universidade da Coruña,
Campus de Elviña S/N, 15071 A Coruña, Spain
e-mail: nbarreira@udc.es

L. Sánchez-Brea
e-mail: luisa.brea@udc.es

L. Ramos
e-mail: l.ramos@udc.es

A. Mosquera
Departamento de Electrónica y Computación, Universidade de Santiago de Compostela,
Campus Universitario Sur, 15782 Santiago de Compostela, Spain
e-mail: antonio.mosquera@usc.es

© Springer International Publishing AG 2018
D.E. Holmes and L.C. Jain (eds.), *Advances in Biomedical Informatics*,
Intelligent Systems Reference Library 137,
https://doi.org/10.1007/978-3-319-67513-8_7

123

7.1 Introduction

Optometry is a health care profession concerned with examination, diagnosis and treatment of the human eye. It provides a wide range of services, from diagnosis and management of several ocular diseases, injuries and disorders of the visual system to identification of related systemic conditions affecting the eye. Although Optometry is usually confused with Ophthalmology, they are complementary since the former provides primary health care whereas the later is a branch of medicine specialized in eye diseases and surgery. This way, optometrists treat common vision disorders associated with refractive problems such as myopia, hyperopia, presbyopia or astigmatism that can be corrected with eyeglasses or contact lenses. Moreover, they can diagnose anomalies in the eye fundus related to the retina, the macula or the blood vessels (Fig. 7.1, left) such as diabetic retinopathy, macular degeneration or ocular hypertension as well as alterations on the eye surface (Fig. 7.1, right) such as conjunctivitis or dry eye.

Dry eye syndrome (DES) occurs when the eyes do not produce enough tears, or when the tears are of poor quality and evaporate too quickly. Most people has experienced dry eye in numerous situations. Common symptoms are stinging or burning of the eye, a gritty feeling as if something is in the eye, pain, redness, uncomfortable contact lenses, and eye fatigue. Moreover, some people also experience wet eyes since extra tearing is a natural reflex to alleviate the eye pain. These symptoms can be temporary or chronic so that if they persist, professional advice is required in order to avoid permanent damage in the surface of the eye. Even though dry eye cannot be cured, symptoms can be alleviated in order to keep the eyes healthy. The most common treatment is the use of artificial tears or substances that increase the quality of tears. Ointments or gels are prescribed in severe cases as well as medicines in order to reduce the inflammation of the eye.

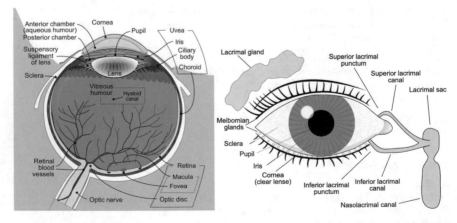

Fig. 7.1 Human eye. *Left* Main parts of the eye. *Right* Surface of the eye and tear system. *Source* Wikimedia Commons

DES can be associated with several factors such as the inflammation of the surface of the eye, the lacrimal gland or the conjunctival tissue, any disease process that alters the chemical composition of the tears, problems with normal blinking, normal aging process, exposure to environmental conditions such as air conditioners, heating or smoke, LASIK surgery, allergies, computer use, chemical and thermal burns in the eye or side effects of several medications (antihistamines, nasal decongestants, oral contraceptives, antidepressants). For this reason, its prevalence is increasing around the world, mainly in developed countries. In the Unite States, the Beaver Dam Offspring Study found that its prevalence rate is 14.5% in adults and that DES affects more women (17.9%) than men (10.5%) [1]. Worldwide epidemiological studies reveal a prevalence of 7.4% in Australia, 25% in Canada, 27.5% in Indonesia, 33% in Japan and 33.7% in Taiwan [2]. In 2011, the associated costs of DES in the United States was estimated in 3.84 billions of dollars for the health care system and in 55.4 billions for the society overall [3]. In summary, the growing prevalence, the chronic pain that affects the quality of life and the economic burden force researchers to pay more attention to this chronic disease.

In 2007, the International Dry Eye Workshop formally defined DES as *a multifactorial disease of the tears and ocular surface that results in symptoms of discomfort, visual disturbance and tear film instability with potential damage to the ocular surface. It is accompanied by increased osmolarity of the tear film and inflammation of the ocular surface* [4]. Moreover, DES was classified into two subgroups: *aqueous deficient* and *evaporative*. The former is caused by a low tear production whereas the later is related to anomalies in the tear composition that lead to tear instability and evaporation. Thus, the quantity and quality of tears are key issues in DES. In turn, *aqueous-deficient* DES may depend on *Sjögren Syndrome*[1] or not (lacrimal gland obstruction, systemic drugs) and the *evaporative* DES may be *intrinsic*, if it depends on tear film features or lid dynamics, or *extrinsic*, if external agents affect the tear film such as contact lenses or allergies. Regarding the tear film features, a normal tear film consists of three layers as Fig. 7.2 shows. Alterations in the composition of these layers can produce DES.

Due to its multifactorial character, there is not a single clinical test to diagnose DES but a set of procedures to assess different factors related with the dry eye:

- **Symptoms and comfort**. Patients fill standardized questionnaires (OSDI, McMonnies, VAS, SPEED,...) about their perception of dryness, which is, of course, very subjective.
- **External signs**. Optometrists perform a preliminary biomicroscopic examination to evaluate different symptoms such as hyperemia or ocular damage using different dyes to reveal the underneath stainings or epithelial damages.
- **Tear secretion**. Schirmer's and phenol red thread tests evaluate the tear production by means of a reactive strip placed on the eye. Usually these tests are uncomfortable for the patient and require anesthesia.

---

[1]Autoimmune disorder that affects the glands that secrete fluids.

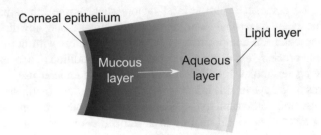

Fig. 7.2 Tear film layers. From *left* to *right*, the inner mucous layer, produced by globet cells in the conjunctiva, that adheres tears to the eyes. Second, the aqueous layer, produced by lacrimal glands, that nourishes and protects the cornea. Finally, the outer lipid layer, produced by the meibomian glands in the eyelids, that lubricates and prevents evaporation

- **Tear volume**. The tear meniscus height observed through a slit-lamp biomicroscopy is a more patient-friendly tear volume estimation.
- **Tear quality**. There are several tests that evaluate different aspects of the tear film quality, such as stability (break-up time, non-invasive break-up time), presence of lipids (lipid layer pattern assessment), osmolarity or chemical composition (lysozyme and lactoferrin laboratory tests).

Even though some tests are completely objective (laboratory tests, Schirmer's and phenol red thread tests, osmolarity measures), most of the tests used for DES diagnosis are subjective since they depend on the patient's opinions or the clinician observations. In this chapter we propose several approaches to apply machine learning techniques to several DES tests based on direct observation in order to reduce the subjectivity of the assessment.

First, we deal with the automatic classification of the lipid layer patterns. Interferometry techniques combined with a slit lamp and a camera produce images of the tear film that show different patterns (waves, fringes, marble-like patterns) depending on the quality of the lipid layer. Features extracted from these images and expert annotations are the input to train classifier algorithms in order to produce an automatic output. The tool presented provides an objective and repeatable assessment of the tear quality.

Conjunctival hyperemia is another medical condition linked to dry eye, among other diseases, in which the sclera is characterized by redness due to blood vessel engorgement. It is diagnosed by direct observation of the eye with a slit lamp. A value in a scale is assigned to the redness where the higher the value, the worst the patient's condition. The scales are a collection of pictures of different levels of redness so the grading is highly subjective and requires training. Regression techniques are machine learning methods that can be useful to translate the quantity of red in an image to a value in the given scale. This way, the proposed methodology simplifies and puts in objective terms the image assessment.

Finally, the break-up time (BUT) test is a clinical procedure to measure the stability of the tear film through time. Sodium fluorescein is instilled in the eye and the patient blinks several times in order to spread the dye. Then, the clinician observes with a slit lamp the evolution of the tear film, from its generation just after the blink up to its evaporation, this is, when the tear film is broken and dark spots appear within the dye. The location, size and shape of these spots are also related with different qualities of tear film. This test is also very subjective since it requires lots of attention to identify quickly the tear film breakups. In this chapter a methodology is proposed to analyze the BUT videos and detect the breakup areas. From these areas, shape descriptors are extracted from the images to feed classification algorithms and provide a more accurate description of the tear film quality. The proposed methodology is fully automatic and provides accurate results in short time.

7.2   Classification of the Lipid Layer Patterns

The assessment of the lipid layer patterns observed in the tear film is usually the first clinical test made by dry eye specialists [5]. This test consists in acquiring an image of the tear film using the Tearscope Plus [6], a non-invasive instrument to assess the lipid layer thickness, and categorizing it using a grading scale defined by Guillon [6]. This scale is composed of five lipid layer patterns, based on morphological and color features, which in increasing thickness are: open meshwork, closed meshwork, wave, amorphous and color fringe.

This clinical test is a useful method to evaluate the quality and the structure of the tear film in a non-invasive way. However, it is affected by the subjective interpretation of the observer, specially with thin layers which are more difficult to interpret due to the lack of color fringes and other distinct morphological features [7]. Training may also affect this test according to the learning curve for lipid layer assessment established by Nichols et al. [8]. Summarizing, the manual classification of the lipid layer patterns is a difficult and time-consuming task, making the use of a computer-based analysis system highly desirable.

First automatic approaches to the automatic classification of the tear film lipid layer can be found in [9, 10], which demonstrated how the lipid layer patterns can be characterized by means of color and texture properties. The results provided by these approaches were later improved by applying five popular different classifiers to the problem at hand [11], analyzing a wide set of texture analysis techniques and color spaces [12], and using three feature selection filters to reduce the computational requirements [13].

Multiple patterns may be observed in the tear film lipid layer, a fact that motivated the development of tear film maps [14, 15] to complement the information provided by previous approaches which classify a tear film image into a single category. Both automatic tools, tear film classification and tear film maps, were included in a web-based system for dry eye assessment known as iDEAS [16]. Furthermore, parallel computing was used to reduce the runtime of these approaches and increase their acceptance among practitioners [17, 18].

Regarding machine learning, which is the main topic of this chapter, two different works should be highlighted for tear film classification. A systematic procedure was presented in [19], which include class binarization, feature selection and artificial neural networks. This procedure was later improved [20], in which a pipeline of processes was proposed to optimize and evaluate our classification problem by means of multiple-criteria decision-making methods and rank correlation.

Section 7.2.1 presents the image dataset and Sect. 7.2.2 the methods used to automatically classify the tear film lipid layer patterns into one of the target categories. These methods include image analysis techniques as well as machine learning methods, such as class binarization, feature selection and decision-making algorithms [20]. Finally, the results and discussion of the evaluation process are included in Sect. 7.2.3.

7.2.1 Image Dataset

The VOPTICAL_I1 dataset [21] contains images of the preocular tear film in which the lipid layer patterns can be observed. All the images were acquired and annotated by specialists from the Optometry Service of the University of Santiago de Compostela (Spain). The image acquisition procedure was carried out with the Tearscope Plus (Keeler, Windsor, UK) attached to a Topcon SL-D4 slit lamp and a Topcon DV-3 digital video camera. The slit lamp's magnification was set at 200X and the images were stored via the Topcon IMAGEnet i-base at a spatial resolution of 1024×768 pixels in RGB. For the annotation procedure they used the Guillon's grading scale [6], and so each image was labeled with its corresponding lipid layer pattern.

The dataset contains 105 images taken over optimum illumination conditions, and acquired from healthy subjects aged from 19 to 33 years. The lipid layer patterns are distributed as follows: 27.62% of the images correspond to the open meshwork pattern, 27.62% to the closed meshwork pattern, 23.81% to the wave pattern, and 20.95% to the color fringe pattern. Notice that images within the amorphous category have not been included in the dataset since it rarely appears isolated. Figure 7.3 includes four representative images of the VOPTICAL_I1 dataset.

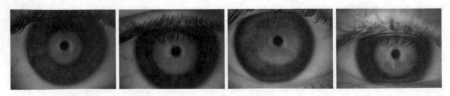

Fig. 7.3 Representative images from the dataset, from *left* to *right* open meshwork, closed meshwork, wave and color fringe

7.2.2   Methods

Figure 7.4 illustrates the main steps to carry out the automatic classification of the lipid layer patterns by means of image analysis and machine learning techniques [20]. Given a tear film image, its region of interest is located and a set of color and texture features are computed from it. The feature set obtained is then converted into new sets using class binarization and feature selections techniques. Next, the classification step is performed and some metrics are calculated, which are subsequently evaluated by means of decision-making methods. Finally, a conflict handling mechanism based on the Spearman's coefficient is applied to obtain a single ranking with the best solutions for the problem at hand.

7.2.2.1   Image Analysis

Given an input image acquired with the Tearscope Plus, the first step entails the localization of its region of interest, followed by the extraction of a set of relevant features from it. The methods used for image analysis are as follows [12]:

- **Extraction of the region of interest** (ROI). In order to analyze this kind of images, practitioners focus their attention on the bottom part of the iris, in which the tear film can be perceived with higher contrast. Therefore, the image analysis step takes place in this area, knows as ROI, which can be selected as following described [9]: the input image in RGB is transformed to the L*a*b* color space and only the component luminance L* is here considered; next, a set of ring-shaped templates previously generated, which covers the different shapes of the ROI, is match with the L* component using the normalized cross-correlation technique [22]; then, the region with the maximum correlation value is selected and, as the ROI is located at the bottom part of the template, the top area is rejected; finally, the rectangle of maximum area inside this bottom part is located and so the ROI of the input image is obtained (see Fig. 7.5).

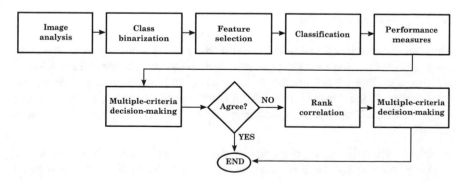

Fig. 7.4 Main steps for the automatic classification of the lipid layer patterns

Fig. 7.5 Extraction of the region of interest (ROI), from *left* to *right* input image, illustrative ring-shaped templates, and ROI

- **Color analysis**. Color features are extracted from the ROIs by using the L*a*b* color space [23], a 3D model where its three components represent: the luminance of the color $L*$, its position between magenta and green $a*$, and its position between yellow and blue $b*$. The use of the L*a*b* color space entails converting the three channels of the ROI image in RGB into the three components of L*a*b* [24]. Then, each component is analyzed separately and its texture descriptor is obtained as following described. Therefore, the final descriptor is the concatenation of the three individual descriptors.
- **Texture analysis**. Texture features are extracted from the ROIs by applying the co-occurrence features method [25]. This technique generates a set of gray level co-occurrence matrices for a specified distance, which determines a neighborhood of analysis, and extracts 14 several statistical measures from their elements. Then, the mean and the range of these measures are calculated across matrices, and so the texture descriptor for a particular distance contains 28 features.

Distances from one to seven were used in the co-occurrence features method [12], and three components of color are considered due to the use of L*a*b*. Thus, the size of the final descriptor obtained from an input image is: 28 features × 7 distances × 3 components = 588 features.

7.2.2.2 Class Binarization

Class binarization techniques simplify multiclass problems by converting them into a set of binary problems, i.e. problems with only two classes, which allow a sensible decoding of the prediction [26]. This simplification may improve the performance of the approach [27] and, for this reason, it has been applied to our 4-class problem. The most common strategies for class binarization are described as follows:

- **One-vs-all**. It divides a c-class problem into c binary problems. Each problem is solved by a binary classifier which has to distinguish one of the classes from all other classes.

- **One-vs-one**. It divides a c-class problem into $c(c-1)/2$ binary problems. Each problem is solved by a binary classifier which has to distinguish between a pair of classes.

Decoding methods have to be used after training the binary classifiers in order to obtain the predictions. If the algorithms are soft, they compute the "likelihood" of classes for a given input. That is, they obtain a confidence p for the *positive* class and a confidence $1 - p$ for the *negative* class. In the *one-vs-all* technique, if we assume the *one*-part as the positive class and the *all*-part as the negative class, the decoding method is done according to the maximum probability p among classes. This method is not valid for *one-vs-one* techniques, so three different decoding methods were considered in this case:

- **Hamming decoding** [27]. This method uses a matrix $M \in \{-1, 1\}^{N \times F}$, where N is the number of classes and F is the number of binary classifiers, i.e. $F = N \times (N-1)/2$. It induces a partition of the classes into two "metaclasses", where a sample is placed in the positive metaclass for the j-th classifier if and only if $M_{y_i j} = 1$, where y_i stands for the desired class of the sample.
- **Loss-based decoding** [28]. The use of the loss function L instead of the Hamming distance is suggested in order to take into account the significance of the predictions, which can be interpreted as a measure of confidence. In this research, the most appropriate loss function is the logistic regression $L(z) = log(1 + e^{-2z})$.
- **Accumulative probability with threshold** [28]. It extends the Hamming matrix to $M \in \{-1, 0, 1\}^{N \times F}$. It ignores binary classifiers if the difference between the confidence for the positive and negative classes is under a threshold.

7.2.2.3   Feature Selection

Feature selection can be defined as the process of selecting a subset of relevant features from the original set, i.e. optimally reducing the feature space according to an evaluation criterion. The main benefits of feature selection include [29, 30]: making the training step more efficient (the size of the data is reduced), increasing the classification accuracy (noise features are eliminated), and avoiding overfitting.

Feature selection techniques can be divided into three main groups: filters, wrappers and embedded methods [31]. Filters rely on general characteristics of the data to select feature subsets without involving any learning model. In addition, they allow for reducing the dimensionality of the data without compromising the time and memory requirements. Both wrappers and embedded methods have the risk of overfitting when having more features than samples [30], as it is the case in the dataset here considered. Consequently, the following three filters were chosen:

- **Correlation-based feature selection** (CFS) [32]. It is a multivariate filter that ranks feature subsets according to a correlation based evaluation function. The bias of the function is toward subsets that contain features that are highly correlated with the class and uncorrelated with each other.

- **Consistency-based filter** [33]. This algorithm evaluates the worth of a subset of features by the level of consistency in the class values when the samples are projected onto the subset of attributes.
- **INTERACT** [34]. It is a subset filter based on symmetrical uncertainty, which is defined as the ratio between the information gain and the entropy of two features. It also includes the consistency contribution of a feature, an indicator about how the elimination of that feature will affect consistency.

7.2.2.4 Classification and Performance Measures

Supervised learning entails inferring a function from input features and output labels, in order to predict the class label of new samples [35]. With the aim of providing different approaches to the learning process, the following classifiers were used in the problem at hand:

- **Fisher's linear discriminant** [36]. It is a simple method used to find the linear combination of features which best separate two or more classes.
- **Naive Bayes** [37]. It is an statistical learning algorithm based on the Bayesian theorem that can predict class membership probabilities.
- **Decision tree** [38]. It is a logic-based algorithm which classifies samples by sorting them based on feature values.
- **Support vector machine** (SVM) [39]. It is based on the statistical learning theory and revolves around a hyperplane that separates two classes.
- **Multilayer perceptron** (MLP) [40]. It is a feedforward neural network which consists of a set of units, joined together in a pattern of connections.

In order to quantify the behavior of the different classifiers considered, a wide set of performance measures were used. For all of them, we considered the worst of the individual per-class performance as a lower bound estimation [41]. These metrics are defined as follows:

- **Accuracy**. The percentage of correctly classified instances.
- **True positive rate** (TPR). The proportion of positives which are correctly classified, also known as sensitivity.
- **True negative rate** (TNR). The proportion of negatives which are correctly classified, also known as specificity.
- **Precision**. The proportion of the true positives against all the positive results.
- **F-measure**. The harmonic mean of precision and sensitivity.
- **Area under the curve** (AUC). The area under the receiver operating characteristic curve.
- **Training time**. The time elapsed for training a learning model, which comprises training a set a classifiers when class binarization techniques are used.

Note that the training time is not as relevant as the other measures since the training process is executed off-line. However, it may be helpful when other measures are very similar. Notice also that the testing time is negligible, thus it was not considered.

7.2.2.5 Multiple-Criteria Decision-Making and Rank Correlation

Multi-criteria problems can be formulated using a set of alternatives (classifiers) and criteria (performance measures), and a decision matrix where x_{ij} is the value of the j-th criterion when using the i-th alternative. In decision-making processes, both alternatives and criteria are identified, and weights are given to each criterion to reflect its relevance. As a result, a ranking list of alternatives is obtained.

Three multiple-criteria decision-making (MCDM) methods were used to analyze the algorithms based on the aforementioned performance measures:

- **Technique for order preference by similarity to ideal solution** (TOPSIS) [42]. It is based on the idea of finding the best alternatives by minimizing the distance to the ideal solution whilst maximizing the distance to the negative-ideal solution.
- **Gray relational analysis** (GRA) [43]. It is based on the degree of similarity or difference of development trends between an alternative and the ideal alternative.
- **VIKOR** [44]. It is a method which provides maximum group utility for the majority and minimum individual regret for the opponent.

When using several MCDM methods, as in this case, different ranking lists of alternatives can be obtained. Therefore, we need a mechanism to handle this type of conflicts and to merge the different rankings into a single one.

The Spearman's rank correlation coefficient is a nonparametric technique for evaluating the degree of correlation between two independent variables [45], and was used for the task of conflict handling. It can be calculated as:

$$\rho = 1 - \frac{6 \sum_{i=1}^{m} d_i^2}{m(m^2 - 1)} \tag{1}$$

where d_i is the difference between ranks for each alternative-criterion pair and m is the number of pairs.

The procedure is described as follows:

1. Apply the MCDM methods using the different configurations of machine learning methods as alternatives and the performance measures as criteria. The weights are arbitrarily given to each criterion in order to reflect its importance.
2. Compute the average similarities between the k-th method and the other methods:

$$\rho_k = \frac{1}{q-1} \sum_{i=1, i \neq k}^{q} \rho_{ki}, k = 1, 2, \ldots, q \qquad (2)$$

where q is the number of MCDM methods and ρ_{ki} is the Spearman's coefficient between the k-th and i-th MCDM methods.

3. Normalize the ρ_k values:

$$\sum_{k=1}^{q} \rho_k = 1 \qquad (3)$$

4. Apply the MCDM methods, now using the ranking scores previously obtained as criteria and the normalized Spearman's coefficients as weights.

7.2.3 Results

This section presents the evaluation of the methods previously presented using the VOPTICAL_I1 dataset described in Sect. 7.2.1. The experimentation was carried out with the following configurations:

- SVM: radial basis kernel and automatic parameter estimation [46].
- MLP: one single layer and different numbers of hidden units [19].
- Validation: leave-one-out cross-validation for better generalization [47].
- MCDM: the weights were assigned equally but for the training time which was reduced to 0.01.

Firstly, the performance of the different feature selection filters is analyzed. Since class binarization methods changes the input search space, the filters were applied for each one-vs-all and one-vs-one approaches, in addition to the single approach. Figure 7.6 illustrates the number of features selected by each filter from the total of 588 features used, which include color and texture information as explained in

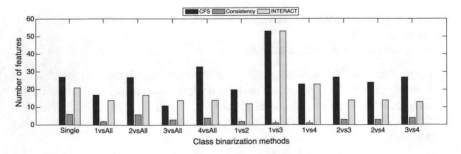

Fig. 7.6 Number of features selected for every filter from a total of 588 features

Sect. 7.2.2.1. As can be seen, the consistency-based filter performs the most aggressive selection and only retains the 0.54% of the original features, on average. In contrast, the other two filters select higher percentages of features: CFS retains a 4.47%, compared to the 3.23% retained by INTERACT. If we analyze the results grouped by class binarization techniques, the filters retain, on average, a 3.06% in the single approach, a 3.00% in the one-vs-one approach and a 2.30% in the one-vs-all. As expected, the percentage of features selected in the single, multiclass approach is larger than one selected when using class binarization methods because they reduce the complexity of the problem. Nevertheless, the number of features selected in one-versus-one is larger than in one-versus-all despite the fact that the complexity of the former is less. This fact can be explained due to the lack of relevant knowledge in the smaller datasets obtained when only two classes are involved.

Table 7.1 presents the ranking lists provided by every MCDM method, and the machine learning techniques (class binarization, feature selection, and classification) associated to each ranked alternative. As can be observed, the one-vs-one approach, with different decoding methods, appear in the 60% (50% in GRA) of the top 10 alternatives. The 30% of the positions correspond to the multiclass approach, and the remaining 10% (20% in GRA) to the one-vs-all. However, the one-vs-all approach ranks first regardless the MCDM method considered. If we focus our attention in the feature selection filters, 40% of the alternatives include them (30% in GRA). These percentages reflect the adequacy of using class binarization and feature selection in the problem at hand. Finally, if we analyze the results in terms of the classifier used, the MLP appear in the 80% of the alternatives (70% in GRA).

Due to the high number of alternatives here considered (96 combinations of the different machine learning methods), only the performance measures of the top 10 alternatives are shown in Fig. 7.7. As can be observed, the top 10 configurations provide very competitive results with values over the 88% at any case, and even

Table 7.1 Machine learning (ML) methods of the top 10 alternatives ranked by the three MCDM methods. Decoding methods in one-vs-one approach are labeled as: [H]Hamming, [L]Loss based, and [T]Threshold

TOPSIS		GRA		VIKOR	
Rank	ML methods	Rank	ML methods	Rank	ML methods
1	1vsAll, No FS, MLP	1	1vsAll, No FS, MLP	1	1vsAll, No FS, MLP
2	Single, No FS, MLP	2	1vs1[H], No FS, SVM	2	1vs1[H], No FS, SVM
3	1vs1[H], No FS, SVM	3	Single, No FS, MLP	3	Single, No FS, MLP
4	Single, CFS, MLP	4	1vs1[H], No FS, MLP	4	Single, CFS, MLP
5	1vs1[L], CFS, MLP	5	1vs1[L], No FS, MLP	5	1vs1[L], CFS, MLP
6	1vs1[H], No FS, MLP	6	1vs1[L], CFS, MLP	6	Single, No FS, Fisher
7	1vs1[L], No FS, MLP	7	Single, CFS, MLP	7	1vs1[H], No FS, MLP
8	Single, No FS, Fisher	8	Single, No FS, Fisher	8	1vs1[L], No FS, MLP
9	1vs1[H], CFS, MLP	9	1vs1[H], CFS, MLP	9	1vs1[H], CFS, MLP
10	1vs1[L], CFS, MLP	10	1vsAll, No FS, Fisher	10	1vs1[L], CFS, MLP

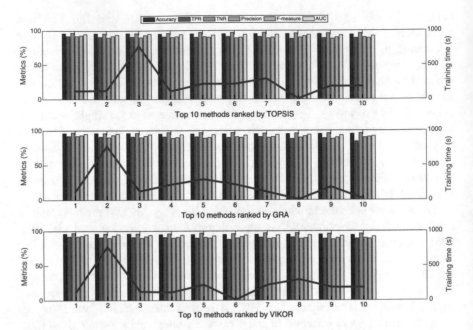

Fig. 7.7 Performance measures of the top 10 alternatives ranked by the three MCDM methods. Note that the right y axis corresponds to the training time in seconds, whilst the left y axis corresponds to the other measures represented in percentage

reaching the 95% in the case of the accuracy, the TNR and the AUC. Regarding the times, the simplest methods as Fisher provide a trained model in only a few seconds whilst both SVM and MLP need a few minutes.

Back on the ranking lists generated by the MCDM methods, they agree on the best alternative but the general level of agreement is only 20%. Thus, it was necessary to apply the rank correlation step in order to solve this conflict. In this sense, the Spearman's coefficient was calculated for the three methods to determine their optimal weight and merge their results into a single ranking.

According to the procedure explained in Sect. 7.2.2.5, the weights obtained are 0.9896 (TOPSIS), 0.9910 (GRA), and 0.9853 (VIKOR). After normalized them, the weights used for re-rank the alternatives are, respectively: 0.3337, 0.3341, and 0.3322. As a result, the three MCDM methods provide exactly the same ranking list of alternatives (see Table 7.2). Thus, it was demonstrated how the conflict handling step is able to change the global agreement between methods, from two to ten.

Summarizing, both class binarization and feature selection are useful to improve the performance of the classifiers in the automatic classification of lipid layer patterns, and they provide a good trade-off between classification performance and training time. Regarding the MCDM methods, they are also effective in our problem in which the performance of the methods was evaluated by means of different metrics. Moreover, the Spearman's coefficient used for conflict handling plays an important role to obtain a single ranking when several MCDM methods are

Table 7.2 One single ranking provided by the MCDM methods after applying the conflict handling by rank correlation, and performance measures of each alternative. Decoding methods in one-vs-one approach are labeled as: [H]Hamming, [L]Loss based, and [T]Threshold

Rank	ML methods	Acc	TPR	TNR	Prec	F	AUC	Time
1	1vsAll, No FS, MLP	0.96	0.92	0.97	0.92	0.93	0.95	118.12
2	1vs1[H], No FS, SVM	0.96	0.91	0.96	0.90	0.93	0.95	773.42
3	Single, No FS, MLP	0.96	0.91	0.96	0.90	0.92	0.94	125.83
4	Single, CFS, MLP	0.95	0.91	0.96	0.90	0.91	0.94	116.18
5	1vs1[L], CFS, MLP	0.95	0.90	0.97	0.91	0.90	0.93	223.71
6	1vs1[H], No FS, MLP	0.95	0.91	0.97	0.89	0.90	0.94	221.36
7	1vs1[L], No FS, MLP	0.95	0.91	0.97	0.89	0.90	0.94	298.69
8	Single, No FS, Fisher	0.95	0.88	0.96	0.90	0.92	0.94	6.70
9	1vs1[H], CFS, MLP	0.95	0.89	0.96	0.88	0.90	0.93	187.08
10	1vs1[L], CFS, MLP	0.94	0.89	0.96	0.90	0.89	0.92	185.72

considered. And, as a conclusion, we can emphasize the combination of the MLP with the one-vs-all approach which provides the most competitive results with an accuracy over 96%, being the agreement between subjective observers in the range from 91% to 100%.

7.3 Hyperemia Grading in the Bulbar Conjunctiva

Bulbar hyperemia is one of the first symptoms that appear in an unhealthy conjunctiva. In order to evaluate it, optometrists typically start by obtaining images of the patient's eyes. Then, they analyze the image by searching for indicators of hyperemia, such as the general hue of the conjunctiva or the vessel quantity. Finally, they decide the grade of hyperemia that the patient has by comparing the photography with a given grading scale.

The manual process presents several drawbacks. First, the observation of the patient's images is tedious and time-consuming. Also, the final evaluations present a high subjectivity both intra- and inter-expert. Moreover, the image features that are involved in the grading are difficult to define. Finally, these issues of the manual process worsen when the specialist has to review several patients on a row, which is a common occurrence. Therefore, there is a clear need for the development of a fully automatic methodology for bulbar hyperemia grading.

To the best of our knowledge, there are no other frameworks that compute the bulbar hyperemia in a fully automatic manner. There are, however, semi-automatic approaches such as [48, 49], that perform a manual selection of the region of interest. Moreover, only a few image characteristics are computed, and the transformation from these values to the grading scale is not described.

In [50], the input of the system are images taken by a keratograph. The main focus is to compare a new corneal topographer to three subjective grading scales in order to asses its validity and reliability.

Both [51] and [52] are focused on the study of allergic conjunctivitis and its effect on hyperemia level. They take into account only information about the vessels morphology, such as width, density or tortuosity. They use each feature separately to obtain the final output of the system.

Also, in [53] a simple image feature, the percentage of red in RGB color space, is used to compare with the evaluations of a group of optometrists and with the automatic evaluation provided by a keratograph. As this work is focused on how a simple objective measure can be compared with the manual approach, it does not include an automatic segmentation of the region of interest, nor the computation and combination of several features. A similar approach is taken in [54], in which a framework is also proposed, but the human operator must intervene to adjust the level of white in the image and the region of interest. They also propose a computationally simple redness measure based on HSV color space, and then remap it to the selected scale. In fact, regarding the computation of image features, several works have proposed new approaches [55, 56].

Finally, the construction and validation of grading scales has been the subject of several studies [57, 58], and their comparison with objective image features [59, 60].

We developed a fully automatic methodology for bulbar hyperemia grading, which receives an image as input and returns the hyperemia level in a given grading scale. First, the region of interest is segmented, removing eyelids, eyelashes and other spurious information [61]. Next, several image features are computed, involving both vessel disposition and color information [62]. Finally, these features are combined by means of machine learning techniques. Therefore, the output of the system is the value in the chosen grading scale.

This section is structured as follows. Section 7.3.1 describes the image set that we use as the input of the system. Then, Sect. 7.3.2 details each step of the methodology. Finally, Sect. 7.3.3 depicts and discusses the obtained results.

7.3.1 Image Dataset

The input of the system are images that show a side view of the patient's eye. The images belong to both eyes and both sides of the eye, from the pupil to the lacrimal or the corner of the eye.

The photographs were obtained and evaluated by specialists from the Optometry Service of the University of Santiago de Compostela (Spain). Two optometrists graded the images with no information regarding the patient identity, and no communication with each other during the process. Moreover, the two specialists performed a second evaluation, months apart. To that end, two widely-known scales were selected: the Efron and the CCLRU grading scales (see Fig. 7.8).

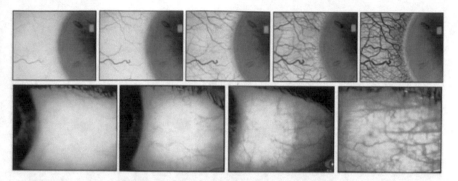

Fig. 7.8 Efron (drawings, levels 0–4) and CCLRU (photographies, levels 1–4) bulbar hyperemia grading scales

At the end of the manual labeling, 163 images of 1024 × 768 pixels were obtained with four evaluations for each image in each scale (i.e. eight evaluations per image). Because of the high subjectivity of the process, we analyzed these values in order to ensure that they were similar. To that end, we had to define a threshold for the difference between two gradings that can still be considered acceptable. We decided to discard those images that had at least two evaluations that differ in more than 0.5 in the same scale. Thus, our data set includes the 114 images that fulfilled this condition (70% of the original set). As ground truth for our system, we used the mean of the four evaluations for each scale. If we divide the values with a step of half point in the scale, their distributions are the following: in the Efron scale, 8% of the images are labeled ≈ 1.0, 29% ≈ 1.5, 40% ≈ 2.0, 18% ≈ 2.5 and 5% ≈ 3.0; and in the CCLRU scale, 24% of the images are labeled ≈ 2.0, 47% ≈ 2.5, 26% ≈ 3.0 and 3% ≈ 3.5.

The images present a high variability in illumination near the edges of the conjunctiva, as can be observed in Fig. 7.9 (top row). The area of conjunctiva depicted also shows a certain level of variation depending on the proximity of the camera to the eye or the position of the eyelids (Fig. 7.9, bottom row). However, as the specialists give more importance to the central part of the image than the borders, the evaluations are unaffected [63].

7.3.2 Methods

The main steps of the automatic methodology for hyperemia grading are depicted in Fig. 7.10. The system receives an image of the bulbar conjunctiva as input. First, the conjunctiva is identified and separated from the regions that do not provide useful information, such as eyelashes or eyelids. Then, several image features are computed in the segmented region. Next, these features are analyzed by means of feature selection techniques. Finally, the chosen features are used as input for machine learning techniques in order to obtain a value in the selected grading scale.

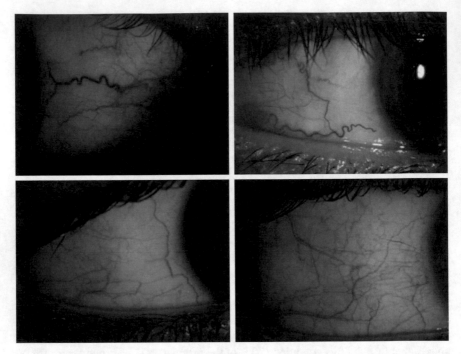

Fig. 7.9 Variability of the conditions of the image set: differences in illumination (*top*), and differences in area of the conjunctiva depicted (*bottom*)

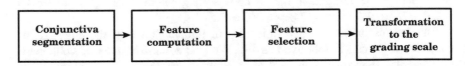

Fig. 7.10 Main steps for the automatic hyperemia grading in the bulbar conjunctiva

7.3.2.1 Conjunctiva Segmentation

The images of our data set depict not only the bulbar conjunctiva, but a number of close regions, such as the pupil, the iris, the eyelashes and, frequently, fragments of the eyelids. These regions do not provide useful information and can be a source of noise. Therefore, the first step was to isolate the conjunctiva region in an automatic and accurate manner [61]. As our image set was obtained from a real world environment, illumination and blurriness issues are fairly common, which increases the complexity of this step.

In order to isolate the conjunctiva, we took into account the general shape of the area by creating an elliptical mask. The horizontal axis starts at the center of the pupil and ends in the lacrimal/corner of the eye, depending on which side view of

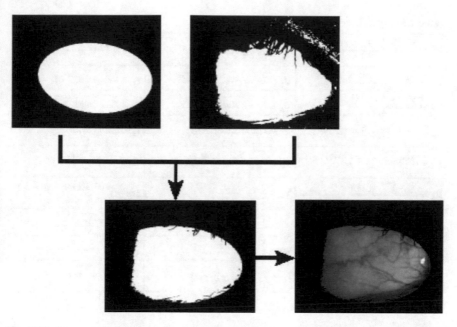

Fig. 7.11 Segmentation of the bulbar conjunctiva by means of the combination of an elliptical mask and a binary threshold

the eye we are processing. The vertical axis goes through the middle point of the horizontal axis, and is perpendicular to it. The starting and end points were obtained by marking this line in a binary threshold of the green channel of the image, and selecting the first points where the transition from white to black happens, starting from the center of the image towards the upper and lower borders.

In order to improve the representation of the iris side, we moved the pupil point in the axis to the exterior of the eye. However, this fact has the drawback of including the pupil and iris areas inside the segmented region. Thus, we used a thresholding step in the green channel of the image to get rid of the pupil and iris, as the area is significantly darker than the conjunctiva. The complete process is depicted in Fig. 7.11.

7.3.2.2 Feature Computation

In order to evaluate the hyperemia level, specialists analyze the bulbar conjunctiva searching for certain characteristics. With the aim of automating this stage of the process, we implemented the 25 image features that are depicted in Table 7.3. Some of the formulas were obtained by researching the literature [55], while others were suggested by the optometrists.

Table 7.3 Implemented hyperemia features

$B_1 = \frac{\sum_{i=1}^{n}\sum_{j=1}^{m}\left((R_{ij}+G_{ij})\overline{VE}_{ij}\right)}{nm}$	$V_1 = \sum_{i=1}^{n}\sum_{j=1}^{m}\left(\frac{R_{ij}VE_{ij}}{R_{ij}+G_{ij}+B_{ij}}\right)$	$I_2 = \frac{\sum_{i=1}^{n}\sum_{j=1}^{m}(R_{ij}-G_{ij})}{nm}$
$B_2 = \frac{\sum_{i=1}^{n}\sum_{j=1}^{m}\left(\lvert 240-H_{ij}\rvert\overline{VE}_{ij}\right)}{nm}$	$V_2 = \frac{\sum_{i=1}^{n}\sum_{j=1}^{m}((R_{ij}-G_{ij})VE_{ij})}{nm}$	$I_3 = \frac{\sum_{i=1}^{n}\sum_{j=1}^{m}(R_{ij}-B_{ij})}{nm}$
$B_3 = \frac{\sum_{i=1}^{n}\sum_{j=1}^{m}(b_{ij}\overline{VE}_{ij})}{nm}$	$V_3 = \frac{\sum_{i=1}^{n}\sum_{j=1}^{m}((R_{ij}-B_{ij})VE_{ij})}{nm}$	$I_4 = \frac{\sum_{i=1}^{n}\sum_{j=1}^{m}\lvert 128-H_{ij}\rvert}{nm}$
$B_4 = \frac{\sum_{i=1}^{n}\sum_{j=1}^{m}(R_{ij}\overline{VE}_{ij})}{nm}$	$V_4 = \frac{\sum_{i=1}^{n}\sum_{j=1}^{m}R_{ij}VE_{ij}}{\sum_{i=1}^{n}\sum_{j=1}^{m}VE_{ij}}100$	$I_5 = \frac{\sum_{i=1}^{n}\sum_{j=1}^{m}a_{ij}}{nm}$
$B_5 = \frac{\sum_{i=1}^{n}\sum_{j=1}^{m}\left(\lvert 128-H_{ij}\rvert\overline{VE}_{ij}\right)}{nm}$	$V_5 = \frac{\sum_{i=1}^{n}\sum_{j=1}^{m}H_{ij}VE_{ij}}{\sum_{i=1}^{n}\sum_{j=1}^{m}VE_{ij}}100$	$C_v = \frac{\sum_{i=1}^{n}\sum_{j=1}^{m}E_{ij}M_{ij}}{n_r}$
$B_6 = \frac{\sum_{i=1}^{n}\sum_{j=1}^{m}(a_{ij}\overline{VE}_{ij})}{nm}$	$V_6 = \sum_{i=1}^{n}\sum_{j=1}^{m}\frac{H_{ij}VE_{ij}}{\mu_{ij}}$	$M_{ij} = \begin{cases} 0 & i \bmod step \neq 0 \\ 1 & i \bmod step = 0 \end{cases}$
$B_7 = \frac{\sum_{i=1}^{n}\sum_{j=1}^{m}((R_{ij}+G_{ij}+B_{ij})\overline{VE}_{ij})}{nm}$	$\mu_{ij} = \frac{\sum_{k=-s/2}^{s/2}\sum_{l=-s/2}^{s/2}\overline{VE}_{ij}H_{i+k,j+l}}{s^2}$	$A_v = \frac{\sum_{i=1}^{n}\sum_{j=1}^{m}VE_{ij}}{nm}$
$B_8 = \frac{\sum_{i=1}^{n}\sum_{j=1}^{m}((V_{ij}+S_{ij})\overline{VE}_{ij})}{nm}$	$V_7 = \frac{\sum_{i=1}^{n}\sum_{j=1}^{m}(a_{ij}VE_{ij})}{nm}$	$P_v = \frac{\sum_{i=1}^{n}\sum_{j=1}^{m}VE_{ij}}{nm}100$
$B_9 = \frac{\sum_{i=1}^{n}\sum_{j=1}^{m}(L_{ij}\overline{VE}_{ij})}{nm}$	$I_1 = \sum_{i=1}^{n}\sum_{j=1}^{m}\left(\frac{R_{ij}}{R_{ij}+G_{ij}+B_{ij}}\right)$	$W_v = \frac{\sum_{r=1}^{\rho}\sum_{c=1}^{\kappa}W_{rc}}{\rho\kappa}$

In order to compute the features, our input is only the conjunctiva region of the image. We use n and m to refer to the size of this area, and i and j to represent a position within it.

The features can be divided in two main groups: those that use only information of the vessel quantity and disposition, and those that need information regarding color. The former are named by a capital letter and the subscript v, while the latter are named by one capital letter among $[B, I, V]$ and an integer subscript. The features that use color information can be computed in the whole conjunctiva (labeled with I), only in the vessels (labeled with V) and only in the background of the conjunctiva (labeled with B).

In order to distinguish the vessels from the background, we used the Canny edge detection algorithm to obtain an edge image E. In the formulas, VE represents the pixels that belong to a vessel in the conjunctiva and $\overline{VE}$ represents the pixels that belong to the background of the conjunctiva.

We computed features in different color spaces, and their channels are represented by the corresponding letter in the formulas: L^*, a^* and b^* for L*a*b*; R, G and B for RGB; and H, S and V for HSV.

The feature V_6 computes the value of the H channel from HSV in each pixel p_{ij}, but taking into account the values of the neighboring pixels μ_{ij} in the surrounding window of size s. The feature C_v measures the quantity of vessels in a number of image rows n_r by applying the mask M. And finally, the feature W_v defines a set of κ circumferences of radius $\rho = [n/2 * \kappa...n/2]$. These circumferences will define cut points with the vessels, where the width of the latter W_{rc} is computed by means of an active contour algorithm [64].

7.3.2.3   Feature Selection

To grade hyperemia, optometrists take into account several parameters. Our main concern when implementing our automatic set of features was to be able to reflect all these parameters in our system, in order to prevent any information loss. However, this creates a new drawback, as we computed several similar features and the same 25 features in three areas of the image. Therefore, some redundancy is expected to appear. Moreover, even if specialists use all this information in order to evaluate the hyperemia level, they do not give the same importance to all of them, although separating the clear indicators from the less relevant ones is a far from straightforward task. Because of these two main reasons, we decided to apply a procedure in order to determine which image features constitute truthful indicators of hyperemia.

To that end, we used different feature selection methods from the three groups of techniques previously mentioned in Sect. 7.2.2.3: filters, wrappers and embedded methods. Filters are already defined in Sect. 7.2.2.3. Wrappers evaluate each feature subset by creating a predictive model and analyzing the accuracy obtained by using the current features. Embedded methods combine both the feature selection and the training process of the predictive model, performing both stages together and providing more accurate results but at a higher computational cost. Among all the algorithms found in the literature, we tested the following five methods:

- **Correlation based feature selection** (CFS). Described in Sect. 7.2.2.3.
- **Relief** [65]. A filter that creates a ranking with all the features sorted by relevance.
- **M5**. A wrapper that uses the M5 algorithm [66], that generates decision trees which have linear regression functions in their nodes. The wrapper uses the separate-and-conquer approach to build a M5 model tree in each iteration, and making the best leaf into a rule. The search strategy is best first.
- **SMOReg**. A wrapper that uses a support vector machine for regression (SVR), with the improvements proposed in [67] for the sequential minimal optimization (SMO) method.
- **SVR-RFE**. An embedded method that applies recursive feature elimination with support vector regression [68]. It starts with the full set of features and removes the worst ones in each iteration, until a given minimum set size. In each iteration, it assigns a weight to each feature and then removes the ones with the smallest absolute value.

7.3.2.4   Hyperemia Evaluation

Once we have a satisfactory subset of features, we obtain a feature vector for each image, where each element can present a different range of values, and none of those elements is clearly related with the grade of hyperemia [69]. Thus, our objective is to find the complex relationship between the features and a given grading scale.

Although hyperemia grading scales consist of a discrete number of prototypes, specialists usually assign a decimal value to their evaluations. This practice allows them to represent the distance from a prototype to the patient's eye, as there is a wide spectrum of situations that take place between two grades of a scale. Because optometrists do not use more than one decimal value for their evaluations, the number of values is still finite, and we could tackle the problem by applying classification methods. However, the number of classes is simply too large in comparison with the number of images, and the approaches that were tested did not provide acceptable results [70].

Therefore, we applied regression methods in order to transform each feature vector to the corresponding value in the grading scale. We tested several machine learning algorithms, that can be divided in two large groups: artificial neural networks (ANNs) and classifiers.

Regarding the ANNs, we tested the following approaches:

- **Multi-layer perceptron** (MLP) [71]. Described in Sect. 7.2.2.4.
- **Radial-basis function network** (RBFN) [72]. An approach that uses radial basis functions as activation functions. It is a feedforward ANN that outputs the linear combination of the computed activation values and the neuron parameters.
- **Self-organizing map** (SOM) [73]. It applies competitive learning and takes into account neighborhood relationships.
- **Learning vector quantization** (LVQ) [74]. It consists of two layers: a competitive one, which works in a similar fashion as a SOM; and a linear one, which transforms the competitive output into target classes defined by the user.

Regarding the classifiers, we tested the following approaches:

- **Decision trees** (DT) [75]. Described in Sect. 7.2.2.4.
- **Random forests** (RF) [76]. An ensemble method that creates several decision trees and outputs their mean prediction.
- **Support vector regression** (SVR) [77]. A variant of SVM (described in Sect. 7.2.2.4) adapted to regression problems.
- **k-nearest neighbors** (KNN) [78]. An algorithm that takes into account the values of the k closest neighbors to the analyzed sample.
- **Partial least squares** (PLS) [79]. A regression technique that creates a linear regression model by projecting the predicted variables to a new space.
- **Naive Bayes** (NB) [80]. Described in Sect. 7.2.2.4.

7.3.3 Results

This section depicts the results obtained after processing the data set described in Sect. 3.1. The values of the parameters for the features (Table 7.3) were set as follows: $s = 3 (V_6)$, $n_r = 10$ and $step = 10 (C_v)$, and $\kappa = 10 (W_v)$.

As we computed the 25 features in three different areas, we assigned an additional subscript to each feature. Thus, subscripts 1–25 refer to features obtained from both sides of the conjunctiva, subscripts 26–50 refer to the ones computed in the pupil side, and subscripts 51–75 refer to the ones computed in the corner of the eye side.

First, the results of the feature selection methods are shown. The feature selection approaches were computed with a 10-fold cross-validation. Therefore, each method did output a different set of features for each fold, and they were combined in order to obtain a unique feature set. To that end, we defined a minimum number of apparitions for a feature to be taken into account. We only included in the feature set for a given method those features that were selected in at least 7 out of 10 folds.

Additionally, Relief is a ranker method which outputs the whole feature set sorted in order of relevance. To obtain a subset, we defined a threshold on the relevance measure. We removed all the features that did not achieve a relevance value of at least half of the higher relevance value in a given fold.

For the embedded method, SVR-RFE, we defined a minimum set size of 5, a value that was determined empirically. Finally, as the subsets of features selected by each method are significantly different, we decided to combine these outputs and create a feature subset as the union of all the former subsets.

Table 7.4 shows the features selected by the feature selection approaches. We can observe how some features were frequently selected, such as V_1 that appears in 4 out of 5 approaches for both scales, or V_7 that appears in 3 out of 5. We can also observe how most of the selected features were computed in the whole conjunctiva. However, half of the features selected by both SVR-RFE and CFS were computed in one of the sides of the conjunctiva, which highlights the importance of the region of computation. Regarding the type of features, the ones that take color into account

Table 7.4 Selected features for each method

Method	Efron
CFS	$V_{1_3}, V_{7_{15}}, V_{1_{28}}, I_{5_{39}}$
Relief	$C_{v_1}, A_{v_2}, V_{1_3}, P_{v_{10}}, V_{6_{13}}, A_{v_{27}}, V_{1_{28}}, P_{v_{35}}$
M5	$V_{1_3}, V_{7_{15}}, I_{5_{39}}$
SMOReg	$V_{1_3}, V_{7_{15}}, V_{5_{37}}$
SVR-RFE	$C_{v_1}, I_{4_9}, W_{v_{25}}, V_{5_{37}}, W_{v_{50}}, B_{8_{73}}$
Union	$C_{v_1}, A_{v_2}, V_{1_3}, I_{4_9}, P_{v_{10}}, V_{6_{13}}, V_{7_{15}}, W_{v_{25}}, A_{v_{27}}, V_{1_{28}}, P_{v_{35}}, V_{5_{37}}, I_{5_{39}}, W_{v_{50}}, B_{8_{73}}$
Method	CCLRU
CFS	$V_{1_3}, V_{7_{15}}, I_{5_{39}}, I_{1_{54}}$
Relief	$C_{v_1}, A_{v_2}, V_{1_3}, P_{v_{10}}, V_{6_{13}}, V_{1_{28}}$
M5	$V_{1_3}, V_{7_{15}}$
SMOReg	$V_{1_3}, V_{7_{15}}$
SVR-RFE	$C_{v_1}, I_{4_9}, W_{v_{25}}, V_{5_{37}}, W_{v_{50}}, V_{7_{65}}$
Union	$C_{v_1}, A_{v_2}, V_{1_3}, I_{4_9}, P_{v_{10}}, V_{6_{13}}, V_{7_{15}}, W_{v_{25}}, V_{1_{28}}, V_{5_{37}}, I_{5_{39}}, W_{v_{50}}, I_{1_{54}}, V_{7_{65}}$

and, specially, those that take into account the hue in the conjunctival vessels (V_{n_m}), seem to be the most relevant. Finally, regarding the size of the subsets, CFS, M5, and SMOReg provide smaller subsets than the other methods in both scales.

After analyzing the importance of some features over the others, the following step is the transformation of the images' feature vectors to the value in a given grading scale. Each image feature has its own range of values and distribution. Moreover, there is not a direct relationship between the feature values and the grading scale ones [69]. Therefore, we need to model a complex relationship and, to that end, we tested several machine learning techniques.

Table 7.5 lists the parameters of each method. We must note that some of the techniques are usually employed in classification environments and, therefore, we needed to adapt them to the particularities of our problem.

The LVQ method processes each input in two stages. The first layer transforms the input (the feature vectors) into clusters. These clusters are not defined by the user, but they are found by the network during the training. Then, the second layer combines these clusters of the previous layer in order to obtain the target classes. These targets are defined by the user. However, in our case we have a potentially infinite number of target classes and, therefore, we started the process creating clusters in the output with KNN. The same situation occurs when applying the NB classifier.

In the SOM we have a different situation, as the output values are the position of the winning neuron, and not the expert value. Therefore, we used a MLP in order to map those neurons (the SOM output) to the desired output (the experts' evaluations).

Table 7.5 Parameters of the regression methods

Method	Parameters
DT	minimum leaf size = 3, minimum parent size = 6
KNN	number of neighbors = 1, distance = cosine
LVQ	number of neighbors for the classification stage = 3, dimensions = 8 (Efron) and 6 (CCLRU)
MLP	configuration = 16, 40, activation function = hyperbolic tangent sigmoid training function = Bayesian regularization backpropagation based on Levenberg-Marquardt optimization
NB	determination of prototypes using a KNN model with number of neighbors = 3 distribution type = mvmn (Efron) and kernel with normal kernel (CCLRU)
PLS	number of components = min(number of features, 8)
RBFN	spread = 0.4, error goal = 0.03
RF	number of trees = 60 (Efron) and 40 (CCLRU), minimum leaf size = 10
SOM	competitive layer size = 8 (Efron) and 6 (CCLRU) number of neighbors = 3, topology function = one-dimensional random pattern distance function = Manhattan, configuration of the MLP = 10
SVR	type = v-SVR (Efron) and ε-SVR (CCLRU), kernel = sigmoid (Efron) and radial basis function (CCLRU) $\gamma = 2^{-12}$ (Efron) and 2^{-10} (CCLRU), C = 2^8 (Efron) and 2^4 (CCLRU)

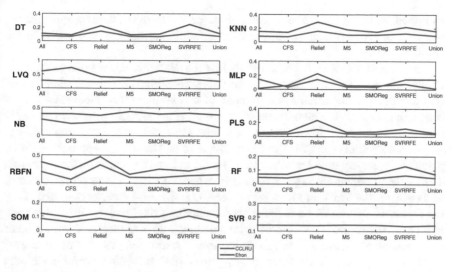

Fig. 7.12 MSE values for all the feature selection options and regression method combinations

We also used a 10-fold cross-validation during this stage. In order to compare the results for the different approaches, we computed the minimum squared error (MSE) for the validation set in each cross-validation iteration, and averaged them. Figure 7.12 depicts the MSE for the different regression methods and feature selection approaches, for both the Efron and the CCLRU scales.

The MLP obtains the lowest error values in both scales. For the Efron scale, the best combination of features is the full set ($MSE = 0.009$), whilst in the CCLRU scale the best results were provided by the CFS subset ($MSE = 0.032$). PLS and RF obtained good results in both scales, while the DT and the SOM approaches obtained good results in CCLRU too, although they are slightly worse than the aforementioned methods. Despite the fact that the lowest MSE value was obtained in the Efron scale, the CCLRU scale obtained lower values in most of the cases.

As we mentioned in Sect. 7.3.1, we selected the images where the experts' evaluations differ in a maximum of 0.5. Therefore, an acceptable system will be one that achieves an MSE lower than 0.25. As there are several approaches that fulfill this condition, it can be said that our system behaves as a human expert.

7.4 Tear Film Break-up Characterization

Break-up time (BUT) test is one of the most commonly tests used in clinical practice to assess tear film stability [81]. It consists in measuring the time that the tear film remains stable without blinking. Besides the time, break-up properties are related to specific aspects of the tear film that could affect dry eye symptoms [82]. Thus, the characterization of break-up areas according to different rupture patterns identified in the literature as streaks, dots, or pools [83] allows a quantitative, objective analysis of tear stability providing additional information useful to the clinical practice.

The main drawback of this test is its low repeatability, mainly due to the variability of the tear film and the subjective identification of the dark areas. Furthermore, the characterization by hand of break-up areas is a tedious and time consuming task so the automation of break-up characterization would reduce its subjective character, allowing a more accurate evaluation of the tear film stability.

The automation of the BUT test is a little explored field. First approaches for an automatic break-up time measure can be found in [84–86]. These methods provide a BUT measurement through an algorithm which consists in locating the different sequences in the tear film video where the BUT test can be performed, extracting the region of interest in each frame, and computing the break-up time over each sequence. The methodology for the automatic break-up time measurement was later improved by including an analysis of break-up location [87] as well as an assessment of the evolution of the break-up areas [88] providing useful additional information of tear film stability. Regarding the break-up classification into the different rupture patterns, a methodology which computes the timing of the break-up and the characterization of the size and morphology of the segmented break-up areas is exposed in [89].

A fully automatic methodology to analyze tear film videos and classify the detected break-up areas into one of the tear film rupture patterns is subsequently presented. Section 7.4.1 describes the tear film video dataset, then Sect. 7.4.2 details the methods used for the automatic break-up characterization, and finally, Sect. 7.4.3 includes the results and discussion of the evaluation process.

7.4.1 Video Dataset

The video dataset contains tear film videos in which the BUT test can be performed. These videos have been acquired and manually annotated by specialists from the Optometry Service of the University of Santiago de Compostela (Spain). For the video acquisition process sodium fluorescein is instilled into the eye using a micro-pipette and the tear film is recorded with a Topcon DV-3 camera with the help of a cobalt-blue and a yellow filters attached to a Topcon SL-D4 slit lamp biomicroscope to improve the visibility of the fluorescein emission. The frame rate of these videos is 30*fps* with a spatial resolution of 1024×768 pixels in RGB. The annotation procedure includes the detection of the sequences of interest for BUT measurement, the break-up time for each of them as well as the labeled of the break-up areas with its corresponding rupture pattern [83].

The dataset consists of 20 videos from healthy patients with ages ranging from 19 to 33, varying from very dry eye to no visible dryness. Each video has a duration of several minutes and contains different sequences of interest for BUT analysis. From these sequences, 96 segmented break-up areas have been manually labeled,

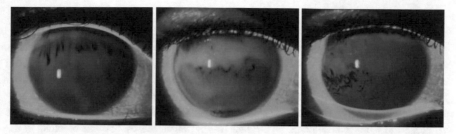

Fig. 7.13 Representative images of the dataset, from *left* to *right* streak, dot, and pool rupture patterns

containing 26.04% dots, 34.37% pools and 39.58% streaks. Figure 7.13 includes some representative images for the different rupture patterns.

7.4.2   Methods

Figure 7.14 shows the main steps to perform the automatic classification of the tear film rupture patterns. Each tear film video is divided in several BUT sequences, denoted as sequences of interest (SOIs), which consists of a set of frames delimited by blinks in which the BUT test can be performed. In order to characterize the tear film break-up, the videos were preprocessed to locate each SOI in the sequence and then, to segment the break-up areas in the frame before the last blink of each SOI, defined as the end reference frame (ERF). For this reason, the region of interest (ROI) corresponding to the visible part of the iris was extracted for each ERF and after that a correction process was applied to normalize the contrast and illumination variability. Then, a break up threshold was computed to determine if a pixel corresponds to a break-up area or not. Once the break-up threshold was applied to the ERF, a shape descriptor containing morphological features was built for each segmented break-up area, and then, different classifiers were applied to these descriptors to decide their tear film rupture patterns.

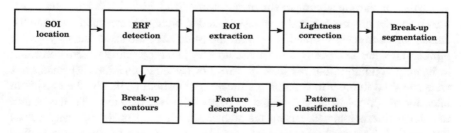

Fig. 7.14 Main steps for the automatic tear film break-up characterization

7.4.2.1 Tear Film Video Preprocessing

The different SOIs contained in a tear film video were detected as the sequences located between two consecutive blinks. Since the frames in which the eye is closed have a higher mean gray value than the frames in which the eye is open, blink detection was based on calculating the finite differences of mean values of gray between consecutive frames, and then applying a threshold. For each sequence the interest, the ERF was selected as the last frame before the eye begins to close at the end of the SOI because it represents the maximum expansion of the break-up in the SOI.

Once the ERF is selected in each SOI, the next step was to extract the ROI in order to discard regions without relevant information. The ROI corresponds to the visible part of the iris which may vary slightly throughout the sequence depending on the eye aperture and the appearance of shadows due to outer parts of the eye like eyelids or eyelashes. Assuming the iris has an approximately circular shape, its initial segmentation was carried out by correlation in the frequency domain with a set of masks formed by circumferences with different radii covering typical eye sizes. Since the eye does not remain fixed in the same position throughout the video, the ROI was registered by aligning each frame with the immediately preceding frame. This way, the methodology is independent of slight motions of the ROI. In some cases, the opening degree of the eye varies slightly throughout the sequence and the ROI contains outer parts like eyelids or eyelashes. These elements can disrupt the results so an adjustment was performed according to the eye features in order to discard these outer regions. For this adjustment, the mask radius was slightly reduced and the ROI was cropped at the top and at the bottom.

The video acquisition processes and the spherical surface of the eye cause luminosity and contrast heterogeneity within the tear film frames. This could affect the break-up characterization, mistaking poorly illuminated areas with real break-up areas. In order to overcome this problem, a lighting correction based on the approach proposed by Foracchia et al. [90] was performed. This process consists in normalizing the lightness and contrast variability in images, based on estimating both features in background small areas, spreading to the whole image and then removing from it.

After the luminosity correction, the break-up areas were segmented from the intensities of the green component of the normalized EFR. Each tear film video presents variations in color and lightness related to biological characteristics and the amount of fluorescein instilled, so not all the SOIs have the same intensity levels. Furthermore, the dark pixels at the break-up vary in a range close to zero according to lighting conditions, but not exactly zero. For these reasons, the ERF intensities were analyzed in order to determine a break-up threshold t_b, that is, the maximum intensity of a pixel to be considered as part of the break-up area. To this end, a multilevel thresholding [91] from the statistic distribution of the intensity values was applied. This method uses the mean and the variance of the image to find optimum thresholds for segmenting the image into multiple levels. The threshold t_b was computed from one of the lowest levels obtained in the multilevel thresholding, since they represent the darkest areas of the frame. The break-up threshold t_b was applied to the ERF in order to segment the break-up areas.

7.4.2.2   Image Features

Once the break-up threshold was applied to the ERF, the shape of the segmented break-up areas was analyzed in order to classify them into the different rupture patterns. Figure 7.15 shows an example of each pattern. The streak rupture pattern presents a linear shape while the dots have a circular morphology. Pools are characterized as disturbances of the tear film which conforms irregular regions that have neither a linear nor a circular shape.

First, the break-up areas were smoothed by applying morphological operations of opening and closing in order to discard noise [92]. For analyzing the shape of the break-up areas, the break-up contours were computed [93]. The algorithm for extracting the contours takes as input the thresholded ERF in which the break-up areas are represented by pixels with value 1 and the rest of the image by pixels with value 0. Starting with the upper left corner, the algorithm scans the image until it finds the first pixel with value 1, defined as starting point. Candidates to contour points are those pixels with value 1 and which have some pixel with value 0 in its neighborhood. Thus, the connected pixels which satisfies the condition of contour points are followed until the starting point is reached. This set of pixels with value 1 defines the contour of a break-up area. After this, the scan resumes, looking for new starting points. When the scan reaches the lower right corner of the frame, the algorithm stops.

Break-ups with the same rupture pattern show a high variability so there is no general models to characterize them. For this reason, the break-up areas were classified by analyzing their morphological features. To this end, spatial and central moments were computed for each contour [94]. Moments provide a geometrical meaning by several parameters which allows to extract different features for each contour [95, 96].

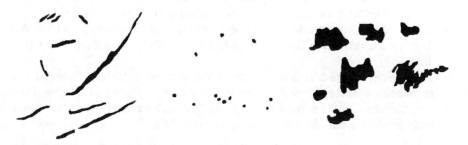

Fig. 7.15 Rupture patterns, from *left* to *right* streak, dot, and pool

The spatial moments are calculated from the following equation:

$$m_{p,q} = \sum_{i,j} i^p j^q T_k(i,j) \tag{4}$$

where $T_k(i,j)$ corresponds to the thresholded image, in which the background pixels are 0 and the pixels belonging to the contour are 1.

The moments are usually classified by the order, which depends on the indexes p and q. Thus, the sum $p + q$ is the order of the moment $m_{p,q}$. The zero order moment $m_{0,0}$ describes the area A delimited by the contour. The first order moments $m_{1,0}$, and $m_{0,1}$, contain information about the center of gravity of the contours $(\bar{i}, \bar{j})$, from which can be derived the central moments:

$$mu_{p,q} = \sum_{i,j} (i - \bar{i})^p (j - \bar{j})^q I(i,j) \tag{5}$$

In this manner, a shape descriptor was built with the features extracted from the moments of each contour. According to the shape of the different rupture patterns, the features should distinguish between linear and circular morphologies. Therefore, dot and streak patterns can be discriminated. For this purpose, the following features were computed:

- **Axis ratio** r_{ab}. The main inertial axes of the object correspond to the semi-major and semi-minor axes a and b of the ellipse which can be used as a approximation of the considered contour. The main inertial axis are those axis around which the contour can be rotated with minimal (major semi-axis a) or maximal (minor semi-axis b) inertia. The main inertial axis a and b can be derived from the second central moments

$$a, b = \sqrt{\frac{1}{2}\left(mu_{2,0} + mu_{0,2} \pm \sqrt{(mu_{2,0} - mu_{0,2})^2 + 4mu_{1,1}^2}\right)} \tag{6}$$

  r_{ab} is the ratio between the semi-major and the semi-minor axis. This ratio gives an idea whether the contour is more or less elongated (see Table 7.6).
- **Roundness** κ. The roundness κ is computed from the perimeter ρ and the area A. If the contour is a circle κ is equal 1, for other objects is greater than 1 (see Table 7.6).
- **Eccentricity** ε. The eccentricity ε can directly derived from the semi-major and semi-minor axes a and b of the contour. The eccentricity ε can have values from 0 to 1. Values of 0 are related to a perfectly rounded contour and values of 1 correspond to a line shaped contour (see Table 7.6).

However, pool patterns are regions with variable morphology, which can be small or large, with irregular boundaries, and can present an elongated or circular

Table 7.6 Features extracted from contours to build a descriptor for break-up classification

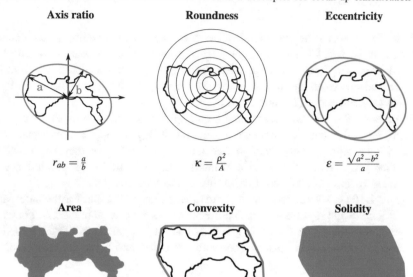

Axis ratio	Roundness	Eccentricity
$r_{ab} = \frac{a}{b}$	$\kappa = \frac{\rho^2}{A}$	$\varepsilon = \frac{\sqrt{a^2-b^2}}{a}$
Area	**Convexity**	**Solidity**
$A = m_{0,0}$	$C = \frac{\rho_h}{\rho}$	$S = \frac{A}{H}$

global distribution. Thus, the area was used as a feature to distinguish between dots and small circular pools.

- **Area** A. The number of pixels of the object bounded by the contour, which is defined by the spatial moment of zero order (see Table 7.6).

On the other hand, the convex hull area is related to the number or size of concavities in the contour and can be combined with the perimeter or area to get an indicator of the contour roughness. Thus, the next features were added to the shape descriptor:

- **Convexity** C. It is defined as the ratio of perimeters of the convex hull ρ_h over the perimeter of the original contour ρ (see Table 7.6).
- **Solidity** S. It describes the extent to which the shape is convex or concave, and it is defined by the relation between the area A of the break-up region and the corresponding convex hull area H. The solidity of a convex shape is always 1 (see Table 7.6).

All these metrics form up a descriptor that was used as the input of a classifier to decide the final type of each break-up area.

7.4.3   Results

This section presents the evaluation of the break-up characterization in the dataset
described in Sect. 7.4.1. Therefore, the break-up areas were segmented in each ERF
of the dataset, and then, a descriptor for each break-up area was computed from its
features: ratio between and major and minor axes (r_{ab}), roundness (κ), eccentricity
(ε), area (A), convexity (C), and solidity (S).

In order to classify each segmented break-up area in dot, pool or streak rupture
patterns, the following classifiers were used: naive Bayes, support vector machine,
multilayer perceptron, and decision tree (see Sect. 7.2.2.4 for more information
about them). Due to the limited size of the data set, a 10-fold cross-validation was
used to assess the generalization capability of the methods [97].

Figure 7.16 shows the performance provided by each classifier for the automatic
classification of the rupture patterns compared to the experts' annotations. For each
rupture pattern the graph shows in green the rate of true positives (TPR), which is
related to the percentage of patterns correctly classified; and in red the false positive

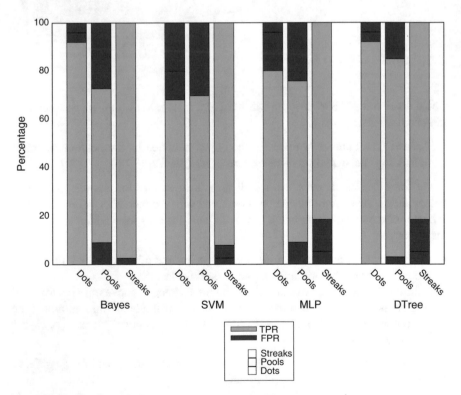

Fig. 7.16 Performance obtained with the different classifiers when comparing the automatic and
manual annotations. Each column represents, from top to bottom, the rates for streak, pool and dot
rupture patterns, as depicted in the legend

rate (FPR), which corresponds to the percentage of confusion with each of the other rupture patterns. This way, the best results are related to the classifier which maximizes the green area for each rupture pattern.

The decision tree classifier got the best results achieving a 84.37% of global accuracy for all rupture patterns. In some cases, very small pools were confused with dots as well as dots quite large were classified as pools. Additionally, some poorly defined streaks were classified as pools while streaks too small where mistaken with dots.

In order to verify the independence of the features in the descriptor and to extract the most relevant ones, principal component analysis (PCA) [98] was applied to the decision tree classifier. The extracted features by applying PCA are the axis ratio, the roundness, the eccentricity, the area, and the solidity. These features cover the main discriminant factors between the different patterns. The convexity is discarded by PCA because it provides similar information to the solidity since both are computed from the convex hull area, which may be redundant. The decision tree applied after discarding this feature achieved the results showed in Fig. 7.17. The dot and streak patterns have increased their sensitivity whereas the sensitivity of

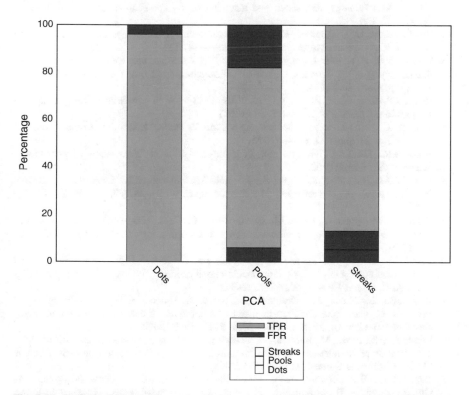

Fig. 7.17 Performance of break-up classification with the descriptors provided by PCA. Each column represents, from top to bottom, the rates for streak, pool and dot rupture patterns, as depicted in the legend

pool pattern has decreased. The global accuracy was 85.42%, which represents a slight increment with respect to the results obtained from the whole descriptor.

The proposed methods allow a quantitative, qualitative analysis of tear film instability, as an extension of BUT measurement, which is focused only on time. The automated system for break-up characterization provides fast, objective results, saving on effort, time and costs.

Acknowledgements This work has been partially funded by the Ministerio de Economía y Competitividad of Spain (project DPI2015-69948-R). Beatriz Remeseiro acknowledges the support of the Ministerio de Economía y Competitividad of the Spanish Government under *Juan de la Cierva* Program (ref. FJCI-2014-21194).

References

1. Paulsen, A.J., Cruickshanks, K.J., Fischer, M.E., Huang, G., Klein, B.E.K., Klein, R., Dalton, D.S.: Dry eye in the beaver dam offspring study: prevalence, risk factors, and health-related quality of life. Am. J. Ophthalmol. **157**(4), 799–806 (2014)
2. Gayton, J.L.: Etiology, prevalence, and treatment of dry eye disease. Clin. Ophthalmol. **3**, 405–412 (2009)
3. Yu, J., Asche, C.V., Fairchild, C.J.: The economic burden of dry eye disease in the united states: a decision tree analysis. Cornea **30**(4), 379–387 (2011)
4. Lemp, M.A., Baudouin, C., Baum, J., et al.: The definition and classification of dry eye disease: report of the definition and classification subcommittee of the international dry eye workshop. Ocul. Surf. **5**(2), B75–B92 (2007)
5. Craig, J.P., Tomlinson, A.: Importance of the lipid layer in human tear film stability and evaporation. Optom. Vis. Sci. **74**, 8–13 (1997)
6. Guillon, J.P.: Non-invasive tearscope plus routine for contact lens fitting. Contact Lens and Anterior Eye **21**(Suppl 1), 31–40 (1998)
7. Korb, D.R.: The Tear Film: Structure, Function, and Clinical Examination. Elsevier Health Sciences, Amsterdam (2002)
8. Nichols, J.J., Nichols, K.K., Puent, B., Saracino, M., Mitchell, G.L.: Evaluation of tear film interference patterns and measures of tear break-up time. Optom. Vis. Sci. **79**(6), 363–369 (2002)
9. Calvo, D., Mosquera, A., Penas, M., García-Resúa, C., Remeseiro, B.: Color texture analysis for tear film classification: a preliminary study. Int. Conf. Image Anal. Recogn. **6112**, 388–397 (2010)
10. Ramos, L., Penas, M., Remeseiro, B., Mosquera, A., Barreira, N., Yebra-Pimentel, E.: Texture and color analysis for the automatic classification of the eye lipid layer. Int. Work Conf. Artif. Neural Netw. **6692**, 66–73 (2011)
11. Remeseiro, B., Penas, M., Mosquera, A., Novo, J., Penedo, M.G., Yebra-Pimentel, E.: Statistical comparison of classifiers applied to the interferential tear film lipid layer automatic classification. Comput. Math. Methods Med. **2012**, 1–10 (2012)
12. Remeseiro, B., Penas, M., Barreira, N., Mosquera, A., Novo, J., García-Resúa, C.: Automatic classification of the interferential tear film lipid layer using colour texture analysis. Comput. Methods Programs Biomed. **111**, 93–103 (2013)
13. Remeseiro, B., Bolón-Canedo, V., Peteiro-Barral, D., Alonso-Betanzos, A., Guijarro-Berdinas, B., Mosquera, A., Penedo, M.G., Sánchez-Marono, N.: A methodology for improving tear film lipid layer classification. IEEE J. Biomed. Health Inf. **18**(4), 1485–1493 (2014)

14. Remeseiro, B., Mosquera, A., Penedo, M.G., Garca-Resúa, C.: Tear film maps based on the lipid interference patterns. 6th Int. Conf. Agents Artif. Int. **1**, 732–739 (2014)
15. Remeseiro, B., Mosquera, A., Penedo, M.G.: CASDES: a computer-aided system to support dry eye diagnosis based on tear film maps. IEEE J. Biomed. Health Inf. **20**(3), 936–943 (2016)
16. Remeseiro, B., Barreira, N., Garca-Resúa, C., Lira, M., Giráldez, M.J., Yebra-Pimentel, E., Penedo, M.G.: iDEAS: a web-based system for dry eye assessment. Comput. Methods Prog. Biomed. **130**, 186–197 (2016)
17. González-Domínguez, J., Remeseiro, B., Martín, M.J.: Acceleration of tear film map definition on multicore systems. Procedia Comput. Sci. **80**, 41–51 (2016)
18. González-Domínguez, J., Remeseiro, B., Martín, M.J.: Parallel definition of tear film maps on distributed-memory clusters for the support of dry eye diagnosis. Comput. Methods Programs Biomed. **139**, 51–60 (2017)
19. Méndez, R., Remeseiro, B., Peteiro-Barral, D., Penedo, M.G.: Evaluation of class binarization and feature selection in tear film classification using topsis. Agents Artif. Intell. Revised Selected Papers ICAART 2013 **2014**(449), 179–193 (2014)
20. Peteiro-Barral, D., Remeseiro, B., Méndez, R., Penedo, M.G.: Evaluation of an automatic dry eye test using MCDM methods and rank correlation. Med. Biol. Eng. Comput. **55**(4), 527–536 (2017)
21. VOPTICAL_I1, VARPA optical dataset acquired and annotated by optometrists from the Optometry Service of the University of Santiago de Compostela, Spain (2012)
22. Russ, J.C.: The Image Processing Handbook, 3rd edn. CRC Press Inc, Boca Raton, FL, USA (1999)
23. McLaren, K.: The development of the CIE 1976 (L*a*b) uniform colour-space and colour-difference formula. J. Soc. Dyers Colour. **92**(9), 338–341 (1976)
24. Bradski, G.: OpenCV. Dr. Dobb's J Softw. Tools **25**, 120–126 (2000)
25. Haralick, R.M., Shanmugam, K., Dinstein, I.: Texture features for image classification. IEEE Trans. Sys. Man Cybern. **3**, 610–621 (1973)
26. Furnkranz, J.: Round robin ensembles. Int. Data Anal. **7**(5), 385–403 (2003)
27. Dietterich, T.G., Bakiri, G.: Solving multiclass learning problems via error-correcting output codes. J. Artif. Intell. Res. **2**, 263–286 (1995)
28. Allwein, E.L., Schapire, R.E., Singer, Y.: Reducing multiclass to binary: a unifying approach for margin classifiers. J. Mach. Learn. Res. **1**, 113–141 (2001)
29. Manning, C.D., Raghavan, P., Schütze, H.: Introduction to Information Retrieval, vol. 1. Cambridge University Press, Cambridge (2008)
30. Loughrey, J., Cunningham, P.: Overfitting in wrapper-based feature subset selection: the harder you try the worse it gets. Res. Dev. Intell. Sys. XXI, **2005**, 33–43 (2005)
31. Guyon, I., Gunn, S., Nikravesh, M., Zadeh, L.: Feature Extraction: Foundations and Applications. Springer Verlag, Berlin (2006)
32. Hall, M.A.: Correlation-based feature selection for machine learning. PhD thesis, The University of Waikato (1999)
33. Dash, M., Liu, H.: Consistency-based search in feature selection. Artif. Intell. **151**(1–2), 155–176 (2003)
34. Zhao, Z., Liu, H.: Searching for interacting features. Proceedings of the 20th international joint conference on Artificial intelligence, 1156–1161 (2007)
35. Mitchell, T.M.: Machine Learning. McGraw-Hill, Boston (1995)
36. Friedman, J.H.: Regularized discriminant analysis. J. Am. Stat. Assoc. **84**, 165–175 (1989)
37. Jensen, F.V.: An Introduction to Bayesian Networks, vol. 210. UCL press, London (1996)
38. Murthy, S.K.: Automatic construction of decision trees from data: a multi-disciplinary survey. Data Min. Knowl. Disc. **2**, 345–389 (1998)
39. Burges, C.J.C.: A tutorial on support vector machines for pattern recognition. Data Min. Knowl. Disc. **2**, 121–167 (1998)
40. Rosenblatt, F.: The perceptron: a probabilistic model for information storage and organization in the brain. Psychol. Rev. **65**(6), 386 (1958)

41. Fernandez-Caballero, J.C., Martnez, F.J., Hervás, C., Gutiérrez, P.A.: Sensitivity versus accuracy in multiclass problems using memetic pareto evolutionary neural networks. IEEE Trans. Neural Networks 21(5), 750–770 (2010)
42. Hwang, C.L., Yoon, K.: Multiple Attribute Decision Making: Methods and Applications, vol. 13. Springer-Verlag, New York (1981)
43. Kuo, Y., Yang, T., Huang, G.W.: The use of grey relational analysis in solving multiple attribute decision-making problems. Comput. Ind. Eng. 55(1), 80–93 (2008)
44. Opricovic, S.: Multicriteria optimization of civil engineering systems. Fac. Civil Eng. Belgrade 2(1), 5–21 (1998)
45. Gautheir, T.D.: Detecting trends using spearman's rank correlation coefficient. Environ. Forensics 2(4), 359–362 (2001)
46. Chang, C., Lin, C.: LIBSVM: A library for support vector machines. ACM Trans. on Intell. Sys. Tech. 2, 1–27, http://www.csie.ntu.edu.tw/cjlin/libsvm (2011)
47. Bramer, M.: Principles of Data Mining, vol. 180. Springer, London (2007)
48. Yoneda, T., Sumi, T., Takahashi, A., Hoshikawa, Y., Kobayashi, M., Fukushima, A.: Automated hyperemia analysis software: reliability and reproducibility in healthy subjects. Jpn. J. Ophthalmol. 56(1), 1–7 (2012)
49. Rodriguez, J.D., Johnston, P.R., Ousler, G.W., Smith, L.M., Abelson, M.B.: Automated grading system for evaluation of ocular redness associated with dry eye. Clin. Ophthalmol. 7, 1197 (2013)
50. Wu, S., Hong, J., Tian, L., Cui, X., Sun, X., Xu, J.: Assessment of bulbar redness with a newly developed keratograph. Optom. Vis. Sci. 92(8), 892–899 (2015)
51. Tort, M., Ornberg, R., Lay, B., Danno, R., Soong, F., Salapatek, A.: Development of an objective assessment of conjunctival hyperemia elicited via Conjunctival Allergen Provocation Testing (CAPT) and Environmental Exposure Chamber (EEC) testing. EEC (N = 13) 2, 5 (2012)
52. Wald, M.J., Lay, B., Danno, R., Grosskreutz, C.L., Chandra, S.: Performance of automated hyperemia assessment in allergic conjunctivitis interventional study. Invest. Ophthalmol. Vis. Sci. 56, 12300 (2015)
53. Downie, L.E., Keller, P.R., Vingrys, A.J.: Assessing ocular bulbar redness: a comparison of methods. Ophthalmic Physiol. Opt. 36(2), 132–139 (2016)
54. Amparo, F., Wang, H., Emami-Naeini, P., Karimian, P., Dana, R.: The ocular redness index: a novel automated method for measuring ocular injectiona novel automated system to measure redness. Invest. Ophthalmol. Vis. Sci. 54(7), 4821–4826 (2013)
55. Papas, E.B.: Key factors in the subjective and objective assessment of conjunctival erythema. Invest. Ophthalmol. Vis. Sci. 41(3), 687–691 (2000)
56. Wolffsohn, J.S., Purslow, C.: Clinical monitoring of ocular physiology using digital image analysis. Contact Lens and Anterior Eye 26(1), 27–35 (2003)
57. Efron, N., Morgan, P.B., Katsara, S.S.: Validation of grading scales for contact lens complications. Ophthalmic Physiol. Opt. 21(1), 17–29 (2001)
58. Fieguth, P., Simpson, T.: Automated measurement of bulbar redness. Invest. Ophthalmol. Vis. Sci. 43(2), 340–347 (2002)
59. Murphy, P.J., Lau, J.S.C., Sim, M.M.L., Woods, R.L.: How red is a white eye? Clinical grading of normal conjunctival hyperemia. Eye 21(5), 633–638 (2007)
60. Wolffsohn, J.S.: Incremental nature of anterior eye grading scales determined by objective image analysis. Br. J. Ophthalmol. 88(11), 1434–1438 (2004)
61. Sánchez, L., Barreira, N., Pena-Verdeal, H., Yebra-Pimentel, E.: A Novel Framework for Hyperemia Grading Based on Artificial Neural Networks, pp. 263–275. Springer, Heidelberg (2015)
62. Sánchez, L., Barreira, N., Sánchez, N., Mosquera, A., Pena-Verdeal, H., Yebra-Pimentel, E.: On the analysis of local and global features for hyperemia grading. Ninth Int. Conf. Mach. Vis. 10341, 103411T–103411T (2017)

63. Sánchez-Brea, M.L., Barreira-Rodrguez, N., Mosquera-González, A., Evans, K., Pena-Verdeal, H.: Defining the optimal region of interest for hyperemia grading in the bulbar conjunctiva. Comput. Math. Methods Med. **2016**, 1–9 (2016)
64. Vázquez, S.G., Barreira, N., Penedo, M.G., Pena-Seijo, M., Gómez-Ulla, F.: Evaluation of SIRIUS retinal vessel width measurement in REVIEW dataset. IEEE 26th Int. Symp. Comp. Med. Syst. **2013**, 71–76 (2013)
65. Robnik-Šikonja, M., Kononenko, I.: An adaptation of Relief for attribute estimation in regression. In Machine Learning: Proceedings of the Fourteenth International Conference, 296–304 (1997)
66. Quinlan, J.R.: Learning with continuous classes. Aust. Jt Conf. Artif. Intell. **92**, 343–348 (1992)
67. Shevade, S.K., Keerthi, S.S., Bhattacharyya, C., Murthy, K.R.K.: Improvements to the SMO algorithm for SVM regression. IEEE Trans. Neural Netw. **11**(5), 1188–1193 (2000)
68. Guyon, I., Weston, J., Barnhill, S., Vapnik, V.: Gene selection for cancer classification using support vector machines. Mach. Learn. **46**(1–3), 389–422 (2002)
69. Sánchez-Brea, M.L., Barreira, N., Sánchez-Maroño, N., Mosquera, A., García-Resúa, C., Giráldez-Fernández, M.J.: On the development of conjunctival hyperemia computer-assisted diagnosis tools: Influence of feature selection and class imbalance in automatic gradings. Artif. Intell. Med. **71**, 30–42 (2016)
70. Sanchez, L., Barreira, N., Mosquera, A., Pena-Verdeal, H., Yebra-Pimentel, E.: Comparing machine learning techniques in a hyperemia grading framework. Int. Conf. Agents Artif. Intell. **2**, 423–429 (2016)
71. Baum, E.B.: On the capabilities of multilayer perceptrons. J Complexity **4**(3), 193–215 (1988)
72. Park, J., Sandberg, I.W.: Universal approximation using radial-basis-function networks. Neural Comput. **3**(2), 246–257 (1991)
73. Kohonen, T.: The self-organizing map Neurocomputing **21**(1–3), 1–6 (1998)
74. Kohonen, T.: Improved versions of learning vector quantization. Int. Jt Conf. Neural Netw. **1990**, 545–550 (1990)
75. Quinlan, J.R.: Induction of decision trees. Mach. Learn. **1**(1), 81–106 (1986)
76. Breiman, L.: Random forests. Mach. Learn. **45**(1), 5–32 (2001)
77. Smola, A.J., Schölkopf, B.: A tutorial on support vector regression. Stat. Comput. **14**(3), 199–222 (2004)
78. Dudani, S.A.: The distance-weighted k-nearest-neighbor rule. IEEE Trans. Syst. Man Cybern. **6**(4), 325–327 (1976)
79. Abdi, H.: Partial least square regression (PLS regression). Encycl. Res. Methods Soc. Sci. **6**(4), 792–795, 2003
80. John, G.H., Langley, P.: Estimating continuous distributions in bayesian classifiers.Conf. Uncertainty Artif. Intell. **1995**, 338–345, (1995)
81. Abelson, M.B., Ousler, G.W., Nally, L.A., Welch, D., Krenzer, K.: Alternative reference values for tear film break up time in normal and dry eye Populations. In Lacrimal Gland, Tear Film, and Dry Eye Syndromes 3, pp. 1121–1125. Springer, New York (2002)
82. King-Smith, P.E., Fink, B.A., Nichol J.J., Braun, R.J., McFadden, G.B.: The contribution of lipid layer movement to tear film thinning and breakup. Invest. Opthalmol. Vis. Sci. **50**, 2747–2756 (2009)
83. Bitton, E., Lovasik, J. V.: Longitudinal analysis of precorneal tear film rupture patterns. In Lacrimal Gland, Tear Film, and Dry Eye Syndromes 2, pp. 381–389. Springer, New York (1998)
84. Yedidya,T., Hartley, R., Guillon, J.P.: Automatic detection of pre-ocular tear film break-up sequence in dry eyes. Digit. Image Comput. Tech. and Appl., **2008**, 442–448 (2008)
85. Cebreiro, E., Ramos, L., Mosquera, A., Barreira, N., Penedo, M.G.: Automation of the tear film break-up time test. Proceedings of the 4th International Symposium on Applied Sciences in Biomedical and Communication Technologies, 123 (2011)
86. Ramos, L., Barreira, N., Mosquera, A., Currás, M., Pena-Verdeal, H. Giráldez, M.J., Penedo, M.G: Adaptive parameter computation for the automatic measure of the Tear Break-Up Time.

16th International Conference on Knowledge-Based and Intelligent Information & Engineering Systems, 243, 1370–1379 (2012)

87. Ramos, L., Barreira, N., Mosquera, A., Penedo, M.G., Yebra-Pimentel, E., García-Resúa, C.: Analysis of parameters for the automatic computation of the tear film break-up time test based on cclru standards. Comput. Methods Programs Biomed. **113**(3), 715–724 (2014)

88. Ramos, L., Barreira, N., Pena-Verdeal, H., Giráldez, M.J., Yebra-Pimentel, E.: Computational approach for tear film assessment based on break-up dynamics. Biosys. Eng. **138**, 90–103 (2015)

89. Ramos, L., Barreira, N., Mosquera, A., Pena-Verdeal, H., Yebra-Pimentel, E.: Break-up analysis of the tear film based on time, location, size and shape of the rupture area. International Conference Image Analysis and Recognition, 695–702 (2013)

90. Foracchia, M., Grisan, E., Ruggeri, A.: Luminosity and contrast normalization in retinal images. Med. Image Anal. **9**(3), 179–190 (2005)

91. Arora, S., Acharya, J., Verma, A., Prasanta, K.: Panigrahi. Multilevel thresholding for image segmentation through a fast statistical recursive algorithm. Pattern Recogn. Lett. **29**(2), 119–125 (2008)

92. Dougherty, E.R.: An introduction to morphological image processing. SPIE Optical Engineering Press, Tutorial texts in optical engineering (1992)

93. Suzuki, S., Abe, K.: Topological structural analysis of digitized binary images by border following. Comput. Vis. Graph. Image Process. **30**(1), 32–46 (1985)

94. Hu, M.: Visual pattern recognition by moment invariants, computer methods in image analysis. IRE Trans. Inf. Theory **8**, 179–187 (1962)

95. Reed-Teague, M.: Image analysis via the general theory of moments. J. Opt. Soc. Am. **70**(8), 920–930 (1980)

96. Nunes, J.F., Moreira, P.M., Tavares, J.M.R.S: Shape based image retrieval and classification. 5th Iberian Conference on Information Systems and Technologies (2010)

97. Rodriguez, J., Perez, A., Lozano, J.: Sensitivity analysis of k-fold cross validation in prediction error estimation. IEEE Trans. Pattern Anal. Mach. Intell. **32**(3), 569–575 (2010)

98. Jolliffe, I.T.: Principal Component Analysis. Springer Verlag, New York (1986)

Chapter 8
Intelligent Decision Support Systems in Automated Medical Diagnosis

Florin Gorunescu and Smaranda Belciug

Abstract The Intelligent Decision Support Systems (IDSSs) represent an inter-disciplinary research domain bringing together Artificial Intelligence/Machine Learning (AI/ML), Decision Science (DS), and Information Systems (IS). IDSS refers to the use of AI/ML techniques in decision support systems. In this context, it should be emphasized the special role of statistical learning (SL) in the process of training algorithms from data. The purpose of this chapter is to provide a short review of some of the state-of-the-art AI/ML algorithms, seen as intelligent tools used in the medical decision-making, along with some important applications in the automated medical diagnosis of some major chronic diseases (MCDs). In addition, we aim to present an interesting approach to develop novel IDSS inspired by the evolutionary paradigm.

Keywords Intelligent decision support system · Neural networks · Support vector machines · Evolutionary computation · Computer-aided medical diagnosis

8.1 Introduction

The explosive growth of medical databases derived from a wide range of diseases, and the widespread development of efficient AI/ML algorithms, led to the development of high performance IDSSs aiming to support the computer-aided medical diagnosis (CAMD). From a medical perspective, a differential diagnosis is the complex process of differentiating among different possible diseases presenting

F. Gorunescu (✉)
Chair of Mathematics, Biostatistics and Computer Science, University of Medicine and Pharmacy of Craiova, Petru Rares 2-4, 200349 Craiova, Romania
e-mail: gorunef@gmail.com; florin.gorunescu@webmail.umfcv.ro

S. Belciug
Department of Computer Science, Faculty of Sciences, University of Craiova, A.I. Cuza 13, 200585 Craiova, Romania
e-mail: smaranda.belciug@gmail.com; sbelciug@inf.ucv.ro

© Springer International Publishing AG 2018
D.E. Holmes and L.C. Jain (eds.), *Advances in Biomedical Informatics*,
Intelligent Systems Reference Library 137,
https://doi.org/10.1007/978-3-319-67513-8_8

161

more or less similar symptoms. This process consists of highlighting the connection between a certain disease and the patient's history recorded in the patient's file, physical examinations, tests, clinical, radiological and laboratory data, etc. From a biomedical informatics point of view, the differential diagnosis is perceived as a classification procedure involving a decision-making process based on the available medical data and processed by an "intelligent" system (IS) built from AI/ML algorithms. Since these medical databases are processed by means of statistical and AI/ML techniques, the process of developing IDSSs is naturally embedded in the data mining (DM) field [1]. Directly helped by IDSSs, physicians have thus access to a wide range of DM techniques such as (artificial) neural networks (NNs), genetic/evolutionary algorithms (GAs/EAs), support vector machines (SVMs), naïve Bayes (nB), random forests (RFs), and hybrid algorithms using particle swarm optimization (PSO), ant colony optimization (ACO), etc., from which to choose the most appropriate one to the situation at hand. Up-to-date online medical databases can be used with the efficient aid of IDSSs to support clinical decision-making, offering thus direct access to medical evidence.

The paradigm underlying the functionality of an IDSS can be summarized as follows. The input consists of different symptoms (attributes), while the output consists of possible diseases caused by them (decision classes). After comparing a certain data corresponding to a yet undiagnosed patient with the observations and corresponding diagnoses contained in the medical database (medical data of patients), IDSS will provide the most probable diagnosis based on the human knowledge embedded in that database.

It has been more than 10 years since the U.S. Food and Drug Administration (FDA) has approved the first computer-aided medical diagnosis (CAMD) system, the R2 ImageChecker (http://investors.hologic.com/news?item=233), using film-based mammography. Since then, various advanced DM techniques have been developed and applied on high dimensional medical databases, also including specific medical data obtained from high performance medical imaging devices. A simple search on Medline (http://www.ncbi.nlm.nih.gov/pubmed) regarding terms such as "artificial intelligence", "machine learning algorithms", "data mining", etc., reveals a rich literature regarding the use of IDSSs in various pathologies, such as different types of cancer (breast, pancreatic, liver, lung, thyroid, gastric), neurological diseases, heart diseases, etc.

The aim of this chapter is to review some of the most efficient IDSSs used in CAMD, since nowadays there is a tremendous opportunity for AI/ML algorithms to assist the physician deal with a flood of patient information and scientific knowledge. Its goal is to draw attention to the applications and challenges faced by the use of IDSSs in CAMD, and to highlight the idea that, although they are underused, they have a strong potential in helping to efficiently solve medical decision problems. The chapter was written for all those working in the medical field, willing to use the state-of-the-art computerized techniques, and for all researchers in the health informatics domain, eager to develop such "intelligent" tools.

8.2   State-of-the-Art IDSSs for CAMD

The use of medical datasets belonging to large data repository, such as UCI Machine Learning Repository (http://archive.ics.uci.edu/ml/), requires state-of-the-art ML techniques. A variety of IDSSs have been developed focusing on a specific medical problem. There are two approaches in this research field. On the one hand, IDSS is reduced to one algorithm, its target being just a specific disease of a specific organ. On the other hand, to overcome this situation and get more efficient IDSSs, one can consider instead the paradigms underlying both the committees of machines and parallel computing/distributed systems. In such a way, a more reliable decision-making process is expected, based on the synergism of different algorithms brought together to solve the same problem.

NNs are among the most used AI/ML algorithms in medical decision-making, based on their undoubted efficiency in solving classification issues. Highly accurate classification of biomedical images is an essential task in the clinical diagnosis. A deep convolutional NN has been proposed to use raw pixels of original biomedical images instead of using the manual design of the feature space, or to seek an effective feature vector classifier, or to segment specific detection object and image patches [2]. The problem of cell segmentation in histopathological images has been solved using convolutional NNs, stacked autoencoders, and deep belief networks [3]. Coronary artery is one of the widespread cardiovascular disease, being associated with high costs and major side effects. A hybrid NN-genetic algorithm has been proposed for computer-aided decision-making to detect it [4]. Since the prevalence of high hyperlipidemia is increasing around the world nowadays, a model based on multiple linear regression and NN algorithm has been proposed to analyze the relationship of triglyceride (TG) and cholesterol (TC) with indexes of liver function and kidney function, and to develop a prediction model of TG, TC in overweight people [5]. The problem of the diagnostic accuracy of mammography in breast cancer detection has been approached with the aid of NNs within a multipurpose image analysis software [6].

SVMs are efficient ML algorithms used mostly in classification problems. Among different ML approach in breast cancer diagnosis, a SVM along with SVM ensembles have been proposed [7] for small and large scale breast cancer datasets. The identification of a subset of genes having the ability to capture the necessary information to distinguish classes of patients has been solved based on ensemble SVM-recursive feature elimination and applied to childhood leukemia [8]. Pancreatic cancer classification has been approached using a SVM optimized by an improved fly optimal algorithm [9]. Some quantum mechanics features and attributes from circular fingerprints are computed and classified using the SVM for predicting some potential sites of metabolism on a series of drugs that can be metabolized by flavin-containing monooxygenase enzymes [10]. A quantitative non-invasive *in vivo* cell localization method using contrast enhanced multiparametric MRI and SVM-based post-processing has been developed for cell cancer [11].

An nB classifier assumes independent predictors, being useful for very large datasets, often outperforming more sophisticated classification methods. A Bayesian approach for classification of proteomic profiles generated by liquid chromatography-mass spectrometry used in cancer treatment has been proposed [12]. An nB classifier was trained to investigate candidate biomarkers with new modified aptamer-based proteomic technology, developing afterwards a 7-protein panel that discriminates lung cancer from controls [13]. An nBs classifier has been built to calculate pretest probabilities adjusted for age, tumor localization and sex for primary bone tumors [14]. A Bayesian network classifier has been proposed in order to increase the ability to integrate diverse information from different sources, to effectively predict protein-protein interactions, to infer aberrant networks with scale-free and small-world properties, and to group molecules into functional modules or pathways based on the primary function and biological features in hepatocellular carcinoma [15]. A method to enhance a Bayesian classifier performance by exploiting the correlations between the class-discriminating miRNA and the expression of an additional normalized miRNA, has been proposed to discriminates samples into two classes of pulmonary tumors, adenocarcinomas and squamous cell carcinomas [16].

A RF consists of a collection or ensemble of simple tree predictors, each capable of producing a response when presented with a set of predictor values, leading to significant improvement in prediction accuracy. RF has been used to evaluate candidate region descriptors for lesion detection in breast cancer screening adjunct technology (automated breast ultrasound) [17]. RF models with variable selection, using both genetic and clinical variables in order to predict the response of a patient using pathological complete response as the measure of response, have been developed for breast cancer [18]. RF regressors have been developed to map parameters of the vertical, anterior-posterior, and medio-lateral ground reaction forces (i.e., mean value, push-off time, and slope) via rule induction to the degree of knee osteoarthritis [19]. A random survival forest (RSF) has been proposed to determine the key factors affecting survival in gastric cancer [20]. A new feature selection strategy, named genetic algorithm based on RF, has been proposed in order to choose the most predictive subset of features in oesophageal cancer [21].

A suggestive illustration of the way IDSSs work in CAMD is displayed in Fig. 8.1 below.

8.2.1 Neural Networks-Based IDSSs

The research field of NNs at more than 70 years from their occurrence -the first artificial neuron was designed in 1943 by the neurophysiologist Warren McCulloch and the logician Walter Pitts- is still under a continuous development, new algorithms, increasingly sophisticated, being proposed [1]. In recent decades, NNs appear as a practical technology, designed to successfully solve many problems in various fields, with particular emphasis on CAMD.

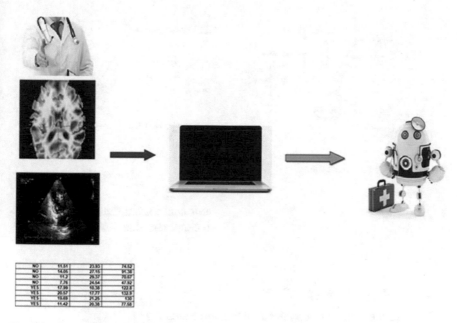

Fig. 8.1 IDSS applied to CAMD

8.2.1.1 Traditional NNs for CAMD

Basically, NNs represent non-programmed (non-algorithmic) adaptive information processing systems. They can be considered as a massively parallel distributed computing structure, inspired by the way the human brain processes information. They learn from examples and behave like the biological brain, the way of processing the available information being sketchy for nonprofessionals. The similarity between NNs and human brain's functionality resides in two main aspects:

- The available knowledge is acquired by the network through the *learning (training)* process;
- The intensities of the inter-neuron connections, known in the neural computing domain as *(synaptic) weights,* are used to store acquired knowledge.

The architecture of an artificial neuron (perceptron) can be illustrated graphically by the figure below (Fig. 8.2).

The learning paradigm underlying the perceptron classification process for a simple two-class decision problem (C_1, C_2) can be summarize in the following algorithm (more details in [1]).

Input: $\mathbf{x}(n) = (1, x_1(n), x_2(n), \ldots, x_p(n))$ – input training vector
$\mathbf{w}(n) = (b(n), w_1(n), w_2(n), \ldots, w_p(n))$ – (synaptic) weight vector
$y(n)$ – actual response
$d(n)$ – desired response

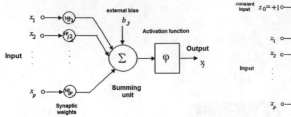 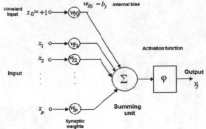

Fig. 8.2 Perceptron architecture

η – constant learning-rate parameter (positive and less than unity)

1. *Initialization.* Set $\mathbf{w}(0) = \mathbf{0}$.

   Perform the following computations for time step $n = 1, 2, \ldots$.

2. *Activation.* At time step n, activate the perceptron by applying continuous-valued input training vector $\mathbf{x}(n)$ and desired response $d(n)$.

3. *Computation of actual response.* Compute the actual response of the perceptron, given by:

$$y(n) = sign\left[\mathbf{w} \cdot \mathbf{x}(n)^T\right],$$

where *sign* represents the signum function.

4. *Adaptation of the weight vector.* Update the weight vector of the perceptron:

$$\mathbf{w}(n+1) = \mathbf{w}(n) + \eta \cdot [d(n) - y(n)] \cdot \mathbf{x}(n),$$

where:

$$d(n) = \begin{cases} +1, & \mathbf{x}(n) \in C_1 \\ -1 & \mathbf{x}(n) \in C_2 \end{cases}$$

5. *Continuation.* Increment time step n by one and GOTO step 2.

   Output: After a finite number of time steps n, the rule for adapting the synaptic weights of the perceptron must terminate

NN is composed of a large number of highly interconnected processing elements -artificial neurons/perceptrons- working in parallel to solve a specific problem. NN consists of: *a*) the *input* (units) fed with information from the environment, *b*) the *hidden neurons* controlling the actions in the network, and *c*) the *output* (units) synthesizing the network response. All the computing units must be interconnected in order that the network becomes fully functional. Much of the current NNs have a *feedforward* type structure. The signal moves from input to output, passing through all the network's hidden units, so the outputs of neurons are connected just to the next layer. The best-known type of such network is the *multilayer feedforward network*, also known as *multi-layer perceptron* (MLP). Briefly, the information from the environment enters the network by the input layer, representing the inputs of neurons in the second layer (i.e., the first hidden layer), then, after it is processed by them, it will become the input of the next layer (i.e., the second hidden layer), and so on.

The most popular (supervised) learning paradigm for MLP is the well-known *back-propagation* (BP) algorithm. In essence, the network output values are compared with the actual values and the error is computed based on a predefined error function E depending on the weights w_i. Then, according to the result thus obtained, one acts backward through the network to adjust the weights in order to minimize the error E by the iterative process of *gradient descent* using the error gradient ∇E.

Unlike MLP, where the computing units use a non-linear activation function, based on the dot product between the input vector and the weight vector, in the *radial basis function* network (RBF) case the activation of a (hidden) unit is based on the distance between the input vector and a prototype vector (center). Instead of using hyper-planes to divide the problem space as in the MLP case, the RBF divides up the space by using hyper-spheres characterised by centres and radii. A two-stage learning paradigm is used: *a*) in the first stage, the input dataset is used to determine the parameters of the basis function (usually Gaussian), and *b*) in the second stage the weights are optimized.

Another interesting interpretation of the network outputs is to estimate the probability of class membership, the network attempting to learn an approximation of the probability density function (pdf) corresponding to a certain decision class. This particular NN type, known as *probabilistic neural network* (PNN), replaces the activation function (usually, sigmoid) with an exponential function, based on the Bayesian decision theory. PNN uses the training dataset in order to estimate the pdfs corresponding to the decision classes, and requires a single pass over all the training patterns. In the training process, the only adjustable parameter (weight) is the *smoothing/scaling* parameter σ-the standard deviation of a sum of small multivariate Gaussian distributions (*Parzen window*), centered at each training sample.

Unlike the above (supervised) networks, the *Kohonen self-organizing map* – SOM represents an unsupervised NN trained to obtain a transformation of an incoming signal pattern of arbitrary dimension into a one- or two -dimensional discrete map, performing this transformation adaptively in a topologically order way. SOM has a feedforward structure, with one layer devoted to computation, and the weights being first initialized by assigning them small random values. The (unsupervised) learning process continues with three steps: *a*) *competition* (a particular neuron with the largest value for a discriminant function is declared as *winning neuron*), *b*) *cooperation* (the winning neuron will determine the spatial location of a topological neighborhood of excited neurons, providing the basis for cooperation among the neighboring neurons), and *c*) *synaptic adaptation* (the excited neurons are enabled to increase their individual values of the discriminant function in relation to the input patterns through suitable adjustments applied to their synaptic weights).

A lesser-known network is represented by the cluster network (CN). CN consists of a number of class-labeled samples, each represented by a radial neuron. The samples are assigned centers by a traditional cluster technique, such as the *k*-means algorithm. The labeling process is based on nearby cases, after that the centers are fine-tuned using Learned-Vector-Quantization (LVQ) method.

More details about NNs and their applications in diverse fields can be found in [1].

8.2.1.2 Applications in CAMD

In [22], inspired by the stimulus-sampling paradigm, a novel learning technique is proposed, based on the enhancement of the standard backpropagation algorithm performance with the aid of a (Markovian) stimulus-sampling procedure applied to the output neurons. According to the stimulus-sampling theory, the learning process is gradual and cumulative, and environmental changes cause variability for learning. Based on this paradigm, the model uses the observable behavior that varies throughout the training process by stimulating the correct answers through corresponding rewards/penalties assigned to the output neurons. Different from other approaches trying to improve the rate of convergence of MLP trained with the BP algorithm, the underlying idea in this paper is to consider the learning process as a mix BP/stimulus-sampling. Thus, while the BP algorithm has the task of updating weights in a standard manner, the stimulus-sampling-based approach strengthens or weakens the responses of the neurons belonging to the output layer at a certain step depending on the correctness of the answers to the previous step. To simplify the computations, the memoryless/Markov property has been considered. Thus, the occurrence at a certain step of the learning process of a strengthening stimulus (reward), conditioned by the entire previous evolution depends only on the preceding answer of an output neuron. Technically, a counter c_j has been assigned to each output neuron, counting how many cases were classified correctly (initially, all counters were set to zero). At the end of an epoch -after all the samples have been presented to the network- each of the output neurons received a stimulus s_j in concordance to its performance (the percentage of cases correctly classified). In this way, each output neuron has been proportionally rewarded with its performance at each epoch.

The algorithm behind this model consists of the following steps:

- *Forward pass*: the signal of each input vector **x** is propagated forward through the network. The normalized output values are computed, and the *winner-take-all* procedure is applied.
- *Tuning counters*: each counter c_j, $j = 1, 2,..., q$ is updated according to the response of the corresponding output neuron.
- *Backward pass*: the error signal propagates backward in the standard manner.
- *End of the first epoch*: all training samples $\mathbf{x}_k$, $k = 1, 2,..., n$, are presented to the network.
- *Assigning stimuli*: each of the output neurons receives a stimulus s_j, $j = 1, 2,...,$ q. The values of the output neurons for the next epoch are multiplied with the value of the corresponding stimuli, rewarding the answers proportionally to the degree of correctness.

The above steps are repeated until the predefined number of iterations (epochs) is reached.

In order to assess the efficacy of this model, it has been applied to six real-life breast cancer, lung cancer, colon cancer, heart attack, and diabetes databases. The statistical comparison to well-established machine learning algorithms (standard 3-layer MLP, RBF, PNN, SVM, nB, and k-nearest neighbors (k-NN)) proved beyond doubt its efficiency and robustness.

The idea to use the stimulus-sampling paradigm to enhance the BP algorithm is straightforward, advantageous and handy in several aspects. Firstly, it is easy to understand and implement, and secondly, the model proved to be adaptable and robust when applied to both numerical and categorical data.

Paper [23] proposes a novel training technique replacing the well-known BP algorithm, by gathering together the error-correction learning, the posterior probability distribution of weights given the error function, and the Goodman-Kruskal Gamma rank correlation in order to assembly them in a Bayesian learning strategy. The underlying idea behind this approach is to use the error-correction learning and the posterior probability distribution of weights given the error function. The synaptic weights belonging to the unique hidden layer have been adjusted inspired by the Bayes' theorem. They were considered as posterior probabilities estimated using priors and likelihoods expressing only the natural association between object's attributes and the network output, or the error function, respectively, through the non-parametric Goodman-Kruskal Gamma rank correlation.

The architecture of the algorithm involved one hidden layer with the number of hidden units equaling the number of decision classes, the hyperbolic tangent as activation function, the initialization based on the Goodman-Kruskal Gamma rank correlation between attributes and decision classes, and the network output computed using the *winner-takes-all* paradigm. In order to use the Bayes model, the corresponding synaptic weight between an input attribute and a certain hidden neuron has been considered as the values of a certain random variable, as well as the network error. Since a synaptic weight refers to the strength of a connection between two units, perceived as a measure of this strength, then, assuming that the real-valued synaptic weights belong to the interval [0, 1], they might be interpreted as a probability-like measure encoding the strength of a connection. Thus, the Bayes rule can be used in the updating process, by considering the synaptic weights as *posterior* (*probabilities*).

Technically, for each decision class C_j, $j = 1, 2,..., q$, and for each attribute A_i, $i = 1, 2,..., p$, the corresponding mean attribute value m_i^j has been computed. Then, for each hidden neuron, the initialization of the corresponding synaptic weights w_{ij} has been done using the Goodman-Kruskal Gamma rank correlation Γ between attributes and related out.

The updating formula connecting $w_{ij}(n+1)$ and $w_{ij}(n)$ has been expressed using again the Goodman-Kruskal Gamma rank correlation Γ between attributes, decision classes and the network error E.

The model ability in medical decision-making has been assessed using real-life breast and lung cancer, diabetes, and heart attack medical databases. The experimental results showed that this model provided performance equaling or exceeding the results reported in literature.

Paper [24] proposes the hybridization of NN with GA in order to enhance the classification performance of the NN in the case of breast cancer detection and recurrence. A MLP has been designed for this purpose, using a GA routine to set weights. Traditionally, MLPs have been successfully trained with the popular BP algorithm. Training MLP with the BP algorithm consists in adjusting the synaptic weights in a way to produce the optimal solution to specific problems, practically by regulating them to minimize an error function, usually the sum of squared errors (SSE). Since the learning paradigm of MLP appeals to the process of adapting synaptic weights, this problem can be solved using GAs, seen as adaptive algorithms. GAs are built inspired by the biological processes of life, reproduction and death. They consist of a populations of chromosomes, selection according to fitness, crossover operator to produce new offspring, and (random) mutation of new offspring. At the beginning of a GA run, a large population of random chromosomes is created. At each iteration, all of the existing chromosomes are evaluated by means of a cost function and the best chromosomes are kept. Then they are mated with each other to replenish the population, and evaluated again by the aid of the cost function. Some of them are deliberately mutated to help in getting the search out of local optima. The chromosomes are again evaluated for fitness and the process is repeated until a given convergence criterion is met.

The hybrid model consisted of two components:

- A multi-layered feedforward network architecture – the MLP component;
- A special designed GA model to achieve the MLP weights optimization – the GA component.

The MLP consisted of a number n of input units, representing the number of attributes in the database, and one layer with two hidden units corresponding to the two decision classes. The network output was computed using the *winner-takes-all* rule. Regarding the GA structure, the weight vector in MLP has been encoded as a vector of real numbers (network synaptic weights). In such a way, a synaptic weight vector has been represented through a chromosome, which contained a number of genes equaling the number of neurons from the input layer multiplied by the number of neurons from the hidden layer. The synaptic weights between input units and hidden units have been read off the network from top to bottom, representing the components of the weight vector. The model hyper-parameters, consisting of population size, number of generations, mutation rate, etc., have been set heuristically, in order to obtain optimal performance. The classical sigmoid $f(x) = \frac{1}{(1+e^{-x})}$ has been chosen as activation function. As crossover parameters, the *arithmetic crossover/BLX-α*, the *linear BGA crossover*, the *Wright's heuristic crossover*, and the *uniform crossover* have been considered. The process of mutation consisted of two steps. A number between 0 and 1 has been randomly generated. If the number was smaller than the standard default threshold, a subtraction has been made, otherwise an addition. Then, using the chromosome's error, each gene has been mutated according to the previous step.

The hybrid model has been benchmarked against other NN algorithms (MLP, RBF, PNN) using four publically available datasets concerning breast cancer. In comparison to its competitors, the performance of this hybrid model proved to be superior. Moreover, this hybrid MLP/GA model was very flexible in terms of providing accurate classification, even with different types of attributes (numerical, categorical and mixed).

Paper [25] deals with a simpler variant of MLP, namely partially connected NN (PCNN). Although fully connected feed-forward NNs are used in almost all NNs applications since such architecture provides the best generalization power, however, they need large computing resources and have low speed when they are applied to large databases. In order to counteract this disadvantage, partially NNs succeeded in obtaining very good accuracy in comparison with fully connected NNs by reducing the computing resource consumption during the classification process, and increasing the speed as well.

PCNN can be considered as a pruned NN, containing only a subset of the entire set of possible connections between the networks neurons. The novel approach proposed an artificial replica of the way the human brain works. Thus, if a signal is being processed by the brain, only certain neurons participate to that course of action, those who have been excited. PCNN is in fact a MLP with some deactivated synaptic connections. PCNN has been trained using the traditional BP algorithm, and the classical sigmoid function has been considered as activation function. The key idea behind the partially connected NN model was that the synaptic weights that did not suffer major modifications, i.e., they did not surpass a certain threshold after a certain number of training samples presented to the network, have been erased from the network's architecture, being inhibited.

This new algorithm has been tested on four different publically available datasets concerning breast cancer detection and recurrence, with different data types in order to validate our conclusions. The computational results strongly suggested the suitability of the PCNN methodology for the classification/prediction of cancer and recurrent events in breast cancer

In paper [26], a tandem model consisting of a feature selection (FS) mechanism and an evolutionary-driven neural network (MLP/GA) has been proposed to support the liver fibrosis stadialization in chronic hepatitis C using the Fibroscan device.

FS is considered a key factor in designing an IDSS, since even the best models will perform poorly if the features are not chosen well. The FS mechanism has been based on both specific statistical tools, such as discriminant function analysis, multiple linear regression (both forward and backward stepwise approaches), analysis of correlation matrix, and the subsequent sensitivity analysis provided by the use of some traditional NNs (linear NN, PNN, RBF, 3-MLP, and 4-MLP). Each of the FS techniques used in the first place gave its own hierarchy, preserving only the truly important attributes in the decision-making process. Next, in order to create a unique hierarchy based on the contribution of each technique, a specific procedure has been proposed. Thus, a score has been assigned to each feature, calculated as the sum of the values that represents its rank in each hierarchy. Next, all features have been ordered from the most important to the least important, obtaining thus a "core" dataset containing the features with the hierarchy level exceeding a certain threshold.

The evolutionary-trained NN followed the development presented in [24]. The binary tournament selection and the total arithmetic recombination have been considered as variation operators.

The hybrid MLP/GA algorithm has been evaluated over two benchmark datasets: a) the complete liver fibrosis dataset containing 25 attributes, and b) the reduced dataset containing only 6 attributes. Taking into account both the computational speed and the ability to detect the most important liver fibrosis stages, i.e., F1 -disease starting stage, and F4 –cirrhosis, experimental results have shown that the proposed tandem system has provided a significantly better classification performance than its competitors (PNN, MLP, RBF, and SVM).

Paper [27] is concerned with the use of a 4-layer MLP (i.e., MLP with two hidden layers) in order to assess the accuracy of real-time Endoscopic Ultrasound Elastography (EUS elastography) in focal pancreatic lesions. As it is well-known, the differential diagnosis of pancreatic cancer and chronic pseudo-tumoral pancreatitis is still difficult and it can be a challenge for the expert gastroenterologist. In this context, EUS elastography images offer complementary information added to conventional EUS with minimal prolongation of the examination time, minimum costs, and no added morbidity or mortality. Unfortunately, the direct visual analysis of elastography recorded movies offered unsatisfactory results concerning sensitivity and specificity for two different observers blinded to clinical data. Accordingly, the use of the "objective" observations provided by NN after the analysis of the digitalized EUS images, improved the classification process.

The diagnosis accuracy of the MLP model designed in the EUS elastography multi-centric study was in accordance with reported standard EUS elastography experience, nevertheless, with an improved robustness, reliability, and reproducibility.

In paper [28], the effectiveness of RBF in the case of evaluation of liver fibrosis has been assessed in comparison with other ML techniques. In this study, RBF had the standard form $y_k(\mathbf{x}) = \sum_{i=1}^{M} w_{ki} \cdot \Phi_i(\|\mathbf{c}_i - \mathbf{x}\|, \sigma_i)$, where $y_k(\mathbf{x})$ is the k-th output, $\mathbf{x}$ is an input vector, w_{ki} is the weight from the i-th kernel node to the k-th output node, $\mathbf{c}_i$ is the centroid of the i-th kernel node, σ_i is the bandwidth of the i-th kernel node, M is the number of kernel nodes, $\Phi_i(\cdot) = \exp(-\|\mathbf{c}_i - \mathbf{x}\|^2/(2\sigma_i^2))$ is a Gaussian RBF with bandwidth σ_i. The RBF diagnosing performances has been considered in two cases: *a*) using the new technique provided by Fibroscan (stiffness), and *b*) excluding the parameter stiffness (classical medical approach). Comparing the results of the two experiments, it was clear that the inclusion of the Fibroscan output in the diagnosis process slightly improved the decision accuracy.

Among other ML techniques (e.g., MLP, SVM, etc.) applied in estimating the liver fibrosis stages, RBF represents a reliable and fast tool, providing good diagnosis results without using large computational resources. Moreover, it assesses without any doubt the importance of the new non-invasive technique represented by Fibroscan in liver fibrosis evaluation.

A comparison between the classification performance regarding breast cancer, obtained by using a SOM neural network and some supervised NNs such as MLP, RBF and PNN, is presented in paper [29]. The idea is to compare the efficacy of supervised learning against unsupervised learning using real data regarding breast cancer.

Training NN using supervised learning can be thought of as learning with a 'teacher'. In this context, there are two phases in neural information processing: the *training* (*learning*) phase and the *testing* phase. The 'teacher' is supposed to have a priori knowledge of the environment (*input-output* couple gathered in the training set). When the 'teacher' and NN are both exposed to a training data (input), the teacher is able to provide NN with the corresponding correct (desired) output (known label). The error value for a given training data is the difference between the correct output (data label) provided by the teacher and the actual output, provided by the network. By decreasing the network error, step-by-step, in the learning phase, a certain level of accuracy is reached by the network.

Training the NN using unsupervised learning is thought of as learning without a 'teacher'. The network does not provide a feedback on the quality of solutions obtained, because there are no labels to be compareed, but a task-independent measure of the quality of representation is provided, and the network is required to learn it. The network parameters are optimized with respect to this measure. The output neurons of the network compete among themselves to be activated, with the result that only one output neuron, or one neuron per group is on at any one time (*winning neuron*) [1].

Both the supervised NNs and the unsupervised SOM have been applied to the Wisconsin Prognostic Breast Cancer (UCI Machine Learning repository). The experiment showed that for breast cancer detection the SOM model performed similarly to standard NN classifiers. As the main advantage of SOM, one can consider its ability of using medical data without previous diagnosis, since it exploits its self-organizing feature in contrast to the traditional supervised NNs, which require the corresponding diagnosis in the training process.

In paper [30] is presented a clustering based approach for breast cancer recurrence in which SOM has been compared with two clustering algorithms, i.e., the classical *k*-means algorithm and a CN. Using the publically available dataset Wisconsin Recurrence Breast Cancer from UCI Machine Learning Repository, the experimental results have shown that the performance of the three ML algorithms were in accordance to the reported modern medical imaging experience, the best performance belonging to CN, followed by SOM and *k*-means algorithm.

8.2.2 Support Vector Machines-Based IDSSs

The philosophy behind SVMs refers to the kernel-based methodology starting in 1963 (Vapnik and Chervonenkis) and finished in the standard form (Cortes and Vapnik) [1], [31]. Generally speaking, SVM can be regarded as a linear feedforward learning machine, incorporating a learning procedure that is independent of the problem dimension. Fitted with special features and based on the *structural risk minimization* (SRM) method, SVM can provide a good generalization performance in pattern recognition problems.

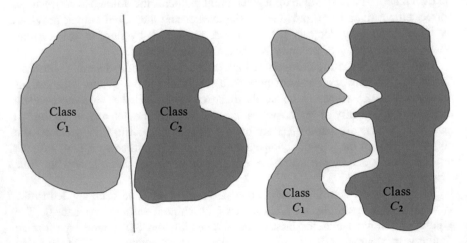

Fig. 8.3 Classes separability

8.2.2.1 SVMs for CAMD

There are two types of cases when we face a classification problem: a) linear separability (linear learning), and b) non-linear separability (non-linear learning), both illustrated in figure below (Fig. 8.3).

Mathematically, if training data are linearly separable, there exists a hyper-plane H, such that $H : w \cdot x - b = 0$, where w is the weight vector and b is the bias. In the case of non-separable data, one can use the *kernel trick*, by mapping the original non-linearly separable points into a higher-dimensional space, where a linear classifier is subsequently used. This technique is illustrated in figure below (Fig. 8.4).

In the case of linearly separable data, the goal of SVM is to find that hyper-plane which maximizes the margin of separation. In this case, the decision surface thus found is called *optimal hyper-plane*. The problem at hand is to estimate the parameters w and b based on the training dataset. The illustration of the linearly separability is synthetically presented in Fig. 8.5.

The particular points belonging to the two margins of separation are called *support vectors*, hence the name "support vector machine".

In the case of non-linearly separability, one considers a set of additional non-negative scalar variables $\{\xi_i, i = 1, 2, \ldots, N\}$, called *slack variables*, measuring the deviation of a point from the ideal condition of pattern separability. In this case, instead of finding the equation of the optimal hyper-plane, one minimizes the

cost function $L(w, \xi) = \frac{\|\vec{w}\|^2}{2} + C\left(\sum_{i=1}^{N} \xi_i^k\right)$.

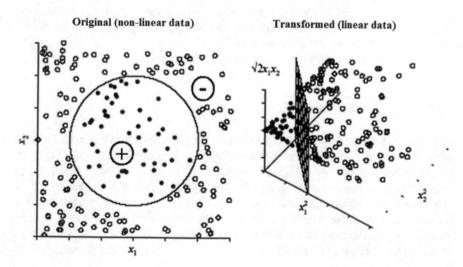

Fig. 8.4 Kernel trick

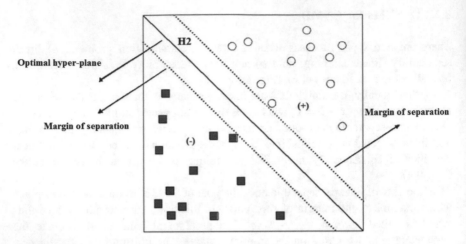

Fig. 8.5 Optimal hyper-plane and margin of separation

Finally, let us mention here three types of SVMs, based on the most frequently kernel types used in practice: a) *polynomial learning machine*, b) *radial-basis function network*, and c) *two-layer perceptron*.

A quite novel approach regarding SVMs is based on the evolutionary paradigm. The framework of the evolutionary-driven SVM (ESVM) is meant both to solve the complexity of modeling the embedded optimization issue, and to broaden the range of maneuvering the kernel machine of SVM, especially for classification problems. Briefly, ESVM inherits the SVM classical approach regarding the learning process, while the estimation of the decision function parameters being performed by using an evolutionary algorithm. In the classification problems, the only parameter to be estimated is the separation hyper-plane. ESVM can always determine the learning function parameters, often impossible in the classical case, obtaining the coefficients directly from the evolutionary algorithm.

More details about NNs can be found in [1] and [31].

8.2.2.2   Applications in CAMD

In paper [32], a novel evolutionary classification technique deriving from the field of cooperative coevolutionary algorithms (CCEAs) is presented, accomplishing the collaboration between evolving rule populations of different outcomes towards improvement in prediction accuracy. In this regard, a potential architecture considers the final decision suite as to contain only one or more randomly selected rules for each outcome of the classification task. In addition, the algorithm is enhanced by an archiving mechanism, preserving the best performing combinations during each evolutionary cycle, and displaying a complete and varied output rule set.

CCEA has been introduced as an alternative evolutionary approach to function optimization, by considering as many populations as the number of variables of the function. In this context, each variable represents a component of the solution vector, and is separately treated using some type of EA. For classification task, the output has been imagined as to be represented by a set of rules that contains at least one prototype for each class, and the decomposition of each potential problem solution into components is performed by assigning to each population the task of building the rule(s) for one certain class. In this way, the number of species equals the number of outcomes of the classification problem. CCEA develops p rules, one for each class, so p populations are considered, each with the purpose of evolving one of the p individuals. Each individual (that is, a rule) in every population follows the same encoding as a sample from the data set, and is represented as a simple IF-THEN rule having the condition part in the attributes space and the conclusion in the classes space. In classification problems, an individual will not encode the class, as all individuals within a population have the same outcome. The values for the genes of all individuals are randomly initialized following a uniform distribution in the definition intervals of the corresponding attributes in the data set. For evaluating an individual from a certain population -that is a rule of a certain outcome- a collaborating rule from each of the other populations is selected according to the collaborator selection pressure choice for a number of times equal to the pool size parameter (cps). From an evolutionary point of view, fitness proportional selection was employed, intermediate recombination was used, and mutation with normal perturbation was considered.

CCEA has been improved in what regards the production of multiple rules for each outcome of the decision problem, since the method must be endowed with a mechanism to conserve a diverse selection of rules. Briefly, an archive population has been created to coexist and update during the evolutionary loop. At the end of each generation, it seizes a fixed number of best collaborators from the current population and the previous archive. These fittest combinations are copied into the new archive such that there are no two rules alike.

Experimentations have been carried on two real-world problems of tumor diagnosis, both having only two classes, one coming from the UCI Repository of Machine Learning Databases (breast cancer), and the other collected from the University Hospital in Craiova, Romania (hepatic cancer). The results obtained during the two experiments indicated that the proposed type of collecting the suites of rules that performed best on the training set indeed led to higher success than the initial CCEA classifier.

Paper [33] presents an automatic tool capable to learn from a medical dataset. Data has been obtained from the University of Medicine and Pharmacy of Cluj-Napoca, Romania, consisting of 24 medical indicators characterizing each sample. The proposed technique used the acquired knowledge to differentiate between five degrees of liver fibrosis. The methodology combined a hill climbing algorithm that selects subsets of important attributes for an accurate classification, and a core represented by a CCEA classifier that builds rules for establishing the diagnosis for every new patient. The hill climbing algorithm was used to dynamically pick the proper attributes from the data set, and then CCEA was employed to classify the selected data into the five different classes. Experimental results have shown that the proposed technique performed better than other state-of-the-art classification techniques like SVMs of NNs, with or without a preprocessing method. In addition, the model can provide prototypes of rules for each fibrosis stage and information about the importance and interaction of the attributes.

In paper [34], a precise estimation of patient length of stay (LoS) is investigated in the context of colorectal cancer, using a discrete dataset. Several classifiers from distinct conceptual families provided an estimation or even further information on the LoS of patients that had been operated of cancer in certain stages and invasion at various parts of the colon or rectum. SVMs and NNs gave a black-box prediction of the hospitalization period, while decision trees (DTs) and EAs additionally offered the underlying rules of decision.

Technically, a linear SVM, a C5.0 DT/rule-based model and a feedforward NN with a single hidden layer have been considered in this study. Experiments have envisaged two black-box predictors, while a rule-based EA has been also introduced to address comprehensibility along with the DT. In addition, a collaborative design between the EA and the SVM has been appointed in order to explore the effects of a degree of interdependence between comprehensibility and accuracy. The experimental results have shown that the prediction accuracy was adequately high, the rules were understandable and useful, with those of the EA being quickly graspable as concerns the more important factors and their values. Finally, the methods were helpful for increasing efficiency in time, and accuracy in discovering the patterns in the large collection of records.

A cooperative coevolution (CC) technique, used to extract discriminative and compact class prototypes following a SVM model, is presented in paper [35]. Three main hypotheses: *a*) fidelity to the SVM prediction, *b*) superior accuracy to the CC classifier alone, and *c*) a compact and comprehensive resulting output achieved through a class-oriented form of feature selection are tested in this study.

The proposed SVM-CC white-box classification framework consisted of a SVM classification model, an indicative and minimal threshold representation for class prototypes, and a CC engine to determine these attribute edge values against the resulting SVM labeled data. Several experiments have been conducted to investigate whether the intermediate use of SVMs conducts to significant enhancements in the classification ability of the CC technique, while the prediction accuracy of the resulting white-box classifier is not inferior to that of the opaque predictor alone. The experiments have envisaged medical data regarding diabetes, breast cancer, transfusion, Ecoli, and different other datasets (Iris, Glass, Ionosphere, Letters) originating from UCI Machine Learning Repository.

The advantage of the proposed SVM-CC model as compared with other ML algorithms, such that genetic programming, ant colony optimization, multi-objective EAs, etc., is the simplicity of the encoding as well as the specificity and compactness of each prototype for a diagnosis outcome. It is noteworthy that, from the practical perspective, such a flexible and efficient knowledge extractor from an accurate opaque engine can be further used at the frontier between science and the real-world problems.

8.2.3 Evolutionary-Inspired IDSSs and Other Approaches

The idea of building a complex IDSS inspired by the evolutionary metaphor is quite simple and straightforward. By combining several algorithms into a composite system, a committee of algorithms is then built, and the overall decision of the committee system is represented by a certain combination of the decisions of its components. On the other hand, one considers in the broadest sense possible the parallel computing as the simultaneous use of multiple algorithms to solve a computational problem, and the distributed systems as a collection of algorithms communicating with each other to achieve a common goal.

Starting from this paradigm, the idea behind the system's functionality is based on the collaborative competition model, inspired by the synergetic evolutionary metaphor. One considers a population of simple potential "intelligent" solutions to the medical decision problem interacting locally with one another and with the environment. They are competing and collaborating with each other based on the evolutionary metaphor, hence making an "intelligent" global decision.

At the beginning, one chooses efficient types of AI/ML algorithms to form the initial population of solutions for a decision problem. Then, they will "evolve" in such a way that:

a. new variants are obtained by 'mutation';
b. hybrids algorithms are obtained by 'crossover'.

Thus, offspring and mutants will form the next population.

For selection, they are iteratively tested based on their fitness given by the individual decision performance, and the best of them are retained in order to replenish the new generation. After a certain number of generations, the most performing algorithms will form the intelligent decision system. The synergism of this complex configuration involves the collaboration between the selected algorithms in making the final decision, technically based on a proportionally weighted voting system (WVS). Inspired again from the evolutionary metaphor, in this context the 'crossover' operator refers to the 'recombination' of the individual 'decision genes' using the fitness-based weights, and the 'offspring' represents the 'joint' final decision.

The fitness measure of selecting the best performing algorithms is generally defined by the *decision accuracy* of each algorithm in the testing phase, along with the corresponding *standard deviation*, measuring the algorithm's stability to multiple independent computer runs. The benchmarking methodology used for selection consists in measuring the fitness-based contrast between the algorithms performance, usually based on the one-way ANOVA technique along with the

Tukey's honestly significant difference (Tukey HSD) *post hoc* test. Based on the fitness measure used in the benchmark process, the initial algorithms are statistically compared and the fittest are chosen to seed the next generations by applying the 'variation' approach.

As mentioned above, the hybridization of two algorithms is considered as the 'recombination' operator, inspired by the idea underlying the classical variation operators from the evolutionary computing field. The development of new variants of algorithms is considered as a 'mutation'. It is worth mentioning that the hybrid NN/GA model presented in [24], as well as the PCNN presented in paper [25] may be considered as algorithms designed using the evolutionary paradigm, hybridization being seen as 'crossover', while the partial connectivity being seen as 'mutation'.

The survivor selection mechanism is based on a steady-state model in which the entire population is not replaced at once, but just a number of individuals are replaced using a rank-based selection, and the end of the evolutionary process is determined by choosing different STOP conditions, such as: flat improvement curve, limited population diversity, etc. After the 'evolutionary' process ended, the best algorithms thus selected to form the IDSS are applied to new data and an overall decision is made based on a certain WVS.

Paper [36] presents a competitive/collaborative neural computing system (CCNCS) designed to support the medical decision process using medical imaging databases. The automated differential diagnosis of chronic pancreatitis and pancreatic cancer has been chosen as concrete application.

Different from other approaches integrating different ML techniques, this study proposes and investigates the effectiveness of a computational system combining a certain number of NNs, which process information in a competitive-collaborative way in order to obtain a global automatic medical diagnosis. The system is designed to use medical imaging data, and consists of two modules:

- The medical imaging information, previously digitalized using a certain image-to-numerical data converter, is provided to CCNCS;
- CCNCS, consisting of a set of n NN algorithms working in both competitive and collaborative way, processes the available information and makes the automated decision-making.

CCNCS works in both competitive and collaborative modes, as follows:

- *Competitive mode* firstly, by choosing the first k best NN models ($k < n$), based on a thorough benchmarking process concerning their classification performances; the number k is chosen by the user, depending on the problem to be solved.
- *Collaborative mode* secondly, by choosing for a new case the diagnosis provided by WVS based on the k best NN individual diagnoses.

In the competitive phase, all algorithms are evaluated based on some statistical measures of performance (e.g., t-test for independent samples, Mann & Whitney U test, two-sided z-test, Cohen's kappa test, area under the ROC curve, etc.)

In the collaborative stage, CCNCS works in a weighted collaborative mode. Each of the k best NN algorithms are applied to new data, and an overall diagnosis will be retained as the final computing system output, based on WVS. By default, the number of votes is directly proportional to the testing performance of each algorithm.

In the concrete application to the automated differentia diagnosis of chronic pancreatitis and pancreatic cancer, the 'competitors' were represented by:

- *Linear neural network* (LNN), represented by a NN having no hidden layers, but an output layer with linear neurons;
- *MLP* with one and two hidden layers trained with the standard BP algorithm;
- *RBF*;
- *PNN*.

Three statistical measures of performance have been chosen for the benchmarking process, namely the training performance, the testing performance, and the confusion matrix.

The statistical analysis focused on the following three aspects:

- *Data screening*, involving the suitability of the data for the type of statistical analysis that is intended;
- *Hypotheses testing*, involving the comparison between the testing performances obtained in the diagnosing process;
- *Over-learning control*, involving the correlation between training and testing phases, related to the generalization ability of each competitor.

Data screening consisted of the classical normality tests (*Kolmogorov-Smirnov & Lilliefors* test and *Shapiro-Wilk W* test), along with the *Levene's test* for verifying the equality of variances assumption. Hypothesis testing consisted of the classical comparison tests (t-test for independent samples & Mann-Whitney U test), two-sided test (z-value), Cohen's *kappa* test). t-test for independent samples and Pearson's r correlation coefficient have been used to investigate the over-learning behavior.

For the final decision, the hierarchy of the three ML algorithms, kept to form CCNCS, consisted of a (four-layer) MLP, a (three-layer) MLP, and a RBF. A concrete example consisting in three different testing cases has been considered for the model assessment. In all cases, CCNCS provided the correct diagnosis, while the component algorithms indicated different decisions. It is noteworthy that CCNCS might be naturally widened by including in its structure different ML algorithms, such as recurrent NNs, SVMs, Bayesian classifiers, k-nearest neighbors, etc.

Paper [37] proposes an ensemble of state-of-the-art classifiers consisting of SVMs, NNs, DTs, and logistic regression (LR) in order to estimate the LoS after surgery in patients diagnosed with colorectal cancer. The experimentations involved a dataset from the University of Medicine and Pharmacy of Craiova, Romania, and concerns a certain hospitalization period for patients diagnosed with colorectal cancer that underwent surgery. The ensemble construction envisaged the following strategy. Once all three classifiers have been trained, a voting mechanism has been used to establish the class for fresh examples. For every new case, one label has been derived from each method and the class most frequently appearing in this triple response has been assigned as the combined vote result. In the case that the three predictions are however all distinct, then priority has been given to the decision of the SVM (as pre-experimentally proven most accurate).

Paper [38] examines the behavior of two evolutionary algorithms that discover prototypes for each possible diagnosis outcome, with the aim of scrutinizing alternative architectures for prototype representation to reach the centroids with desired accuracy and in acceptable time. It is noteworthy that the discovered centroids are used to provide understandable thresholds of differentiation among the decision classes. The methodology has been applied to the Breast Cancer Wisconsin (UCI Machine Learning Repository), and to the hepatitis C dataset from the University of Medicine and Pharmacy, Cluj-Napoca, Romania.

In this study, a solution to a classification task consists of the class prototypes. Basically, a prototype holds the thresholds or coordinates that decide the separation between the current class and the other possible diagnosis outcomes. One way to evolve the k-prototype individuals is the use of standard EA. Once the centroids are discovered, each test sample receives its predicted outcome following the most similar (nearest) prototype from the generated set of k representatives. Alternatively, one can consider CC, so an individual is a prototype for a class distinguished by the subpopulation it belongs to, and multiple subpopulations are thus maintained and communicate towards achieving the common goal of a high prediction accuracy.

This study compares the performance of a standard EA to evolve the k-prototype individuals, and a CC approach for centroids detection. The experimental results for the two datasets have shown an overall advantage of the EA versus the CC.

8.3 How Useful Are IDSSs in CAMD?

ML algorithms have been from the beginning used to analyze medical datasets. IDSS technology is currently well suited for analyzing medical data, and there is a lot of work done in efficiently supporting the medical diagnosis process.

Paper [39] presents a short review regarding the applications of ML algorithms, in other words, IDSSs, in the health care domain. A special attention has been paid to the 'Ups and Downs' of this domain. Thus, it is noteworthy to mention just the following three facts:

- Medical diagnosis is 'subjective' and depends both on available data, and on the experience of the physician;
- Various ML classifiers perform roughly the same. However, there are significant differences between them, mostly depending on the database and data type;
- IDSS-based diagnosis remains just another source of possibly useful information that helps to improve the diagnostic accuracy. The final responsibility whether to accept or reject this information belongs to the physician.

Although diagnosis seems to be a relatively straightforward ML problem, and the existing "intelligent" technology codes both symptoms and conditions using large datasets for training, the clinical diagnosis still rely mostly on doctors' expertise and intuition. Under these circumstances, one may conclude that, despite the great achievements, there is still a reluctance to use IDSS in CAMD. The reasons for this situation may lie in:

a. the doctors are still more effective than algorithms;
b. there could be regulations of liability that prevent hospitals from relying too heavily on IDSSs for diagnosis;
c. doctors resent the idea that computers could replace them.

It is noteworthy that the most misplaced expectation is the idea that IDSS is intended to replace doctors in the medical decision process.

To conclude, the future will show exactly how important is IDSS for CAMD, a straight collaboration between doctors and computers being expected.

8.4 Privacy Issues

In using IDSSs for automated medical diagnosis, it is mandatory to consider possible implications regarding privacy when using medical datasets. Medical records contain personal information that can be used for unethical purposes. For this reason, in all developed countries, the respective governments and other official institutions have considered strict regulations regarding the use of such data.

8.5 Conclusions

Nowadays, researchers have to cope with collections of large and complex datasets, difficult to be processed with traditional tools. In this context, by "big data" we commonly understand the tools, processes, and techniques enabling to create, manipulate and manage very large datasets. In this framework, clinical datasets store large amounts of information of all kinds about patients and their medical conditions. State-of-the-art IDSSs are designed to deal with this information in order to discover relationships and patterns which are helpful in studying the progression and the management of diseases.

To conclude, there are many challenges in dealing with big data and efficient IDSSs. The ultimate goal of future research in bringing together computer technologies and medical decision-making is to bridge physicians, IDSSs and medical informatics communities to foster interdisciplinary studies between this three research groups. The future will show how important is the extended use of IDSSs in modern diagnosis making process.

References

1. Gorunescu, F.: Data Mining Concepts, Models and Techniques. Springer-Verlag, Berlin Heidelberg (2011)
2. Pang, S., Yu, Z., Orgun, M.A.: A novel end-to-end classifier using domain transferred deep convolutional neural networks for biomedical images. Comput Methods programs. Biomed. (2017). Epub 2017 Jan 6. doi:10.1016/j.cmpb.2016.12.019
3. Hatipoglu, N., Bilgin, G.: Cell segmentation in histopathological images with deep learning algorithms by utilizing spatial relationships. Med Biol Eng Comput. [Epub ahead of print] (2017). doi:10.1007/s11517-017-1630-1
4. Arabasadi, Z., Alizadehsani, R., Roshanzamir, M., Moosaei, H., Yarifard, A.A.: Computer aided decision making for heart disease detection using hybrid neural network-Genetic algorithm. Comput Methods programs Biomed. Epub 2017 Jan 18 (2017). doi:10.1016/j.cmpb.2017.01.004
5. Ma, J., Yu, J., Hao, G., Wang, D., Sun, Y., lu, J., Cao, H., Lin, F.: Assessment of triglyceride and cholesterol in overweight people based on multiple linear regression and artificial intelligence model. lipids health Dis. (2017). doi:10.1186/s12944-017-0434-5
6. Becker, A.S., Marcon, M., Ghafoor, S., Wurnig, M.C., Frauenfelder, T, Boss, A.: Deep Learning in Mammography: Diagnostic Accuracy of a Multipurpose Image Analysis Software in the Detection of Breast Cancer. Invest Radiol. [Epub ahead of print] (2017). doi:10.1097/RLI.0000000000000358
7. Huang, M.W., Chen, C.W., Lin, W.C., Shih-Wen Ke, S.W., Tsai C.F.: SVM and SVM Ensembles in Breast Cancer Prediction. PLoS One 12(1), (2017). doi:10.1371/journal.pone.0161501
8. Anaissi, A., Goyal, M., Catchpoole, D., Braytee, A., Kennedy, P.: Ensemble Feature Learning of Genomic Data Using Support Vector Machine. PLoS One 11(6), doi:10.1371/journal.pone.0157330 (2016)
9. Huiyan Jiang, H., Zhao, D., Zheng, R., Ma, X.: Construction of Pancreatic Cancer Classifier Based on SVM Optimized by Improved FOA. Biomed Res Int. (2015). doi:10.1155/2015/781023
10. Fu, C.W., Lin, T.H.: Predicting the Metabolic Sites by Flavin-Containing Monooxygenase on Drug Molecules Using SVM Classification on Computed Quantum Mechanics and Circular Fingerprints Molecular Descriptors. PLoS One 12(1), (2017). doi:10.1371/journal.pone.0169910
11. Weis, C., Hess, A., Budinsky, L., Fabry, B.: In-Vivo Imaging of Cell Migration Using Contrast Enhanced MRI and SVM Based Post-Processing. PLoS One 10(12), (2015). doi:10.1371/journal.pone.0140548
12. Banerjee, U., Braga-Neto, U.M.: Bayesian ABC-MCMC Classification of Liquid Chromatography-Mass Spectrometry Data. Cancer Inform 14(5), 175–182 (2017)

13. Jung, Y.J., Katilius, E., Ostroff, R.M., Kim, Y., Seok, M., Lee, S., Jang, S., Kim, W.S., Choi, C.M.: Development of a Protein Biomarker Panel to Detect Non-Small-Cell Lung Cancer in Korea. Clin Lung Cancer. [Epub ahead of print] (2016). doi:10.1016/j.cllc.2016.09.012

14. Benndorf, M., Neubauer, J., Langer, M., Kotter, E.: Bayesian pretest probability estimation for primary malignant bone tumors based on the Surveillance, Epidemiology and End Results Program (SEER) database. Int. J. Comput. Assist. Radiol. Surg. **12**(3), 485–491 (2017)

15. Wang, J., Zuo, Y., Man, Y., Tadesse, M.G., Ressom, H.W.: Identification of functional modules by integration of multiple data sources using a Bayesian network classifier. Circ Cardiovasc Genet **7**(2), 206–217 (2014)

16. Ricci, L., Del Vescovo, V., Cantaloni, C., Grasso, M., Barbareschi, M., Denti, M.A.: Statistical analysis of a Bayesian classifier based on the expression of miRNAs. BMC Bioinformatics (2015). doi:10.1186/s12859-015-0715-9

17. Sreekumari, A., Shriram, K.S., Vaidya, V.: Breast lesion detection and characterization with 3D features. Proc IEEE Conf Eng Med Biol Soc., pp. 4101–4104 (2016)

18. Yu, K., Sang, Q.A., Lung, P.Y., Tan, W., Lively, T., Sheffield, C., Bou-Dargham, M.J., Liu, J.S., Zhang, J.: Personalized chemotherapy selection for breast cancer using gene expression profiles. Sci Rep. (2017) doi:10.1038/srep43294

19. Kotti, M., Duffell, L.D., Faisal, A.A., McGregor, A.H.: Detecting knee osteoarthritis and its discriminating parameters using random forests. Med Eng Phys. [Epub ahead of print] (2017) doi:10.1016/j.medengphy2017.02.004

20. Adham, D., Abbasgholizadeh, N., Abazari, M.: Prognostic Factors for Survival in Patients with Gastric Cancer using a Random Survival Forest. Asian Pac. J. Cancer Prev **18**(1), 129–134 (2017)

21. Paul, D., Su, R., Romain, M., Sébastien, V., Pierre, V., Isabelle, G.: Feature selection for outcome prediction in oesophageal cancer using genetic algorithm and random forest classifier. Comput Med Imaging Graph. [Epub ahead of print] (2016). doi:10.1016/j.compmedimag.2016.12.002

22. Gorunescu, F., Belciug, S.: Boosting backpropagation algorithm by stimulus-sampling: Application in computer-aided medical diagnosis. J. Biomed. Inform. **63**, 74–81 (2016)

23. Belciug, S., Gorunescu, F.: Error-correction learning for artificial neural networks using the Bayesian paradigm. Application to automated medical diagnosis. J. Biomed. Inform. **52**, 329–337 (2014)

24. Belciug, S., Gorunescu, F.: A hybrid neural network/genetic algorithm system applied to the breast cancer detection and recurrence. Expert Systems **30**(3), 243–254 (2013)

25. Belciug, S., El-Darzi E.: A partially connected neural network-based approach with application to breast cancer detection and recurrence. In: Proc. 5th IEEE conference on intelligent systems-IS, 7–9 July 2010, London, UK. pp. 191–196 (2010)

26. Gorunescu, F., Belciug, S., Gorunescu, M., Badea, R.: Intelligent decision-making for liver fibrosis stadialization based on tandem feature selection and evolutionary-driven neural network. Expert Syst. Appl. **39**(17), 12824–12832 (2012)

27. Saftoiu, A., Vilmann, P., Gorunescu, F., et al.: Efficacy of an Artificial Neural Network-Based Approach to Endoscopic Ultrasound Elastography in Diagnosis of Focal Pancreatic Masses. Clinical Gastroent and Hepatol **10**(1), 84–90 (2012)

28. Gorunescu, F., Belciug, S., Gorunescu, M., Lupsor, M., Badea R., Ştefanescu, H.: Radial basis function network-based diagnosis for liver fibrosis estimation. In Proc. 2nd International Conference on e-Health and Bioengineering-EHB 2009, 17–18th September, 2009, Iaşi-Constanţa, Romania, Ed. UMF "Gr.T. Popa" Iasi. pp. 209–212 (2009)

29. Belciug, S., Gorunescu, F., Gorunescu M., Salem A.B.: Assessing Performances of Unsupervised and Supervised Neural Networks in Breast Cancer Detection. In: Proc. 7th IEEE International Conference on INFOrmatics and Systems-INFOS 2010. Advances in Data Engineering and Management-ADEM, March, 28–30, 2010, Cairo, pp. 80–87 (2010)

30. Belciug, S., Gorunescu, F., Gorunescu, M., Salem, A.B.: Clustering-based approach for detecting breast cancer recurrence. In: Proc. 10th IEEE International Conference on Intelligent Systems Design and Applications-ISDA10, Nov 29 – Dec 1, 2010, Cairo, pp. 533–538 (2010)

31. Stoean, C. Stoean, R.: Support vector machines and evolutionary algorithms for classification. Springer (2014)
32. Stoean, C., Stoean, R.: Evolution of Cooperating Classification Rules with an Archiving Strategy to Underpin Collaboration. Springer (Evolution of Cooperating Classification Rules with an Archiving Strategy to Underpin Collaboration, Intelligent Systems and Technologies-Methods and Applications), pp. 47–65 (2009)
33. Stoean, C., Stoean, R., Lupsor, M., Stefanescu, H., Badea, R.: Feature Selection for a Cooperative Coevolutionary Classifier in Liver Fibrosis Diagnosis. Comput. Biol. Med. **41**(4), 238–246 (2011)
34. Stoean, R., Stoean, C., Sandita, A., Ciobanu, D., Mesina, C.: Interpreting Decision Support from Multiple Classifiers for Predicting Length of Stay in Patients with Colorectal Carcinoma, Neural Processing Letters, pp. 1–17, (2017). doi:10.1007/s11063-017-9585-7
35. Stoean, C., Stoean, R.: Post-evolution of variable-length class prototypes to unlock decision making within support vector machines. Appl. Soft Comput. **25**, 159–173 (2014)
36. Gorunescu, F., Gorunescu, M., Saftoiu, A., Vilmann, P., Belciug, S.: Competitive/ collaborative neural computing system for medical diagnosis in pancreatic cancer detection. Expert Syst **28**(1), 33–44 (2011)
37. Stoean, R., Stoean, C., Sandita, A., Ciobanu, D., Mesina, C.: Ensemble of Classifiers for Length of Stay Prediction in Colorectal Cancer. International Work-Conference on Artificial Neural Networks (IWANN 2015), Advances in Computational Intelligence, Lecture Notes in Computer Science, Springer, Volume 9094, Palma de Mallorca, Spain, 10–12 June, pp. 444–457 (2015)
38. Stoean, C., Stoean, R., Sandita, A.: Investigation of Alternative Evolutionary Prototype Generation in Medical Classification. In: IEEE Post-Proc. 16th International Symposium on Symbolic and Numeric Algorithms for Scientific Computing (SYNASC), September 22 – 25, 2014, Timisoara, Romania, pp. 537–543 (2014)
39. Gorunescu, F.: Intelligent decision systems in Medicine -a short survey on medical diagnosis and patient management (*keynote speech*). In: Proc. 5th IEEE International Conference on "E-Health and Bioengineering"-EHB 2015, 19–21 November 2015, Iasi, Romania, pp. 1-8 (2015)

Chapter 9
On The Automation of Medical Knowledge and Medical Decision Support Systems

Vicente Moret-Bonillo, Isaac Fernández-Varela,
Elena Hernández-Pereira, Diego Alvarez-Estévez and Volker Perlitz

Abstract This chapter follows the steps undertaken by other researchers in the field of knowledge engineering in Medicine. The material here presented is concerned with four key issues: The nature of the medical knowledge, the characteristics of the reasoning processes in medicine, the automatic acquisition of medical knowledge, and the effective handling of uncertain medical knowledge. The chapter reviews first categorical logic models and Bayesian methods. From a first conclusion on inherent uncertainty of medical decision making and reasoning, a vector representation of medical knowledge is proposed. This vector representation facilitates automation of the processes involved in the development of medical decision support systems. The use of contingency tables for automatic and objective knowledge acquisition is also proposed. To facilitate the understanding of the presented material, we have tried to illustrate each statement, idea, proposal or approach, with examples taken from the literature. Finally, the chapter concludes with an analysis of the possibilities of the overall method. Major potential contributions are justified, explained and discussed. The mentioned analysis focus on the application of the proposed approach on a simplified clinical case derived from the experience of the authors in the domain of the Sleep Medicine. Finally, the chapter concludes with a discussion, and with the establishment of the required conclusions.

Keywords Artificial intelligence in medicine · Knowledge engineering · Uncertain reasoning · Knowledge representation · Medical decision support systems

---

V. Moret-Bonillo (✉) · I. Fernández-Varela · E. Hernández-Pereira
Department of Computer Science, Faculty of Informatics, University of A Coruña,
15071 A Coruna, Spain
e-mail: vicente.moret@udc.es

D. Alvarez-Estévez
Sleep Center & Clinical Neurophysiology, Lijnbaan 32, 2512 VA The Hague, Netherlands

V. Perlitz
CEO Simplana GmbH, Aachen, Germany

© Springer International Publishing AG 2018
D.E. Holmes and L.C. Jain (eds.), *Advances in Biomedical Informatics*,
Intelligent Systems Reference Library 137,
https://doi.org/10.1007/978-3-319-67513-8_9

187

9.1 Introduction

In 1959 Ledley and Lusted published an article entitled "Reasoning Foundations of Medical Diagnosis" [1]. In that article they analyzed the inherent difficulty of the problem of medical diagnosis from three points of view: (a) The Symbolic Logic [2], (b) The Theory of Probability [3] and (c) The Decision Theory [4]. The initial hypothesis of Ledley and Lusted is that the rational use of computers could greatly simplify the process of medical diagnosis, or at least, could provide an important resource for the efficient management of the relevant information to obtain a medical diagnosis. However, at the time this important article was written, computers faced with a huge challenge: Its storage capabilities, and computing capabilities, were many orders of magnitude below of today's current computers. This situation allows us to establish a first conclusion: Although the theoretical foundations of the approaches proposed by Ledley and Lusted were reasonably clear, the complexity of the knowledge representations associated with the proposed methods greatly limited the size of the problems that could be handled.

In their approach, Ledley and Lusted identify three factors that play a key role in diagnostic tasks: (a) The medical knowledge that relates sets of symptoms with possible sets of diagnoses, (b) The sets of symptoms observed in a specific case, and (c) The sets of possible diagnoses that are consistent with the observed symptoms and with the knowledge of the physician who is analyzing the case. In this context, it seems clear that this knowledge is mainly heuristic [5].

An important feature of the analysis of Ledley and Lusted is that they deal with the problem of medical diagnosis from the point of view of the Differential Interpretations [6] that, basically, consists in eliminating situations of incompatibility between symptoms and possible diagnoses, given the knowledge domain.

This approach to the problem is very common in medicine, but from a computational perspective, and more specifically from the viewpoint of the Knowledge Engineering [7] and Artificial Intelligence [8] this way to face the problem greatly complicates the knowledge acquisition task, a task that is always subjective, and that usually results in a computational model of apparently intelligent behavior, but often incomplete, sometimes inefficient, and eventually unnatural [9].

The work we present in this chapter follows closely the ideas and approaches proposed by Ledley and Lusted, although we will not discuss the issues related to Decision Theory. We start from the very same example that has been presented by Ledley and Lusted in their article, and we will discuss it from the point of view of the Categorical Logics and of the Theory of Probability to subsequently introduce new approaches and models that could help to improve the understanding of the difficult task of medical diagnosis, while – at the same time – facilitating its computational implementation taking into account the characteristics associated with current technology.

9.2 Ledley and Lusted Approach to Medical Diagnosis

In order to explain the starting point of the analysis of Ledley and Lusted to the medical diagnosis task, which is based – as it has already been mentioned – in the use of Categorical Logics, we will change a little bit the notation originally proposed in the original article. So, we will use 'a_i' to denote specific symptoms, and 'x_j' to denote specific diagnoses or specific diseases. Capital letters indicate 'associations of symptoms or associations of diagnosis'.

- Let "n" be the number of relevant specific symptoms in the domain
- If the symptom "i" is present we say that $a_i = 1$
- If the symptom "i" is not present we say that $a_i = 0$

We are assuming a strictly categorical approach. Thus, in this domain, if "n" is the total number of specific symptoms, we have 2^n possible combinations of symptoms. To avoid ambiguity we adopt the binary numbering system. Thus, if $n = 3$ then we will have $2^3 = 8$ possible associations of symptoms: $A_0 \ldots A_7$. The corresponding values are shown in Table 9.1.

From this strictly categorical perspective, in a particular case, and in terms of associations, the symptoms of a patient are represented by one and only one of the associations of individual symptoms. For this reason, we say that the set of associations of symptoms, $S = \{A_0, A_1 \ldots A_2^n{}_{-1}\}$ is complete and exhaustive, and its elements are mutually exclusive. The same statement is true if we talk about diagnoses.

Assume now that we have two possible symptoms ($S = \{a_1, a_2\}$) and two possible diagnoses ($D = \{x_1, x_2\}$). If we express this in terms of associations, and in accordance with the criteria listed in Table 9.1, we have:

- $S = \{A_0, A_1, A_2, A_3\}$
- $D = \{X_0, X_1, X_2, X_3\}$

At this point, it seems reasonable to assume that there is a relationship between relevant symptoms and possible diagnoses. In this context, and working with the relevant associations of symptoms and diagnoses, we can build what Ledley and Lusted call Expanded Logic Base (ELB) that is built as the Cartesian product $S \times D$. Thus:

$$ELB = S \times D = \{A_0X_0, A_0X_1, A_0X_2, A_0X_3, A_1X_0, A_1X_1, A_1X_2, A_1X_3, A_2X_0, A_2X_1,$$
$$A_2X_2, A_2X_3, A_3X_0, A_3X_1, A_3X_2, A_3X_3\}$$

Table 9.1 Building 'Associations' (*upper case*) from individual events (*lower case*)

	A_0	A_1	A_2	A_3	A_4	A_5	A_6	A_7
a_1	0	0	0	0	1	1	1	1
a_2	0	0	1	1	0	0	1	1
a_3	0	1	0	1	0	1	0	1

ELB is also a complete and exhaustive set and its elements are mutually exclusive. Now, if we make explicit the values of individual symptoms and diagnoses, we can represent ELB as shown in Table 9.2.

Note that in ELB are absolutely all possible combinations of symptoms and diagnoses. Some of such combinations will be out of clinical sense and, therefore, they have to be eliminated. In this regard, Ledley and Lusted propose the use of the knowledge domain to rule out absurd combinations. For this purpose they face the problem from the differential diagnosis perspective. This involves removing from ELB those associations that are incompatible with the medical knowledge. We will try to explain, following the example proposed by Ledley and Lusted, how this works. Suppose that, after a laborious phase of knowledge acquisition, we have been able to identify the following rules:

- R1: If the patient has symptoms, then there must be one or more associated diagnosis.
  (Note that this rule does not preclude asymptomatic problems, only it states that if there are symptoms is because there are problems)
- R2: If the patient has the disease x_2, then he or she must present the symptom a_1.
- R3: If the patient has the disease x_1, but does not have the disease x_2, then he or she must present the symptom a_2.
- R4: If the patient has not the disease x_1, but he or she has the disease x_2 then symptom a_2 must be absent.

This is the knowledge we have about the domain and, with the small variations derived from our notation (already mentioned), Ledley and Lusted represent this knowledge in terms of implications, as follows:

- R1: $(a_1 = 1)$ OR $(a_2 = 1) \rightarrow (x_1 = 1)$ OR $(x_2 = 1)$
- R2: $(x_2 = 1) \rightarrow (a_1 = 1)$
- R3: $(x_1 = 1)$ AND $(x_2 = 0) \rightarrow (a_2 = 1)$
- R4: $(x_1 = 0)$ AND $(x_2 = 1) \rightarrow (a_2 = 0)$

Now, following the differential approach, we must discard those associations that are in ELB, but that also are incompatible with the rules of our knowledge domain. Thus, rule R1 eliminates A_1X_0, A_2X_0 and A_3X_0, since these are the

Table 9.2 Expanded Logic Base, ELB, involving two manifestations of a given problem, a_1 and a_2, and two eventually possible interpretations, x_1 and x_2

	A_0	A_1	A_2	A_3	A_0	A_1	A_2	A_3	A_0	A_1	A_2	A_3	A_0	A_1	A_2	A_3
a_1	0	0	1	1	0	0	1	1	0	0	1	1	0	0	1	1
a_2	0	1	0	1	0	1	0	1	0	1	0	1	0	1	0	1
x_1	0	0	0	0	0	0	0	0	1	1	1	1	1	1	1	1
x_2	0	0	0	0	1	1	1	1	0	0	0	0	1	1	1	1
	X_0				X_1				X_2				X_3			

complexes or associations in which there are symptoms but no associated pathologies. Similarly, rule R2 eliminates A_0X_1, A_1X_1, A_0X_3 and A_1X_3, since both X_1 and X_3 verify that $(x_2 = 1)$, but A_0 and A_1 verify that $(a_1 = 0)$. Similarly, R3 discards A_0X_2 and A_2X_2, since X_2 verifies that $(x_1 = 1)$ and $(x_2 = 0)$, while both A_0 and A_2 verify that $(a_2 = 0)$. Finally, R4 discards A_1X_1 and A_3X_1, because X_1 verifies that $(x_1 = 0)$ and $(x_2 = 1)$, but both A_1 and A_3 verify $(a_2 = 1)$. So, the associations that are removed from ELB are:

$$A_1X_0, A_2X_0, A_3X_0, A_0X_1, A_1X_1, A_0X_3, A_1X_3, A_0X_2, A_2X_2, A_3X_1$$

These associations are incompatible with our knowledge about the domain, and have to be removed from ELB. The result is a Reduced Logic Base (RLB) containing only those associations Symptom-Diagnosis that are possible given the available knowledge. In this case:

$$RLB = \{A_0X_0, A_1X_2, A_2X_1, A_2X_3, A_3X_2, A_3X_3\}$$

Suppose now that we have a clinical case: A patient goes to see the doctor presenting the symptoms $(a_1 = 0)$ and $(a_2 = 1)$. These symptoms responds to the association of symptoms A_1, which is associated, in RLB, with the diagnostic complex $\{X_2 = (x_1 = 1) \text{ AND } (x_2 = 0)\}$. In this case we can say that the patient has the disease x_1 but we can rule out disease x_2.

Suppose now another different case in which the symptoms are $(a_1 = 1)$ and $(a_2 = 0)$. This symptomatology, A_2, is associated in RLB with X_1 and with X_3. But $\{X_1 = (x_1 = 0) \text{ AND } (x_2 = 1)\}$ and, on the other hand, $\{X_3 = (x_1 = 1) \text{ AND } (x_2 = 1)\}$. In this case, we can say that the patient has the disease x_2 with certainty, but we should not rule out the pathology x_1 because there is evidence for and against among which we cannot discriminate. The conclusion is obvious: No matter how much categorical the procedure is -and this procedure is strictly categorical- the uncertainty is inherent to human reasoning, and appears spontaneously and naturally. This fact is very important, and may be due to one, or all, of the following factors:

- The nature of heuristic knowledge.
- The inherently imprecise nature of knowledge.
- The lack of knowledge.
- The subjectivity in the interpretation of the information.

To resolve this problem, Ledley and Lusted propose the use of probabilistic models in order to try to identify which is the most plausible interpretation. Specifically, in the example discussed so far, they posed the following question: Given the symptoms represented by the complex A_2, which one of the diagnostic associations, X_1 or X_3, is more probable? In this context, they suggest the use of conditional probabilities and the Bayesian approach (which can be considered as a form of "a posteriori" reasoning) [10] to find the value of the ratio:

$$\frac{P(X_1/A_2)}{P(X_3/A_2)}$$

In the above expression $P(X_1/A_2)$ is the conditional probability that given A_2 then we can conclude X_1 with a given probability, and $P(X_3/A_2)$ is the conditional probability that given A_2 then we can conclude X_3 with a given probability. Both expressions are linked to the principle of causality.

9.3 Some Thoughts on Medical Reasoning

The purpose of this section is to illustrate how we can obtain the RLB in a way that, in our opinion, could be more natural and transparent. To do this we suggest a change in the way of representing the declarative and procedural knowledge. In this manner, always working with the same example, assume we have two relevant symptoms, a_1 and a_2, which may be present or not in a given case ($a_1 \in \{0, 1\}$ and $a_2 \in \{0, 1\}$). If we establish an ordered set with the individual symptoms such as $S = \{a_1, a_2 \dots a_n\}$ and using the criterion that the value 0 indicates 'absence', and the value 1 indicates 'presence', then we can represent associations as vectors. In the case of our example we get the following result:

$$A_0 = [0\,0] \quad A_1 = [0\,1] \quad A_2 = [1\,0] \quad A_3 = [1\,1]$$

Analogously, for the diagnostic associations of the example, we obtain the following results:

$$X_0 = [0\,0] \quad X_1 = [0\,1] \quad X_2 = [1\,0] \quad X_3 = [1\,1]$$

Accordingly, the representation of the ELB will have the form: ELB $\rightarrow$ [**S, D**], and the elements of this logical base will be of the kind: $[a_1\ a_2\ x_1\ x_2]$. Now we will work with the rule base of the previous example, but considering the logical definition of the implication:

$$\{\text{Antecedent} \rightarrow \text{Consequent}\} \equiv \{\text{NOT Antecedent OR Consequent}\}$$

Thus, the rules:

- R1: $(a_1 = 1)$ OR $(a_2 = 1) \rightarrow (x_1 = 1)$ OR $(x_2 = 1) \equiv (a_1$ OR $a_2) \rightarrow (x_1$ OR $x_2)$
- R2: $(x_2 = 1) \rightarrow (a_1 = 1) \equiv x_2 \rightarrow a_1$
- R3: $(x_1 = 1)$ AND $(x_2 = 0) \rightarrow (a_2 = 1) \equiv (x_1$ AND NOT $x_2) \rightarrow a_2$
- R4: $(x_1 = 0)$ AND $(x_2 = 1) \rightarrow (a_2 = 0) \equiv (\text{NOT } x_1$ AND $x_2) \rightarrow \text{NOT } a_2$

can be written as follows:

- R1: [NOT a_1 AND NOT a_2] OR x_1 OR x_2
- R2: a_1 OR NOT x_2

- R3: a_2 OR NOT x_1 OR x_2
- R4: NOT a_2 OR x_1 OR NOT x_2

We can infer now that:

- R1 is compatible with vectors: $[0\ 0 - -]\ [- - 1 -]\ [- - - 1]$
- R2 is compatible with vectors: $[1 - - -]\ [- - - 0]$
- R3 is compatible with vectors: $[- 1 - -]\ [- - 0 -]\ [- - - 1]$
- R4 is compatible with vectors: $[- 0 - -]\ [- - 1 -]\ [- - - 0]$

In the above vectors symbol (–) indicates that the value 0 or 1 of the corresponding symptom, or of the corresponding diagnosis, is not relevant for the rule to considered 'true'. Now, if we consider the problem globally, only vectors that are compatible with all the rules of our knowledge base will be valid. This can be seen in Table 9.3, in which the corresponding consistency analysis is shown.

Compatible vectors of the original rule base are, therefore:

$$[0\,0\,0\,0] \equiv A_0X_0 \quad [1\,0\,0\,1] \equiv A_2X_1 \quad [0\,1\,1\,0] \equiv A_1X_2$$
$$[1\,1\,1\,0] \equiv A_3X_2 \quad [1\,0\,1\,1] \equiv A_2X_3 \quad [1\,1\,1\,1] \equiv A_3X_3$$

This result is exactly the same to the one obtained by reducing the Expanded Logic Base using the differential method. Furthermore, this procedure allows us the vector representation of the rules, from symptoms to diagnosis which, from a computer-based approach seems to be a more natural way for representing knowledge. Thus, we could now write our four rules as follows:

Table 9.3 Consistency analysis between vectors and rules

Vectors	R1	R2	R3	R4	Consistency
[0 0 0 0]	YES	YES	YES	YES	YES
[0 1 0 0]	NO	YES	YES	YES	NO
[1 0 0 0]	NO	YES	YES	YES	NO
[1 1 0 0]	NO	YES	YES	YES	NO
[0 0 0 1]	YES	NO	YES	YES	NO
[0 1 0 1]	YES	NO	YES	NO	NO
[1 0 0 1]	YES	YES	YES	YES	YES
[1 1 0 1]	YES	YES	YES	NO	NO
[0 0 1 0]	YES	YES	NO	YES	NO
[0 1 1 0]	YES	YES	YES	YES	YES
[1 0 1 0]	YES	YES	NO	YES	NO
[1 1 1 0]	YES	YES	YES	YES	YES
[0 0 1 1]	YES	NO	YES	YES	NO
[0 1 1 1]	YES	NO	YES	YES	NO
[1 0 1 1]	YES	YES	YES	YES	YES
[1 1 1 1]	YES	YES	YES	YES	YES

$$
\begin{array}{llllll}
R1: & IF & S = [0\,0] & THEN & D = [0\,0] \\
R2: & IF & S = [0\,1] & THEN & D = [1\,0] \\
R3: & IF & S = [1\,0] & THEN & D = [0\,1] & OR & D = [1\,1] \\
R4: & IF & S = [1\,1] & THEN & D = [1\,0] & OR & D = [1\,1]
\end{array}
$$

Let us now analyze the particular cases that we raised before, but from this new perspective:

Case 1: The patient has no symptom 1, but symptom 2 is present.

- Associated Symptoms Vector: S = [0 1]
- Rule Activated and Executed: R2
- Associated Diagnostic Vector: D = [1 0]
- Conclusion: The patient has the disease 1 and pathology 2 can be discarded.

Case 2: The patient has symptom 1, but symptom 2 is not present.

- Associated Symptoms Vector: S = [1 0]
- Rule Activated and Executed: R3
- Associated Diagnostic Vectors: D = [0 1] and D = [1 1]
- Conclusion: We can establish with certainty the pathology 2, but the available evidence is contradictory regarding the pathology 1 so that we can neither confirm nor discard it.

Both results coincide again with those obtained by Ledley and Lusted. It could not be otherwise, since the model presented in this section is, from a conceptual perspective, virtually the same as the one proposed by them (although with the nuances already commented). Remember also that these authors propose the use of "a posteriori" probabilistic models to resolve the ambiguity on the second case just analyzed. However, still remain unresolved some knowledge related problems, such as: (a) Subjectivity, (b) Lack of knowledge, or (c) The bottleneck which involves the acquisition of knowledge. We will discuss about these issues in the next section of this paper.

9.4 Analyzing Medical Knowledge Acquisition and Ambiguity

Consider as the starting points of the material to be treated in this section the information shown in Table 9.2 and the RLB resulting from applying on the data of this table the knowledge represented by our four rules:

$$RLB = \{A_0X_0, A_1X_2, A_2X_1, A_2X_3, A_3X_2, A_3X_3\}$$

Note that, for example, A_0X_1 is in ELB but it is not in RLB. The reason is obvious: the A_0X_1 association is not allowed by the knowledge domain, which has been built from a strictly categorical perspective. Now, if we enlarge our point of view, and we perform a probabilistic assessment of the problem, we can say that the joint probability of A_0X_1 is zero $[P(A_0X_1) = 0]$. Following this line of reasoning, the probability of any association of the kind AiXj that is in ELB but not in RLB is zero. But, does this mean that the probability of the associations that are present in RLB is one? The question now is something different, and the answer to this question is ... not necessarily. A different issue is that we were talking in terms of possibility, in which case the answer would be 'yes' because the association to which we refer is allowed by the knowledge domain. Formally:

Let {Pos (AiXj)} be the possibility of an association AiXj

- Pos (AiXj) = 0 ↔ AiXj ∉ BLR
- Pos (AiXj) = 1 ↔ AiXj ∈ BLR

On the other hand, let {P(AiXj)} be the probability of the association AiXj

- P(AiXj) = 0 ↔ AiXj ∉ BLR
- 0 < P(AiXj) ≤ 1 ↔ AiXj ∈ BLR

P(AiXj) can be interpreted as the joint probability of Ai and Xj. If we now analyze the case of the association A_1X_2 we can introduce the principle of causality through conditional probabilities, with which it is related. Conditional probability is defined as the probability that involves two events, in which the occurrence of the second event depends on the occurrence of the first. If Ai is the first event, and Xj is the second event, P(Xj/Ai) it is the conditional probability of Xj given Ai. This can also be represented as follows:

$$A_i \xrightarrow{P(X_j/A_i)} X_j$$

Conditional probabilities are built from the joint probabilities and the total probability of the first event. So:

$$P(X_j/A_i) = \frac{P(A_iX_j)}{P(A_i)}$$

Obviously, P(Xj/Ai) can be interpreted as the 'strength' of the corresponding causal relation, so that, in the case of the association we are studying now, (A_1X_2), the causal relations responds to the expression:

$$A_1 \xrightarrow{P(X_2/A_1)=\frac{P(A_1X_2)}{P(A_1)}} X_2$$

Now consider the case of associations and A_2X_3 A_2X_1. In this regard, following the same approach as before:

$$A_2 \xrightarrow{P(X_1/A_2)} X_1 \quad A_2 \xrightarrow{P(X_3/A_2)} X_3$$

Now, if we remember that:

$$P(X_1/A_2) = \frac{P(A_2X_1)}{P(A_2)} \quad P(X_3/A_2) = \frac{P(A_2X_3)}{P(A_2)}$$

and, since the denominators are the same, we simply divide $P(A_2X_1)$ between $P(A_2X_3)$ to know, given the same symptoms, which of the diagnostic associations is more likely because:

$$\frac{P(X_1/A_2)}{P(X_3/A_2)} = \frac{P(A_2X_1)}{P(A_2X_3)}$$

This is precisely the procedure proposed Ledley and Lusted to discriminate between ambiguous situations. However, this same procedure can be used to acquire medical knowledge in a coherent, complete, and unambiguous manner. Obviously, if in a contingency table, in which we have previously identified symptoms and diagnoses, appear common data, then we can calculate joint probabilities and – since we have also information about total probabilities – we can also calculate conditional probabilities which, as we already know, are related to the principle of causality. In this way we can obtain the corresponding rules in an objective manner. As we shall see shortly, this procedure can be directed from symptoms to diagnosis, or from diagnostics to symptoms, both individually or in the form of associations. To illustrate the method, in this work we will use associations.

An important issue to be considered is to ensure the consistency of the method. So the contingency table should not include data for associations that are not present in RLB. And, similarly, we must have data for other associations that are present in RLB. Not met the criterion of consistency we have to think of some of the following circumstances:

- The data casuistry or the contingency table are incorrect.
- The medical knowledge is incomplete.
- There have been errors during the acquisition of knowledge.

With these considerations in mind, we can return to our example and consider the data shown in Table 9.4, which is nothing else than a contingency table. This table has been built arbitrarily from a presumed statistically and sufficiently

Table 9.4 Contingency table of a hypothetical case

S ↓/D →		X_0 [00]	X_1 [01]	X_2 [10]	X_3 [11]	Number of cases
A_0	[00]	512				**512**
A_1	[01]			75		**75**
A_2	[10]		23		37	**60**
A_3	[11]			5	60	**65**
Number of cases		**512**	**23**	**80**	**97**	**712**

Table 9.5 Joint probabilities and total probabilities of a hypothetical case

S ↓/D →		X_0 [00]	X_1 [01]	X_2 [10]	X_3 [11]	Total probabilities
A_0	[00]	0.719				**0.719**
A_1	[01]			0.105		**0.105**
A_2	[10]		0.032		0.052	**0.084**
A_3	[11]			0.007	0.084	**0.091**
Total probabilities		**0.719**	**0.032**	**0.112**	**0.136**	**0.999**

representative set of data. Also, it has been included the restriction that the symptoms a_1 and a_2 and the diagnoses x_1 and x_2 are potentially relevant. This means that we start from some prior knowledge. Later we will expand the example eliminating this restriction, but this will be once we have discussed some points that we consider of interest. On the other hand, as we have already mentioned, we will work with associations of vectors.

From the data of Table 9.4 we can infer that the unique possible RLB is:

$$BLR = \{A_0X_0, A_1X_2, A_2X_1, A_2X_3, A_3X_2, A_3X_3\}$$

Let us now convert Table 9.4 in a joint probability table with total probabilities (Table 9.5). For that purpose it will be enough to divide the data of each of the cells by the total number of data (712).

With the results of Table 9.5 we can obtain the following conditional probabilities:

$$P(X_0/A_0) = \frac{P(A_0X_0)}{P(A_0)} = \frac{0.719}{0.719} = 1.000$$

$$P(X_1/A_2) = \frac{P(A_2X_1)}{P(A_2)} = \frac{0.032}{0.084} = 0.381$$

$$P(X_2/A_1) = \frac{P(A_1X_2)}{P(A_1)} = \frac{0.105}{0.105} = 1.000$$

$$P(X_2/A_3) = \frac{P(A_3X_2)}{P(A_3)} = \frac{0.007}{0.091} = 0.077$$

$$P(X_3/A_2) = \frac{P(A_2X_3)}{P(A_2)} = \frac{0.052}{0.084} = 0.619$$

$$P(X_3/A_3) = \frac{P(A_3X_3)}{P(A_3)} = \frac{0.084}{0.091} = 0.923$$

Note that in those relations containing the same association in the antecedent, the sum of the conditional probabilities is always 1:

$$A_0 \xrightarrow{1.000} X_0 \Big\} : \sum = 1.000 \quad A_1 \xrightarrow{1.000} X_2 \Big\} : \sum = 1.000$$

$$\left. \begin{array}{c} A_2 \xrightarrow{0.381} X_1 \\ A_2 \xrightarrow{0.619} X_3 \end{array} \right\} : \sum = 1.000 \quad \left. \begin{array}{c} A_3 \xrightarrow{0.077} X_2 \\ A_3 \xrightarrow{0.923} X_3 \end{array} \right\} : \sum = 1.000$$

Now we can write the following rules, which evolve from symptoms to diagnosis:

R1 :	IF	S = [00]	THEN D = [00]	P(D/S) = 1.000
R2 :	IF	S = [01]	THEN D = [10]	P(D/S) = 1.000
R3 :	IF	S = [10]	THEN D = [01]	P(D/S) = 0.381
R4 :	IF	S = [10]	THEN D = [11]	P(D/S) = 0.619
R5 :	IF	S = [11]	THEN D = [10]	P(D/S) = 0.077
R6 :	IF	S = [11]	THEN D = [11]	P(D/S) = 0.923

Let us now apply these results to the two cases that served us as example:

Case 1: The patient has not symptom 1, but he or she presents symptom 2

- Associated Symptoms Vector: S = [0 1]
- Rule Activated and Executed: R2
- Associated Diagnostic Vector : D = [1 0]
- Conclusion: The patient has the disease 1 and pathology 2 can be discarded without uncertainty.

Case 2: The patient has symptom 1, but he or she has does not present symptom 2

- Associated Symptoms Vector: S = [1 0]
- Rules Activated and Executed: R3 and R4
- Associated Diagnostic Vectors: D = [0 1] (P = 0.381) and D = [1 1] (P = 0.619)
- Conclusion: We can establish with certainty the pathology 2, but the available

evidence is contradictory regarding pathology 1 so that we can neither confirm nor discard it. However the presence of pathology 1 is 1.6 times more probable than its absence.

This approach can also be made from a retrospective perspective, which brings us to the differential diagnosis. In this case the process evolves from diagnosis to symptoms and causal relations are represented by the following conditional probabilities:

$$P(A_0/X_0) = \frac{P(A_0X_0)}{P(X_0)} = \frac{0.719}{0.719} = 1.000$$

$$P(A_1/X_2) = \frac{P(A_1X_2)}{P(X_2)} = \frac{0.105}{0.112} = 0.938$$

$$P(A_2/X_1) = \frac{P(A_2X_1)}{P(X_1)} = \frac{0.032}{0.032} = 1.000$$

$$P(A_2/X_3) = \frac{P(A_2X_3)}{P(X_3)} = \frac{0.052}{0.136} = 0.382$$

$$P(A_3/X_2) = \frac{P(A_3X_2)}{P(X_2)} - \frac{0.007}{0.112} = 0.062$$

$$P(A_3/X_3) = \frac{P(A_3X_3)}{P(X_3)} = \frac{0.084}{0.136} = 0.618$$

Note that, once again, the mathematical consistency of the model is fulfilled:

$$\left. X_0 \xrightarrow{1.000} A_0 \right\} : \sum = 1.000 \quad \left. X_1 \xrightarrow{1.000} A_2 \right\} : \sum = 1.000$$

$$\left. \begin{matrix} X_2 \xrightarrow{0.938} A_1 \\ X_2 \xrightarrow{0.062} A_3 \end{matrix} \right\} : \sum = 1.000 \quad \left. \begin{matrix} X_3 \xrightarrow{0.382} A_2 \\ X_3 \xrightarrow{0.618} A_3 \end{matrix} \right\} : \sum = 1.000$$

Now we can write the following rules, which evolve from diagnosis to symptoms:

R1 :	IF	$D = [00]$	THEN $S = [00]$	$P(S/D) = 1.000$
R2 :	IF	$D = [01]$	THEN $S = [10]$	$P(S/D) = 1.000$
R3 :	IF	$D = [10]$	THEN $S = [01]$	$P(S/D) = 0.938$
R4 :	IF	$D = [10]$	THEN $S = [11]$	$P(S/D) = 0.062$
R5 :	IF	$D = [11]$	THEN $S = [10]$	$P(S/D) = 0.382$
R6 :	IF	$D = [11]$	THEN $S = [11]$	$P(S/D) = 0.618$

Table 9.6 Contingency table with asymptomatic cases in a hypothetical example

S ↓/D →		X_0 [00]	X_1 [01]	X_2 [10]	X_3 [11]	Number_of_ cases
A_0	[00]	512		15		527
A_1	[01]			75		75
A_2	[10]		23		37	60
A_3	[11]			5	60	65
Number_of_ cases		512	23	95	97	727

Table 9.7 Joint probabilities and total probabilities with asymptomatic cases in a hypothetical example

S ↓/D →		X_0 [00]	X_1 [01]	X_2 [10]	X_3 [11]	Total probabilities
A_0	[00]	0.704		0.021		0.725
A_1	[01]			0.103		0.103
A_2	[10]		0.032		0.051	0.083
A_3	[11]			0.007	0.083	0.090
Total probabilities		0.704	0.032	0.131	0.134	1.001

Let us now analyze the case of asymptomatic diseases. The idea is as follows: Assume we have a case in which a specific pathology is suspected, but none of the symptoms that may be related is observed. We are certainly facing a difficult case to solve, which could escape the established medical knowledge. However, we can find clues in a well-built database. To illustrate what we mean, we will modify the data in Table 9.4 as shown in Table 9.6, in which have included 15 asymptomatic situations (A_0X_2).

Now the joint probabilities and total probabilities are shown in Table 9.7.

Now we can calculate the conditional probabilities of the causal relations s in which $A_0 = [00]$ is present, either in the antecedent or in the consequent. If we do it, we get the following results:

First Case:

$$A_0 \rightarrow X_0 : \quad P(X_0/A_0) = P(A_0X_0)/P(A_0) = 0.968$$

$$A_0 \rightarrow X_2 : \quad P(X_2/A_0) = P(A_0X_2)/P(A_0) = 0.029$$

Second Case:

$$X_0 \rightarrow A_0 : \quad P(A_0/X_0) = P(A_0X_0)/P(X_0) = 1.000$$

Third Case:

$$X_2 \rightarrow A_0 : \quad P(A_0/X_2) = P(A_0X_2)/P(X_2) = 0.160$$

In this example, the three cases are especially interesting. The first one because, in the absence of symptoms (S = [00]), and in the specific domain considered, and although it is likely that there is no associated pathology, we cannot completely rule out pathology 1 since: $P(D = [10]) \neq 0$.

This can happen, for example, in the following scenario: A patient with a family history of an inherited disease. The patient could have the disease, but the symptoms are not yet manifested. This approach opens a new question in the processes of reasoning in Medicine: Tracking the evolution of a patient, a question on which we will discuss in some depth a little later.

The second case is also interesting, but for other reasons. In this case, we assume that, prior to the application of the proposed approach, we have identified potentially relevant symptoms and diagnoses. Suppose now that using other alternative information we can rule out with absolute certainty relevant pathologies ($D = X_0 = [00]$). In this case we have to expect with complete certainty the absence of symptoms. This situation must be considered in terms of consistency, so that if some symptom appear we may have to revise the knowledge that we have and eventually investigate the reason of the inconsistency. Also we return to this issue later.

Finally, the third case relates to the completeness of the causal relations, which in our example are:

$$X_2 \rightarrow A_0 \quad : \quad P(A_0/X_2) = a$$
$$X_2 \rightarrow A_1 \quad : \quad P(A_1/X_2) = b$$
$$X_2 \rightarrow A_3 \quad : \quad P(A_3/X_2) = c$$

and wherein completeness requires that: a + b + c = 1.

Now we can generalize the process without including any kind of restriction. The only exception is that we must clearly recognize the difference between the concepts of symptom and diagnosis. Suppose that in a particular domain we have a statistically significant number of cases, including diseases and symptoms. With 's' symptoms we have 2^s associations ($A0, A_1, \ldots, A_2{^s} _{-1}$), and with 'd' diagnoses we have 2^d associations ($X_0, X_1, \ldots, X_2{^d} _{-1}$). This involves the construction of a ELB formed by $2^s \times 2^d = 2^{s+d}$ elements. This number can be very large, but do not forget that current technology allows the representation and management of vast amounts of information, so – although there are limits – the problem of space requirements and computing capacity related with actual medical applications is not so important. To give just an indicative example, a simple and conventional spreadsheet, as it can be MS-Excel 2007^{TM}, has a size of 1 048 576 rows and 16 384 columns [11] so it can represent and manipulate around 2^{34} data distributed in 2^{20} rows and 2^{14} columns. This is a significant number that can be handled efficiently with a single desktop computer. Obviously, in this case, due to the limitations of Excel, there are two constraints:

1. IF $14 \leq s \leq 20$ THEN $d \leq 14$
2. IF $14 \leq d \leq 20$ THEN $s \leq 14$

9.5 Methodological Synthesis

Given the above considerations we can – in a given domain – assume the following procedure for the computer generation of medical knowledge. As we will see later, this procedure is able to efficiently handle different kinds of uncertainty. Schematically, the method includes the following steps:

1. We need a statistically significant number of cases.
2. We have to identify individual symptoms and individual diagnoses.
3. We have to generate vectors corresponding associations of symptoms and diagnoses, using the binary numbering criterion.
4. We have to build, with the real cases, the corresponding contingency table.
5. We have to calculate the total and the joint probabilities.
6. We have to calculate the conditional probabilities of the causal relations. This step can be done from symptoms to diagnosis, or from diagnosis to symptoms.
7. We have to check the consistency of the calculations.
8. Finally we can build the relevant rules.

The overall process can also be optimized in many ways. Just to illustrate this idea of optimization with an example, suppose that we incorporate a phase for encoding and decoding numbers of the kind binary-decimal-binary and we apply this to a case in which the number of individual symptoms is $s = 4$, with which we can build $2^4 = 16$ associations of symptoms, which in vector form are the following:

$$
\begin{aligned}
A_0 &= [0000] & A_1 &= [0001] & A_2 &= [0010] \\
A_3 &= [0011] & A_4 &= [0100] & A_5 &= [0101] \\
A_6 &= [0110] & A_7 &= [0111] & A_8 &= [1000] \\
A_9 &= [1001] & A_{10} &= [1010] & A_{11} &= [0011] \\
A_{12} &= [1100] & A_{13} &= [1101] & A_{14} &= [1110] \\
A_{15} &= [1111]
\end{aligned}
$$

Similarly, if $d = 5$ the number of individual diagnoses, we can build $2^5 = 32$ diagnostic associations, which in vector form, are as follows:

$$X_0 = [00000] \quad X_1 = [00001] \quad X_2 = [00010] \quad X_3 = [00011]$$
$$X_4 = [00100] \quad X_5 = [00101] \quad X_6 = [00110] \quad X_7 = [00111]$$
$$X_8 = [01000] \quad X_9 = [01001] \quad X_{10} = [01010] \quad X_{11} = [01011]$$
$$X_{12} = [01100] \quad X_{13} = [01101] \quad X_{14} = [01110] \quad X_{15} = [01111]$$
$$X_{16} = [10000] \quad X_{17} = [10001] \quad X_{18} = [10010] \quad X_{19} = [10011]$$
$$X_{20} = [10100] \quad X_{21} = [10101] \quad X_{22} = [10110] \quad X_{23} = [10111]$$
$$X_{24} = [11000] \quad X_{25} = [11001] \quad X_{26} = [11010] \quad X_{27} = [11011]$$
$$X_{28} = [11100] \quad X_{29} = [11101] \quad X_{30} = [11110] \quad X_{31} = [11111]$$

Let us now consider the following rule:

$$\text{IF } S = [0101] \quad \text{THEN } D = [10100] \quad P(D/S) = a \quad (0 < a \le 1)$$

Note that the vector representation, according to the binary numbering system makes the corresponding vectors match with the sub-index of the associations of symptoms and diagnoses.

Thus, we can easily check that:

- $A5 \rightarrow [0101]_{\text{Base } 2} = [5]_{\text{Base } 10}$
- $X20 \rightarrow [10100]_{\text{Base } 2} = [20]_{\text{Base } 10}$

Therefore we can rewrite the above rule as follows:

$$\text{IF } S = [5] \quad \text{THEN } D = [20] \quad P(D/S) = a \quad (0 < a \le 1)$$

This can be done with arbitrarily large numbers. For example, if we know that $s = 32$ we obtain from A_0 to $A_{4294967295}$ associations of symptoms, and if we have $d = 5$ we obtain from X_0 to X_{32767} associations of diagnoses. In this context, consider now the rule:

$$\text{IF } S = [20251] \quad \text{THEN } D = [4578] \quad P(D/S) = a \quad (0 < a \le 1)$$

In binary terms, and with $s = 32$, the number 20251 represents the symptomatology shown in Table 9.8.

Similarly, with $d = 15$ the number 4578 represents the diagnostic complex shown in Table 9.9.

The interpretation, in accordance with the provisions of the previous rule, is that in this case the patient has symptoms [1, 2, 4, 5, 9–12, 15] that may be associated with pathologies [2, 6–9, 13] with a certain probability—$P(D/S) = a$; $(0 < a \le 1)$.

Table 9.8 Symptoms associated with the decimal number 20251

Position	32	31	30	29	28	27	26	25	24	23	22	21	20	19	18	17	16	15	14	13	12	11	10	09	08	07	06	05	04	03	02	01
Value	0	0	0	0	0	0	0	0	0	0	0	0	0	0	0	0	0	1	0	0	1	1	1	1	0	0	0	1	1	0	1	1

Table 9.9 Diagnosis complex associated with the decimal number 4578

Position	15	14	13	12	11	10	09	08	07	06	05	04	03	02	01
Value	0	0	1	0	0	0	1	1	1	1	0	0	0	1	0

9.6   Imprecision, Uncertainty and Temporal Evolution

What we have discussed up to now has allowed us to see how ambiguous situations appear even in the context of a strictly categorical model. We have also analyzed statistical models based on conditional probabilities for, among other things, try to minimize the ambiguity of the conclusion in a causal relation. This ambiguity is related to the uncertainty of the causal relation itself, which, in turn, generates inaccuracy in the conclusions. In this regard, there is a subtle difference between imprecision and uncertainty. So while imprecision relates to the lack of total certainty in the entities of the domain, either symptoms or diagnoses (declarative knowledge), uncertainty relates to the lack of certainty in the relations between entities (procedural knowledge), and has to do with the strength of the causal relation. Consider the following situations:

(1) $A \xrightarrow{P(B/A)=0} B$

(2) $A \xrightarrow{P(B/A)=1} B$

(3) $A \xrightarrow{P(B/A)=x \in (0,1)} B$

In the first case there is neither uncertainty nor imprecision, just A and B are not connected. In the second case there is neither imprecision nor uncertainty, and given A we can conclude with certainty B. In the third case, however, the uncertainty of the causal relation generates imprecision in the conclusion, although there is no imprecision in the antecedent. But the issue can be further complicated. The reason is that the evidences (symptoms) can:

Table 9.10 Arbitrary weighting of the relevance and severity of symptoms

			Severity				
			None	Light	Moderate	Severe	Very severe
			0.00	0.25	0.50	0.75	1.00
Relevance	None	0.00	0.0000	0.0000	0.0000	0.0000	0.0000
	Small	0.25	0.0000	0.0625	0.1250	0.1875	0.2500
	Significant	0.50	0.0000	0.1250	0.2500	0.3750	0.5000
	Important	0.75	0.0000	0.1875	0.3750	0.5625	0.7500
	Critical	1.00	0.0000	0.2500	0.5000	0.7500	1.0000

V. Moret-Bonillo et al.

1. Have different relevance (i.e., some are more important than others)
2. Have different degrees of severity (i.e., some more than others deviate from the reference values considered normal)
3. Have different relevance and different degrees of severity

These circumstances may modify the results of an interpretation and also as we shall see later, provide valuable information on the evolution of the state of a particular patient, watching how the symptoms evolve. The same argument can be used on the complete set of symptoms. Then we speak of the clinical picture of the patient. In this context we will put an example (arbitrary) based on data (also arbitrary) of Table 9.10 in which we define a number of linguistic labels to ponder numerically according to their relevance or according to their severity (either). Obviously, this is merely an example, and here the clinical work is essential when assessing the situation.

In this context, let us now return to the set of rules seen above that, for clarity we rewrite as follows:

$$R1 : S = [00] \xrightarrow{1.000} D = [00] \quad R2 : S = [01] \xrightarrow{1.000} D = [10]$$
$$R3 : S = [10] \xrightarrow{0.381} D = [01] \quad R4 : S = [10] \xrightarrow{0.619} D = [11]$$
$$R5 : S = [11] \xrightarrow{0.077} D = [10] \quad R6 : S = [11] \xrightarrow{0.923} D = [11]$$

In this case the possible appearance of imprecision that may affect diagnostic associations is the result of the uncertainty of causal relations, since there is no imprecision in the symptoms. The question now is: How can we deal with this problem in the case that the evidence related to the symptoms are also imprecise?

A possible solution to this interesting problem could be based on the ideas expressed by Shortliffe and Buchanan in their model of Certainty Factors [12]. As far as our problem is concerned, we will use of this model the fact that imprecision may be a consequence of the propagation of uncertainty. We will illustrate these ideas with an example.

Let be a specific case in which the symptomatology observed is S = [11]. Both symptoms are critical, and both are presented in a severe degree. According to the criteria in Table 9.10, the imprecision associated with the complex S = [11] is 0.7500

In this example, given that S = [11], rules R5 and R6 are activated. But the imprecision associated with the conclusions may not be the same to the one that can be inferred directly from the activated rules because the antecedent is imprecise. In this regard, we can do the following: If we call E (S) to the evidence associated the complex S = [11] we can write:

$$E(S) \xrightarrow{0.7500} S = [11] \xrightarrow{0.077} D = [10]$$

$$E(S) \xrightarrow{0.7500} S = [11] \xrightarrow{0.923} D = [11]$$

If now we call {Imp(D)} to the imprecision that affects the conclusion D, and which has been obtained by propagation of the uncertainty, we obtain:

- CASE 1: Imp(D = [10]) = 0.7500 × 0.077 = 0.05775
- CASE 2: Imp(D = [11]) = 0.7500 × 0.923 = 0.69225

It is interesting to notice that in this case, the ratio D [11]/D [10] is constant regardless of whether or not there is imprecision in the symptomatology. In this example, D [11] is 12 times more likely than D [10] provided the antecedent is S = [11]:

$$S[11] \rightarrow \frac{P(D = [11])}{P(D = [10])} = \frac{0.923}{0.077} = \frac{0.69225}{0.05775} \approx 12$$

Thus, if we call "Belief" the result of spreading the imprecision by means of conditional probabilities, and depending on the model considered in this particular case, we find the following results:

Categorical Approach:

- S [11] → D [10]: It is possible: Possibility = 1
- S [11] → D [11]: It is possible: Possibility = 1

Probabilistic Approach:

- S [11] → D [10]: It is possible: Probability = 0.077
- S [11] → D [11]: It is possible: Probability = 0.923

Uncertain Approach:

- E (S = [11]) = 0.75 → D [10]: It is possible: Belief = 0.05775
- E (S = [11]) = 0.75 → D [11]: It is possible: Belief = 0.69225

The uncertainty-based approach can also be applied to individual symptoms. In fact, the association S = [11] means that {(a$_1$ = 1) AND (a$_2$ = 1)}. Thus we could build the rules:

$$(a_1 = 1) \, AND \, (a_2 = 1) \xrightarrow{0.077} D = [01]$$

$$(a_1 = 1) \, AND \, (a_2 = 1) \xrightarrow{0.923} D = [11]$$

But now it may happen that not all individual symptoms have neither the same importance nor the same severity. In this context, suppose the symptom a$_1$ is considered 'very relevant' (Relevance [a$_1$] = 0.75 according to the criteria in

Table 9.10), and that in the case considered, manifests 'moderately' (Severity $[a_1]$ = 0.50 also according with the criteria in Table 9.10). So, what we might call 'Causal Relevance' of the symptom a1 – RC $[a_1]$ – would be:

$$RC[a_1] = 0.3750$$

On the other hand, suppose that the symptom a_2 is considered 'important' (Relevance $[a_2]$ = 0.25 according to the criteria in Table 9.10), and that in the case considered is manifested 'light' (Severity $[a_2]$ = 0.25 also according with the criteria in Table 9.10). Therefore, what we might call 'Causal Relevance' of the symptom a_2 – RC $[a_2]$ – would be:

$$RC[a_2] = 0.0625$$

Now we have the following special cases:

CASE 1:

$$\{RC[(a_1 = 1)] = 0.3750\} \text{ AND } \{RC[(a_2 = 1)] = 0.0625\} \xrightarrow{0.077} D = [01]$$

CASE 2:

$$\{RC[(a_1 = 1)] = 0.3750\} \text{ AND } \{RC[(a_2 = 1)] = 0.0625\} \xrightarrow{0.923} D = [11]$$

Once defined the framework the next step is to evaluate the resulting expressions. In the case of AND-rules, Shortliffe and Buchanan propose the use of the clauses with the lowest value. In our example:

CASE 1:

$$\text{Imp}(D = [01]) = \min\{0.3750, 0.0625\} \times 0.077 = 0.0625 \times 0.077 = 0.0048$$

CASE 2:

$$\text{Imp}(D = [11]) = \min\{0.3750, 0.0625\} \times 0.923 = 0.0625 \times 0.923 = 0.0577$$

It is obvious that in all approaches studied in this section (except for the categorical approach), what changes is the absolute value of which, generally speaking, can be called 'plausibility' of the conclusion. This can be checked easily considering that:

$$\frac{D[11]}{D[01]} = \frac{0.923}{0.077} = \frac{0.69225}{0.05775} = \frac{0.0577}{0.0048} \approx 12$$

In relative terms, this ratio is always maintained, what changes is the absolute value of the plausibility of each of the diagnostic associations. But note that if we change the term plausibility by the term 'intensity', what we are actually doing is building a model for the temporal follow-up of a particular patient. To illustrate what we mean imagine the following scenario:

- A patient comes to see de doctor complaining of symptoms a_1 and a_2
- Both symptoms suggest the diagnostic associations [01] or [11]
- We also know that, in the absence of other evidence, the diagnostic association [11] is more frequent than the diagnosis association [01], so the clinician decides to focus on D = [11]
- In any case, both symptoms, a_1 and a_2 are critical for the correct diagnosis
- Following the clinical value of symptoms, both happen to be very severe. This has to be taken into account in order to prescribe appropriate treatment. Therefore, in accordance with Rule 6 of this example:

$$\{RC[(a_1 = 1)] = 1.0000\}\, AND\, \{RC[(a_2 = 1)] = 1.0000\} \xrightarrow{0.923} D = [11]$$

- Imp$(D = [11]) = \min\{1.0000, 1.0000\} \times 0.923 = 1.0000 \times 0.923 = 0.923$
- The Severity Index of D = [11] is around 90%

- In view of these results, the clinician prescribes a treatment, and enjoins the patient to a second visit, which shows that the symptoms have evolved positively from very of severe to severe. Thus:

$$\{RC[(a_1 = 1)] = 0.7500\}\, AND\, \{RC[(a_2 = 1)] = 0.7500\} \xrightarrow{0.923} D = [11]$$

- Imp$(D = [11]) = \min\{0.7500, 0.7500\} \times 0.923 = 0.7500 \times 0.923 = 0.692$
- The Severity Index of D = [11] is around 70%

- Then, the clinician makes adjustments to treatment, and enjoins the patient to a third visit, in which he or she does not observe improvements in the first symptom, but in the second, reduced from severe to moderate. Thus:

$$\{RC[(a_1 = 1)] = 0.7500\}\, AND\, \{RC[(a_2 = 1)] = 0.5000\} \xrightarrow{0.923} D = [11]$$

- I$(D = [11]) = \min\{0.7500, 0.5000\} \times 0.923 = 0.5000 \times 0.923 = 0.462$
- The Severity Index of D = [11] is around 45%

In this way, although the example might seem unrealistic, we intend to illustrate how the model allows to follow the evolution of the pathology of a particular patient in a particular domain.

9.7 Analysis of a Practical Clinical Case

A very common syndrome studied in Sleep Medicine is Apnea Syndrome/hypopnea (SAHS) [13] that affects a range of 3% to 7% of the population, with severe health consequences. Its diagnosis requires analysis of a polysomnography (PSG) [14] in a Sleep Unit of a hospital. Manual analysis of this test is very costly in time and effort to the medical specialist and results in a high economic cost. For a proper diagnosis, analysis of neurophysiological signals related to sleep and analysis of pulmonary signals are performed [15–17]. In our example we will focus on the analysis of pulmonary signals. In order to show how the approach works we will use the data and criteria of the SHHS Case Western Reserve University Program [18]. In this regard, we consider that there are three different types of apneic situations that can be identified in this syndrome:

1. Hypopneas = x_1
2. Central Apneas = x_2
3. Obstructive Apneas = x_3

These situations will be our individual diagnoses. Since there are three different situations, we will have $2^3 = 8$ possible diagnostic complexes. To establish a diagnosis based on respiratory symptoms we can use four key variables, which are:

1. The oxygen desaturation in arterial blood = a_1
2. The thoracic respiratory effort = a_2
3. The abdominal respiratory effort = a_3
4. The respiratory airway flow reduction = a_4

This gives a total of 2^4 complex of symptoms. Therefore, the ELB that, eventually, we could build would consist of $2^{3+4} = 2^7 = 128$ associations of symptom-diagnosis complexes. However, we have said that -not necessarily-, all

Table 9.11 Joint probabilities and total probabilities obtained from a training database with 17 870 real cases epochs

S↓/D→		X_0	X_1	X_2	X_4	Total_probabilities
		[000]	[001]	[010]	[100]	
A_0	[0000]	0.799552				**0.799552**
A_3	[0011]		0.003693			**0.003693**
A_5	[0101]		0.004980			**0.004980**
A_7	[0111]		0.020369			**0.020369**
A_9	[1001]		0.003917	0.000112	0.000783	**0.004812**
A_{11}	[1011]		0.005876		0.000672	**0.006548**
A_{13}	[1101]		0.009737	0.000056	0.001287	**0.011080**
A_{14}	[1110]		0.015725	0.000168	0.000783	**0.016676**
A_{15}	[1111]		0.090935	0.007611	0.033744	**0.132290**
Total_ probabilities		**0.799552**	**0.155232**	**0.007947**	**0.037269**	**1.000000**

Table 9.12 Conditional probabilities obtained from the data of Table 9.11

Conditional_ probabilities		No_event X_0 [000]	Hypopnea X_1 [001]	Central_apnea X_2 [010]	Obstructive_apnea X_4 [100]
A_0	[0000]	1.000000			
A_3	[0011]		1.000000		
A_5	[0101]		1.000000		
A_7	[0111]		1.000000		
A_9	[1001]		0.814007	0.023275	0.162718
A_{11}	[1011]		0.897373		0.102627
A_{13}	[1101]		0.878791	0.005054	0.116155
A_{14}	[1110]		0.942972	0.010074	0.046954
A_{15}	[1111]		0.687391	0.057533	0.255076

the elements of ELB are compatible with the knowledge domain, which, in turn, is necessary to reduce the Expanded Logic Base and turn it into a Reduced Logical Base whose elements do not infringe any of the axioms of the domain. It happens, however, that in this case, we have no such knowledge.

The challenge is therefore to obtain the necessary knowledge. The idea is to build a computational model of intelligent behavior that mimics the way of thinking of the human expert. We will analyze what is known as 'Polysomnographic Register' [14] which can be defined as a collection of signs monitored overnight (7–8 hours of sleep the patient) in 'epochs', that consist of fixed time intervals (typically 30 s duration). In our example, we start from two small databases, the first one for 'training', and the second one for 'testing'.

The training database consists of 15 polysomnographic records from 15 different patients, and includes a total of 17 870 epochs that were scored by experts in sleep medicine, and in which the following associations, which we represent as joint probabilities and total probabilities, were observed (Table 9.11). A detailed explanation of the process of obtaining casuistry can be found in [19].

We are interested now to proceed from symptoms to diagnosis. For this we need the data of Table 9.12, containing the conditional probabilities of the kind $\{P(Xj/Ai)\}$

Now we can obtain the corresponding relations with the antecedent in vector form (three decimals will be more than enough). For example:

$$S = [1101] \rightarrow D = [CENTRAL\ APNEA] \qquad PROBABILITY = 0.005$$
$$S = [1101] \rightarrow D = [OBSTRUCTIVE\ APNEA] \quad PROBABILITY = 0.116$$

With this representation, if for a particular S we have 'n' possible D, then for each of the particular Dj we can define a Normalized Confidence Index, ICN (Dj), as follows:

$$ICN(D_j) = 100 \times Probability(D_j/S)$$

Once we established this index we can construct the following linguistic scale, which is arbitrary and whose elements indicate the plausibility of Dj:

1. CNI (Dj) = 0 → Dj is Impossible
2. 0 < ICN (Dj) ≤ 25 → Dj is very unlikely
3. 25 < ICN (Dj) ≤ 50 → Dj is unlikely
4. 50 < ICN (Dj) < 75 → Dj is Likely
5. 75 ≤ ICN (Dj) < 100 → Dj is very likely
6. ICN (Dj) = 100 → Dj is Confirmed

According to this scheme, one possible rule expressed in natural language, and obtained from the above data in Table 9.12 could be as follows:

RULE [1011]

IF   1. There is reduction in airflow
AND   2. There is no abdominal respiratory effort
AND   3. There is thoracic respiratory effort
AND   4. There is oxygen desaturation in arterial blood

THEN 1. Hypopnea is Very Likely (ICN = 89.7)
AND   2. Central Apnea is Impossible (ICN = 0.0)
AND   3. Obstructive Apnea is Very Unlikely (ICN = 10.2)

To illustrate the consistency and the predictive ability of the proposed approach a simple validation experiment, focusing on the example presented above, was performed. To do this we randomly selected 6 494 epochs corresponding to 5 polysomnographic recordings from 5 different patients, which were scored by experts. All events were accompanied by their associated symptoms, and all of them were represented in the sample. The frequency of apneic events confirmed (Hypopnea, Central Apnea, Obstructive Apnea) was obtained from the corresponding probabilities derived from the training set (Table 9.11) and were expressed in percentage terms. Of all confirmed apneic episodes (Central Apnea, Obstructive Apnea or Hypopnea), the distribution was found to be as follows:

(a) Percentage of Hypopneas = 77%
(b) Percentage of Central Apneas = 4%
(c) Percentage of Obstructive Sleep Apneas = 19%

The knowledge gained from the training set, was applied to the test data set and the results were compared with clinical interpretations obtained from sleep medicine specialists, who were considered as gold standard. Table 9.13 illustrates the results obtained using the data of the test set.

The restrictions, already mentioned, concerning the lack of nuance in the input variables could alter the values of the corresponding ICN, but we have to consider that in this section we only intended to illustrate the problems posed by an example built with real cases, but incomplete. In any case, the results of Table 9.13 seem to

Table 9.13 Results obtained on the test set, considering the clinical expert as the gold standard. $[A_0 \ldots A_{15}]$ = Associations of symptoms. NEP = No_Event predicted by the system. NEC = No_Event identified by the clinician. ANE = Percentage of Total No_Events agreement. HPP = Hypopnea predicted by the system. HPC = Hypopnea identified by the clinician. AHP = Percentage of Total Hyponeas agreement. ACP = Central_Apneas predicted by the system. ACC = Central_Apneas identified by the clinician. AAC = Percent of Total Central_Apneas agreement. AOP = Obstructive_Apneas predicted by the system. AOC = Obstructive_Apneas identified by the clinician. AAO = Percentage of Total Obstructive_Apneas agreement. CPR = Predictive capacity of associations of symptoms. CPE = Predictive capacity for each event. CPSG = predictive capacity of the system as a whole

	NEP	NEC	ANE	HPP	HPC	AHP	ACP	ACC	AAC	AOP	AOC	AAO	CPR
A_0	5255	5255	100	0	0	100	0	0	100	0	0	100	100
A_3	0	0	100	17	17	100	0	0	100	0	0	100	100
A_5	0	0	100	30	30	100	0	0	100	0	0	100	100
A_7	0	0	100	128	128	100	0	0	100	0	0	100	100
A_9	0	0	100	18	17	94	0	0	100	3	4	75	92
A_{11}	0	0	100	32	33	97	0	0	100	4	3	75	93
A_{13}	0	0	100	49	40	82	0	0	100	6	12	50	83
A_{14}	0	0	100	146	147	99	2	0	0	7	8	88	72
A_{15}	0	0	100	550	511	93	46	71	65	204	218	94	88
CPE			100			96			85			87	CPSG = 92

confirm the validity of the analyzed method. It is interesting to notice the predictive capabilities of the rules obtained, directly related to the rules learned from the training set, and the predictive capabilities of the system globally considered in connection with the events listed in the example, depending on the corresponding relative frequencies.

9.8   Discussion

It seems clear that the medical diagnosis is closely linked to the concept of 'inaccuracy' in the broadest sense of the term and nothing pejorative. It also seems clear that this inaccuracy is a direct consequence of the difficulty of the course of medicine. In this regard, any attempt for automation of medical reasoning must be approached under the supervision and advice of the professionals of the medicine, who are the end users of the efforts made by knowledge engineers and specialists in Artificial Intelligence in the development of Intelligent Systems, whose only aim should be to serve as tools to aid decision making. Therefore, it may seem obvious, should be clinicians who determine the true utility of these systems for the diagnosis. And this was precisely the reason why we have not deal with the issues of the Utility Theory in this paper. Another aspect that seems important – again in the context of Medicine – is the choice of the size of the application domain. We will put an example to illustrate why we made this statement: The INTERNIST-1 [20] and CASNET [21] expert systems. INTERNIST-1 is an expert system to aid decision making in the complex field of Internal Medicine, which is an enormous domain, and although INTERNIST-1 proved to be an excellent laboratory for artificial intelligence, the reality was that the practical results obtained were not as expected. By contrast, CASNET, facing specific domain of decision making in the field of glaucoma – a relatively modest and manageable domain – produced more than acceptable practical results.

But we return to the question and return to focus the discussion on the problem of medical diagnosis. In this regard, this work closely approaches what Ledley and Lusted developed in their excellent article, already cited several times. However, that article must be placed at the time it was written and published (1959). Since then, many other approaches have emerged, and the technology available today hardly resembles then. Moreover, many ideas and concepts have changed their perspective. In this context, one of the purposes of our work has been to incorporate issues not treated in the original article, and to adapt the original ideas of Ledley and Lusted to current technology. That said, throughout our work we have also tried to formalize an alternative to the problem of 'inaccuracy'. In this context we have tried to identify some of the possible causes of the occurrence of such inaccuracy. Some of these causes could be:

1. The obvious heuristic component of medical knowledge.
2. The inherent subjectivity of interpretations in Medicine.
3. The lack of knowledge.
4. The way of representing knowledge.
5. The imprecision associated symptoms and diagnoses.
6. The uncertainty associated with causal relationships.
7. The problem of the temporal monitoring of the evolution of patients.

In any case we have not tried to definitively solve all the problems, obviously because the incredibly fast advances in technology, artificial intelligence and medical knowledge, would made the task of questionable utility and -of course- impossible. However, what we have done is a deep thinking on the above problems inherent in the medical reasoning, and have suggested some approaches to try to alleviate as much as possible these problems.

9.9   Conclusions

It is evident that the heuristic knowledge produces excellent results and is indispensable in medical diagnosis, but by definition is not a standard knowledge. This fact, together with the subjectivity of the interpretations and the possible lack of knowledge about the domain, was the reason why we developed an alternative version, though pretty close, to the categorical model proposed by Ledley and Lusted. Perhaps the more important difference has been to provide the procedural knowledge with a computational structure which is from our view point of natural sight. Moreover, the subjectivity of the interpretations and the possible lack of knowledge are approached from the perspective of Bayes, as in the proposal Ledley and Lusted but unlike them, the conditional probabilities are now used to construct rules. The representation of these rules can be improved by defining a vector model, by using the binary numbering system, and optimizing the process by incorporating a phase for encoding, which facilitates the handling of information. Finally the treatment of 'inaccuracy' a term in which we include, among others, concepts like 'imprecision', 'uncertainty', 'confidence', 'Plausibility' and 'time-tracking', is approached from different points of view, being -perhaps- the most notable aspect the treatment of imprecision as a form of propagation of uncertainty. This last point is based on the certainty factor model of Shortliffe and Buchanan.

Acknowledgements This work has-been supported by Spanish regional government of Galicia (Groups of Excellence Program GRC2014/035) and by Spanish MINECO (Research Project TIN2013-40686), co-funded by European ERDF. Special thanks to Case Western Reserve University for permitting us to use Their Sleep Heart Health Study database.

References

1. Ledley, R.S., Lusted, L.B.: Reasoning foundations of medical diagnosis. Science **130**(3366), 9–21 (1959)
2. Ledley, R.S.: Mathematical foundations and computational methods for a digital logic machine. J. Oper. Res. Soc. Am. **2**(3), 249–274 (1954)
3. Uspensky, J.V.: Introduction to Mathematical Probability. McGraw-Hill, New York (1937)
4. Neumann, J. von, Morgenstern, O.: Theory of Games and Economic Behavior. Princeton University Press, Princeton (2007)
5. Gigerenzer, G., Gaissmaier, W.: Heuristic decision making. Annu. Rev. Psychol. **62**, 451–482 (2011)
6. Scott Richardson, W., Glasziou, P., Polashensky, W.A., Wilson, M.C.: A new arrival: Evidence about differential diagnosis. Evid. Based Med., **5**, pp. 164–165 (2000)
7. Shortliffe, E.H., Buchanan, B.G., Feigenbaum, E.A.: Knowledge engineering for medical decision making: A review of computer-based clinical decision aids. Proceedings of the IEEE **67**(9), 1207–1224 (1979)
8. Nilsson, N.: The Quest for Artificial Intelligence: A History of Ideas and Achievements. Cambridge University Press, New York (2009)
9. Wagner, C.: Breaking the knowledge acquisition bottleneck through conversational knowledge management. Infor. Res. Manage. J. **19**(1), 70–83 (2006)
10. Puga, J.L., Kzrywinski, M., Altman, N.: Points of significance: Bayes' theorem. Nat. Methods **12**, 277–278 (2015)
11. Available on Internet. https://support.office.com/es-es/article/Novedades-de-Microsoft-Office-Excel-2007-bcbc55a7-7827-4a01-a872-52c6df64982f
12. Shortliffe, E., Buchanan, B.G.: A model of inexact reasoning in medicine. Math. Biosci. **23**, 351–379 (1975)
13. Alvarez-Estévez, D., Moret-Bonillo, V.: Computer-assisted diagnosis of the sleep Apnea-Hypopnea syndrome: A review. Sleep Disord. **2015**, 1–33 (2015)
14. Berry, R., et al.: The ASSM Manual for the Scoring of Sleep and Associated Events: Rules, Terminology and Technical Specifications. American Academy of Sleep Medicine, USA (2012)
15. Fernández-Varela, I., Alvarez-Estévez, D., Hernández-Pereira, E., Moret-Bonillo, V.: A simple and robust method for the automatic scoring of EEG arousals in polysomnographic recordings. Computers in Biology and Medicine **87**, 77–86 (2017)
16. Fernández-Varela, I., Hernández-Pereira, E., Alvarez-Estévez, D., Moret-Bonillo, V.: Combining machine learning models for the automatic detection of EEG arousals. Neurocomputing 2017, (2017)
17. Hernández-Pereira, E., et al.: A comparison of performance of K-complex classification methods using feature selection. Information Sciences **328**, 1–14 (2016)
18. Quan, S.F., Howard, B.V., Iber, C., et al.: The sleep heart health study: Design, rationale, and methods. Sleep **20**, 1077–1085 (1998)
19. Alvarez-Estévez, D., Moret-Bonillo, V.: Fuzzy reasoning used to detect apneic events in the sleep Apnea-Hypopnea syndrome. Expert Systems with Applications **36**, 7778–7785 (2009)
20. Miller, R.A., Pople, H.E., Myers, J.D.: INTERNIST-1, An experimental computer-based diagnostic consultant for general internal medicine. The New England Journal of Medicine **307**, 468–476 (1982)
21. Weiss, S.M., Kulikowski, C.A., Amarel, S., Safir, A.: A model-based method for computer-aided medical decision-making. Artificial Intelligence **11**, 11145–11172 (1978)

Author Biographies

Vicente Moret-Bonillo (Valencia, Spain, 1962). Degree in Chemistry (B.Sc. Fundamental Chemistry, Major in Physical Chemistry, University of Santiago de Compostela, Spain, 1984). Doctor in Physics (Ph.D. Applied Physics, University of Santiago de Compostela, Spain, 1988, Cum Laude). Post-Doctoral Research Fellow Biomedical Engineering Department (Medical College of Georgia, USA, 1988–1990). Senior Member IEEE (Biomedical Engineering, since 2006). Award of Merit for Significant Contribution in the Field of Clinical Engineering (Quest Publishing Eds, USA, 1990). Principal investigator of more than 30 projects funded on a competitive basis. Director 9 doctoral theses related to various aspects of Artificial Intelligence. Author of more than 145 scientific publications in international journals and congresses (JCR, Scopus, PubMed...). Recognized specialist in Medical Applications of Artificial Intelligence, Intelligent Monitoring Medicine, Imprecise Reasoning Models in Medicine and Intelligent Systems Validation. Areas of current interest: Intelligent Monitoring of Sleep Apnea Syndrome and Quantum Computing.

Isaac Fernández-Varela (Ferrol, Spain, 1988). Degree in Computer Science (University of A Coruña, Spain, 2012). Master in High Performance Computing (University of A Coruña, Spain, 2013). Ph.D. Student in the Laboratory for Research and Development in Artificial Intelligence, Department of Computer Science, University of A Coruña, Spain. Areas of current interest: Machine Learning and Signal Processing.

Elena Hernández-Pereira graduated in computer science from the University of A Coruña (Spain) in 1996. She had a predoctoral fellowship from 1998 to 1999 and a postdoctoral fellowship from 1999 to 2001 in the same university. In 2000, she received her Ph.D. degree working on the area of the application of Artificial Intelligent techniques to sleep apnea diagnosis. She is currently an Associate professor in the Computer Science Department, University of A Coruña. Hers main current research areas are: Machine Learning, Signal Processing and Medical Decision Support Systems. She has published, 14 articles of international scope journals, 10 of them indexed in the JCR and a book chapter. She has also 30 contributions to international conferences and participated in 26 research projects. She has participated in review activities of scientific articles.

Diego Álvarez-Estévez (Ourense, Spain, 1982). Degree in Computer Science Engineering (University of A Coruña, Spain, 2007). Doctor in Computer Sciences and Artificial Intelligence (University of A Coruña, Spain, 2012) with Ph.D. on the application of machine learning and approximate reasoning techniques to the diagnosis of the Sleep Apnea-Hypopnea Syndrome, and awarded with the Special Doctorate Prize. Since 2013 working as Chief Engineer at the Sleep Center and Clinical Neurophysiology department, Medisch Centrum Haaglanden and Bronovo-Nebo, The Hague, The Netherlands. Author of several JCR publications on biomedical signal processing and intelligent data analysis on the field of sleep studies.

Volker Perlitz (Born in 1960, Germany). Doctorate in experimental studies in cardiophysiology (Summa cum Laude, 1984–1986). Post-doctoral fellow Medical College of Georgia, Department of Physiology and Endocrinology, Georgia-USA (1988–1991). Medical resident and researcher university clinic RWTH Aachen, outpatient department for Psychosomatic and Psychoterapeutic Medicine (1991–2008). Specialist for Psychosomatic Medicine (2003). Department for General, Visceral and Transplantation Surgery and Brest center medical facilities RWTH Aachen, Germany (2008). Currently, CEO Simplana GmbH, Aachen, Germany (2015-Present). Areas of Scientific Interest: Nonlinear signal analysis of the Autonomic Nervous System, Bio-Psycho-Social Medicine, Electronic Psychometry, Rehabilitation of Chronic Traumatized Patients.

Chapter 10
Vital Signs Telemonitoring by Using Smart Body Area Networks, Mobile Devices and Advanced Signal Processing

Hariton Costin and Cristian Rotariu

Abstract In the last years the demographic changes and ageing of population increase health care demand. The growing number of chronic patients and elders requires close attention to their health conditions. In this paper we present the realization of a wireless remote monitoring system, based on body area network of medical sensors, capable to measure process and transmit patient's physiologic signals (electrocardiogram, respiratory rhythm, saturation of arterial oxygen, blood pressure and body temperature) to a central medical server. The use of system is suitable for continuous long-time monitoring, as a part of a diagnostic procedure or can achieve medical assistance of a chronic condition. We used custom developed and commercially available devices, low power microcontrollers and RF transceivers that perform measurements and transmit the corresponding numerical values to the Personal Server. The Personal Server, in the form of a personal digital assistant (PDA) or a smartphone, runs a monitor application, receives the physiological signals from the monitoring devices, activates the alarms when the monitored parameters exceed the preset limits, and communicates periodically to the central server by using WiFi or GSM/GPRS connections. The application programs are complemented by automatic atrial fibrillation detection through some advanced signal processing techniques, as well as by patient localization on different maps by means of GPS methodology. Thus, by adding capability of automatic detection of certain heart diseases (for instance), the telemonitoring process becomes "smarter" and more efficient than the traditional approaches based on visualization of raw data.

H. Costin (✉) · C. Rotariu
Faculty of Medical Bioengineering, Grigore T, Popa University of
Medicine and Pharmacy, Iasi, Romania
e-mail: hariton.costin@umfiasi.ro

C. Rotariu
e-mail: cristian.rotariu@umfiasi.ro

H. Costin
Institute of Computer Science of Romanian Academy, Iași Branch, Romania

© Springer International Publishing AG 2018
D.E. Holmes and L.C. Jain (eds.), *Advances in Biomedical Informatics*,
Intelligent Systems Reference Library 137,
https://doi.org/10.1007/978-3-319-67513-8_10

219

Keywords Advanced signal processing · Body area networks · Mobile devices ·
Vital signs telemonitoring · GPS localization · Wireless sensor networks

10.1 Introduction

Telemedicine is part of the expanding use of information and communications
technology (ICT) in health care being used in prevention, disease management,
home health care, long-term care, emergency medicine, gerontology, and other
applications. Nowadays, the diversification of telecommunication networks and
advances in communications technologies, including the Internet, has considerable
potential as a medium for telemedicine applications. In this respect, telemonitoring
and teleconsulting of chronic patients and/or elderly people are priority issues
within modern medicine, confronted with substantial demographic changes.

In the last years, the steady advances in mobile computing, integrated circuits
technology, wireless sensor networks, and medical devices, have opened the way to
miniature, low cost, low power, and multi-functional intelligent monitoring devices,
suitable for many portable medical applications.

Wearable body area networks (WBAN) based on wireless and embedded
monitoring devices allow to continuously monitoring vital physiologic parameters
and provide feedback to help maintain an optimal health status. These networks
make revolutionary changes in health care by allowing continuously long term,
non-invasive, pervasive and ambulatory monitoring of vital signs.

In healthcare domain a numerous integrated systems for remote physiological
signals are developed. In the area of wearable medical systems, the research groups
and commercial vendors has been mainly focused on developing devices for
noninvasive monitoring of physiological signals like the electrocardiogram, blood
pressure, arterial blood oxygen saturation, respiratory rate, or body temperature.

This chapter reviews some of the key aspects of WBANs for medical applications, shows the state of the art in this field and describes the design and implementation issues of the authors' integrated system for remote monitoring of vital
signs by using a smart body area network, smartphones and Internet. The hardware
and software solutions, the power management of wireless devices integrated in
WBAN, and the on-chip bio-signal processing software are also discussed in detail.

Patient monitoring refers to the continuous observation of repeating events of
physiologic function to guide therapy or to monitor the effectiveness of interventions. Historically, these medical instruments are designed to be used by highly
trained personnel, in the intensive care units and operating rooms of hospitals, and
thus such instruments are not applicable for home monitoring.

Successful trauma management requires accurate monitoring of several important physiological parameters, so that proper action can be taken to help maintain
critical functionality.

The main vital signs routinely monitored by medical professionals and health-care providers include the following: heart rate, respiratory rhythm, saturation of oxygen in arterial blood, pulse rate, blood pressure (not considered a true vital sign, but it is measured with other vital signs), and body temperature. Vital signs are useful in detecting or monitoring medical problems and can be measured at hos-pital, at home, at the site of a medical emergency, or elsewhere. Over time, patient monitoring has moved from invasive to non-invasive measurements, which decrease patient's risk of infection.

Recent advances in integrated circuits, wireless communications and physio-logical sensing open the way to miniature, lightweight, low power and intelligent monitoring devices.

Wireless devices had recently a significant development being considered one of the most important technologies of the XXI century. This is due, first of all, to the fact that the mode of transmission (radio waves) completely eliminates data link cables.

Through wireless connectivity it is achieved infrastructure and mobile support for real-time monitoring of the patient regardless of patient location, and tracking system for responding in case of alarm (emergency). Sampling, processing and extracting information from biomedical signals is achieved by appropriate methods, based on performance electronic devices, including computers focused mainly to signal processing.

Wireless monitoring represents a medical practice that involves remotely mon-itoring patients who are not at the same location as the health care provider. Generally, a patient has a number of monitoring devices at home, and the moni-toring results are transmitted to the database server located at the monitoring center. There are several categories of patients who may need continuous monitoring or intensive care. Chronic patients, recovering from serious illness, are often placed in special units, where vital signs can be watched constantly by the use of classic monitoring devices. The computer-assisted measurements involve unwieldy wires between sensors and monitoring devices that are not very comfortable for normal activity [2, 34]. In order to avoid this situation, we used wireless devices, based on low power microcontrollers, having radio micro-transmitters and allowing an autonomic movement of the subject.

Healthcare specialists can receive information regarding patient's condition that has a longer time span than a patient's normal stay in a hospital and this information has great long-term effects on home health care, including reduced expenses for health care. Physicians also have more accessibility to the healthcare specialists, allowing the physician to obtain information on diseases and provide the best health care available. Moreover, patients can thus save time, money and comfort.

Internationally, telemonitoring represents a viable method whose effectiveness has been proven by numerous studies, some of which are still ongoing. Currently there are numerous telemonitoring systems with different complexities. Such sys-tems are deployed mostly in economically and technologically advanced countries because they have the necessary economic and technological potential and, perhaps more importantly, such systems require an increasingly higher percentage of the population with requiring such care.

In the U.S. and Europe there are substantially funded telemonitoring achievements [10, 23,35]. Of these, with a specific complexity needed by the application itself, we can mention the following European projects (within FP5, FP6 and FP7 programs): EPI-MEDICS (for detecting cardiac arrhythmias) [38, 51, 52], AMON (monitoring of vital parameters) [3, 32], MobiHealth (telemonitoring system based on sensor networks), HEALTHSERVICE24 (teleconsulting mobile systems based on computers PDA) [27], INCA (intelligent assisting diabetics), TELECARE (remote monitoring of vital signs using sensors with the help of mobile telephony), THALEA (telemedicine for ICUs). In the U.S., Code Blue (conducted at Harvard University) is a reference project [14, 25, 28, 33].

In Romania there are concerns for the development of the telemonitoring application from research units in Bucharest, Iasi, Timisoara and Cluj. As examples of national telemonitoring systems we can mention: CardioNet ("Integrated system for continuous monitoring in intelligent e-Health network of patients with cardiac disease") [4, 9], BIOMED-TEL ("Biomedical signal acquisition and tele-transmission through mobile equipment") [6], MEDCARE ("cardiac activity monitoring system for the acquisition, transmission and analysis by Internet signal ECG in real time") [16], TELEASIS ("complex system, the support NGN—Next Generation Networking—Telephone helpline for the elderly at home") [54], TELMES ("multimedia platform for complex medical teleservices") [40, 17, 18] and TELEMON ("Integrated Real time telemonitoring of patients and elderly people") [19, 41, 13], http://www.bioinginerie.ro/Cercetare/Contractedecercetare/Telemon%28English%29.aspx, the last three references involving directly the authors of this chapter.

In last years m-health applications related with the use of in-built smartphone sensors (e.g. phone cameras, accelerometers, etc.) have been developed, and already they have millions of users all over the world. For instance, some Android OS and iOS applications for vital signs monitoring and care are as follows.

Application name	Developed by	Function	URL
Instant Heart Rate	Azumio Inc.	Heart rate meter	https://www.azumio.com
Cardiograph	MacroPinch Ltd	Heart rate meter	http://macropinch.com/cardiograph
Handy Logs Heart	HandyLogs	Heart rate meter	http://www.handylogs.com
iBP	Leading Edge Apps LLC	Blood pressure meter	http://leadingedgeapps.com/iBP.html
Breath Biofeedback	Android Research	Respiration biofeedback	http://www.androidresearch.net/breathbiofeedback.html

(continued)

(continued)

Application name	Developed by	Function	URL
Breath Pacer Lite	Android Research	Respiration meter	https://play.google.com/store/apps/details? id=net.androidresearch. BreathPacerLite&hl=en

Every telemonitoring system shown above has its own strengths and weaknesses, due both to the used technology by that time, and to the limited capabilities of the operating and applications software, as well as to the available communications technologies. They can be judged only in relation with the designed technical and economical parameters, as a single telemonitoring system may not have all features at the highest level [5].

However, in this respect we think that the use value of such systems might be increased by adding facilities offered by advanced (bio)signal processing methods, as well as by decision support modules that make use of artificial intelligence paradigms and techniques to implement human medical knowledge for a better decision in health care. This is why our approach was to develop telemonitoring tools for all vital signs encountered in emergency medicine, and in the same time to deploy and fuse the numerical and linguistic information within those complex applications that need such a hybridization of information. In fact, this is the design feature that differentiate our approach from other similar achievements.

10.2 Physiologic Monitored Parameters

Specialized literature proposes for monitoring several physiological parameters, which vary depending on choice of pathologies, technical and financial possibilities [26, 7, 59, 36]. Only the signals considered relevant to common disorders will be taken into account. The term "vital physiological parameters" refers to measuring the heart rate, respiratory rate, blood pressure, body temperature and blood oxygen saturation [56, 15].

Heart rate (HR) is a vital physiological parameter often used in remote monitoring because it allows a rapid assessment of the condition of the patient, cardiac arrhythmias allow prompt recording and variations can be easily differentiated, this giving indications on the cardiovascular function [45, 47, 24]. Usually, the pulse is taken with automatic devices using photoelectric plethysmography to measure the peripheral arteries, but can also be determined through the electrical activity of the heart by automatic analysis of electrocardiographic (ECG) signals [44, 57]. It is also possible to assess the effects of psychological stress on human body using ECG signals analysis [20, 58].

Respiratory rate (RR) One of the vital signs that should be closely monitored is breathing, especially respiratory frequency and its fluctuations [46]. Usually this parameter is shallowly watched, as a doctor who evaluates the breath at a certain moment of time may find a rate of 20 breaths per minute, regardless its evolution in time. However, this parameter can be manifested in various ways, so that at one time it may appear to be normal, but at the immediately following moments it may change its characteristics. In human breath can be seen a series of patterns during its evolution, and they are important and effective to diagnose various diseases. Every breath may differ from the previous one as well as period, amplitude or duration. Respiratory disorders during sleep can induce dysfunctions or diseases that may manifest during the day. Episodes of apnea, hipopnea, partial or total obstruction of the airways, as snoring, lead to a decrease in oxygen saturation of hemoglobin. Thus, these disorders can be associated with tiredness, morning headaches, excessive daytime sleepiness, loss of efficiency of work, sexual dysfunction or even with traffic accidents. According to current studies, some respiratory disorders may be a projection of some disorders of nervous system, specifically the vagus nerve, spinal nucleus or even sinoatrial node (during inspiration, vagus nerve activity is attenuated, leading to an increase heart rate; during expiration, vagus nerve activity is increased, leading to lower heart rate). Also, recent studies show the relationship between respiratory activity and emotional activities. Considering these aspects, it is obvious the need to study more closely this vital parameter, which requires a long-term monitoring in various situations of daily life of the patients. This, as well as current technological advances, have led to the development of miniature portable devices, that do not compromise convenience and freedom of movement of the patient, enhancing thus their quality of life and in case of emergency, to be able of making decisions and to alert the medical staff.

Oxygen saturation (SpO$_2$) is defined as the ratio of oxyhemoglobin to the total concentration of hemoglobin. Along with heart rate, blood pressure, body temperature, and breathing rate, the SpO$_2$ is an important vital parameter, used for detection of hypoxemia. Optimal hemoglobin in O$_2$ is defined by the SpO$_2$ values ranging between 94–100%, slight hypoxemia by saturation values of 88–93%, average hypoxemia by saturation values of 83–88%, and severe hypoxemia by values less than 83%. From medical point of view it is considered that the decrease of the patient's SpO$_2$ below 93% must be followed immediately by compensatory measures.

Blood pressure (BP) is a physiological parameter frequently evaluated and used in monitoring, as sampling is easy, non-invasive, and provides relevant data on cardiac activity. Moreover, this recording allows accurate diagnosis of the type of hypertension, antihypertensive medication titration and control of hypotension with medication, which is useful even in conditions such as preeclampsia, diabetes and / or coronary artery disease. On the market there are numerous devices for measuring blood pressure, many of which can be used at home and can be easily integrated into complex system for telemonitoring. Also known as hypertension, high blood pressure is a growing concern in our society. According to recent studies the number of adults with high blood pressure has increased in the last years. This fact is largely due to an overweight and aging population and leads to an increased risk of stroke, heart attack, heart failure and kidney failure. While the primary cause of hypertension is unknown, high blood pressure is easy to detect, monitor and control. Many people with high blood pressure measure their blood pressure at home, in between visits to their doctor or nurse. Some people may also be asked by their doctor or nurse to take measurements at home for a short period of time to find out whether they have high blood pressure or not.

Body temperature is commonly measured by the patients at home, especially in case of fever or hypothermic state. However, it is rarely measured, and patients do not always know when it is the ideal time to evaluate this parameter. Abnormal body temperature may be affected or disturbed by infection, trauma, tumors, thyroid disease, stroke, autoimmune diseases, metabolic diseases, or exposure to cold or heat. Fever, identified by increased body temperature above the normal range, is a defense mechanism by activating the body's immune system against such variable aggression (e.g.: against infectious pathogens).

10.3   A Concrete Implementation: TELEMON Project

This is a project managed by the authors of this paper and refers to secure acquisition, transmission, analysis and archiving of biomedical vital signs, images and laboratory analyses in order to enhance the telemedical consultancy services. The main objective of the system, named TELEMON, is to enable personalized telemedical services delivery and patient safety enhancement based on (almost) real time electrophysiological signals acquisition and on-line or off-line bio-signal analysis and telediagnosis. The system allows persons having different (chronic) diseases and to elderly/lonely people to be monitored from medical point of view. In this way the medical risks and accidents are significantly diminished. The system acts as a pilot telemedical center, destined to the implementation of a public e-health service, "everywhere and every time (24/7)", in real time, for people being in different hospitals, at home, at work, during the holidays, on the street, etc.

The patients may be located in a limited area or they may be mobile, and a GPS and map-based localization of their position allow immediate intervention of emergency units, when needed.

As for chosen bio-signals, as discussed in previous section, continuous long-time cardiac activity monitoring through *ECG signal analysis* is an important issue. As a part of a diagnostic procedure, it is possible to achieve medical assistance of a chronic condition, or the patient can be supervised during recovery from an acute event or surgical procedure. For instance, we have approached the detection and prediction of atrial fibrillation—an insidious but dangerous cardiac disease. Also, we have used heart rate and morphologic variability of the ECG signal as effective tools for stress detection and classification.

Another vital sign to be monitored is the *respiration*, which is important for a number of medical conditions including rhythm analysis, polysomnography, stress monitoring and ischemic heart disease or heart failure.

The number of deaths from heart conditions caused by high *blood pressure* continues to increase. Early detection and treatment of high blood pressure is critical to prevent future heart problems, and a variety of home monitoring devices are available for this purpose.

A pulsoximeter sensor can be used for *heart-rate* detection and *blood oxygenation* (SpO$_2$) measurement, another two vital signs.

The *body temperature module* allows continuous telemonitoring of this parameter.

A *fall detection module* alerts at a remote site that a monitored people has fallen down and also may monitor continuously the person's activity. The user's activity monitoring allows the characterization of users' mobility for diagnosis purposes.

The TELEMON system is built around a database server, which receives data from *local subsystems* and also from *mobile subsystems*. The transferred information to database server are represented by those data above the limits (eventually alarms), by medical recordings themselves and also by audio/video recordings. The database server stores the recordings and a human operator sends alarms to the ambulance service and patient's doctor. Also, the database server can be connected to another database server, for example a hospital server, in order to send the patient's medical data. The subsystems are connected to the database server through an Internet connection (if it is available) or through GSM/GPRS connection.

The *Local Subsystem* for the monitoring of the patient at home or in hospital may be built around a personal computer (PC) or a smartphone, which wirelessly receive data from the patient.

A *personal server* is made by means of a PDA (Personal Digital Assistant) or a smartphone running a personal health monitor that is responsible for a number of tasks and provides a transparent interface to the wireless medical sensors, an interface to the user, and an interface to the central server by using the GSM connection. If the patient is in a bad condition and he needs an emergency medical assistance, the PDA can send an alarm to the central database server, including the precise coordinates of the patient by using the internal GPS module. A special application allows the precise positioning of patient on an appropriate map.

The *software applications* were developed for:

1. medical data acquisition and processing, and data transmission to the central server, for fixed or mobile patient;
2. interface programs for communication between local sub-system and central server;
3. specific alarms generation and their transmission to the Ambulance, family doctor, to the specialist or to a nurse/paramedic;
4. the electronic health record (EHR) for patients and medical staff involved;
5. specific applications of e-rehabilitation and e-learning;
6. metadata processing programs, at the central server level (medical statistics, researches, etc.);
7. decision aided/expert module, based on medical specific knowledge, e.g. for ECG automatic analysis.

The *wireless body area network* (WBAN) is composed of the different types of devices for continuous measurement of the following physiological signals: electrocardiographic signal (ECG), heart rate (HR), respiratory rhythm (RR), oxygen saturation (SpO$_2$), arterial blood pressure (BP) and body temperature (BT).

The proposed system architecture shown in Fig. 10.1 [11, 43] is centered on a medical database server, which is designed to service registered users and it is connected to medical personnel and healthcare professionals.

The user must wear a WBAN of devices and a *personal digital assistant* (PDA, usually a smartphone) attached to the body. Each wireless device samples the physiologic signal and transfers the raw data to the PDA by using a network protocol (IEEE 802.15.4). The PDA performs the WBAN configuration and management (acts as a coordinator), receives the raw signals from devices, and, after a local signal processing according to the specific monitored feature, the processed data are transmitted periodically (at a preset interval of time), via Internet, to the medical database server. If one or more of the monitored parameters are above the preset limits, an alarm is sent in short time to the monitoring center.

The medical database server stores the medical records of registered users and provides services to the physicians, healthcare specialists or ambulance centers.

The WBAN is built by using custom developed and available commercially devices for physiologic signals measurement and a low power microcontroller board (evaluation module eZ430-RF2500 from Texas Instruments) for wireless communication.

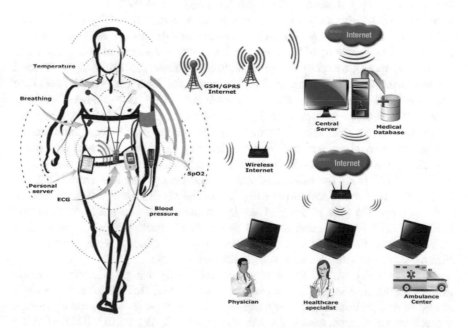

Fig. 10.1 Integrated system for wireless monitoring—network architecture

The eZ430-RF2500 uses the MSP430F2274 16 bit microcontroller, providing all the hardware and software for the MSP430F2274 microcontroller, and Chipcon CC2500 2.4 GHz wireless transceiver, designed for low-power wireless applications. The eZ430-RF2500 is a complete MSP430 wireless development tool providing all the hardware and software for the MSP430F2274 microcontroller and CC2500 2.4 GHz wireless transceiver. Operating on the 2.4 GHz unlicensed industrial, scientific and medical (ISM) bands, the CC2500 provides extensive hardware support for packet handling, data buffering, burst transmissions, authentication, clear channel assessment and link quality. The radio transceiver is interfaced to the MSP430 microcontroller using the serial peripheral interface. The RF transceiver is integrated with a highly configurable baseband modem. The modem supports various modulation formats and has a configurable data rate up to 500 kBaud.

Based on the system architecture illustrated in Fig. 10.1 we have developed a prototype of a WBAN of devices for physiological signal measurement, represented in Fig. 10.2 [43].

The ECG signals were acquired by using a 3-lead custom developed ECG amplifier. The amplifier has for each channel a high gain (500), a high input impedance (>10 MΩ) and common mode rejection (>90 dB), a cut-off frequency around 35 Hz and is AC coupled [13, 41].

For SpO$_2$ and HR measurements we used a commercially available Micro Power Oximeter Board (from Smiths Medical), connected as in Fig. 10.3 [1, 8].

The pulse oximeter is used to collect the SpO$_2$ and HR from the patient and has the following specifications: measurement range of 0–99% SpO$_2$ with ±2% accuracy for 70–99% SpO$_2$, and pulse rate measurement range of 30–254 bpm with ±2 bpm or ±2% accuracy. The pulse oximeter determines the SpO$_2$ and HR by passing two wavelengths of light, one in the visible red spectrum and the other in the infrared spectrum, through body tissue to a photodetector. The SpO$_2$ is obtained by measuring the absorbance of light by blood for each wavelength, and then computing the ratio between these two intensities. The pulse oximeter's microcontroller processes the received light intensities, separates the time invariant parameters (tissue thickness, light intensity, or venous blood) from time variant parameters (arterial volume and SpO$_2$) to identify the signal produced by arterial pulsations, and computes the SpO$_2$ and HR. The pulse oximeter is connected to the MSP430F2274 from the eZ430RF2500 module through an asynchronous serial interface as it is represented in the Fig. 10.3. The serial interface uses microcontroller's serial I/O pins at 3.3 V. Data transmission between pulse oximeter and MSP430F2274 is performed at 4800 bps with 60 packages per second. Each sampled data has 4 byte packets length and includes SpO$_2$ level, HR, Plethysmographic (PPG) signal, and Status bits [45, 24].

Respiratory rhythm measurement was performed by using a piezoelectric respiration transducer. The respiration transducer used, PNEUMOTRACE (from UFI), placed around the thorax generates a high-level linear signal in response to changes in thoracic circumference, associated with respiration [12, 49]. PNEUMOTRACE is a

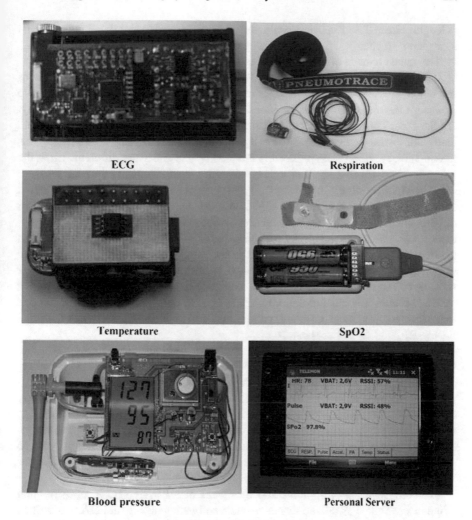

ECG   Respiration

Temperature   SpO2

Blood pressure   Personal Server

Fig. 10.2 The wireless body area network of devices

Fig. 10.3 Micro Power Oximeter Board connected to eZ430RF2500

sturdy piezoelectric respiration transducer, which requires no excitation voltage, the
output voltage level being form 20 to 200 mV into a 1 MΩ load and it has a 122 cm
hook-and-loop strap provided to link ends of transducer package, which has 280 mm

in length. The belt is placed around the thorax and generates a high-level, linear signal in response to changes in thoracic circumference, associated with respiration [50].

The measurement of the blood pressure was performed by using UA-767PC blood pressure monitor (BPM—from A&D) [42]. The method of measurement of the blood pressure is the oscillometric method. It is often used in the automatic device for the measurement of the blood pressure because of its excellent reliability. On the other hand, it is less precise than the auscultatory (Korotkoff) method, but many automatic devices measuring the blood pressure during 24/48 h use this measurement method. The pulsations induced by the artery are different: when the artery is compressed, no pulsation is received by the BPM, then when the pressure decreases in the cuff, the artery starts to emit pulsations: the pressure then measured on the device defines the maximal blood pressure or systolic blood pressure. During the pressure decrease in the cuff, the oscillations will become increasingly significant, until maximum amplitude of these oscillations defines the average blood pressure [29]. The BPM takes simultaneous blood pressure and pulse rate measurements and has the following technical specifications: measurement range for BP: 20–280 mmHg, HR: 40–200 bpm, accuracy measurement for BP: ±3 mmHg or 2% (whichever is greater) and HR: ±5%. It includes a serial port connection that facilitates bi-directional communication at 9600 kbps and is connected to the EZ430-RF2500 as in Fig. 10.4.

For the body temperature measurements we used the TMP275 temperature sensor (from Texas Instruments) directly connected to the eZ430-RF2500 by using the I2C bus [48]. The TMP275 is a 0.5 °C accurate, two-wire, serial output temperature sensor available in an SO8 package. The TMP275 is capable of reading temperatures with a resolution of 0.0625 °C. The TMP275 is directly connected to the eZ430-RF2500 using the I2C bus and requires no external components for operation except for pull-up resistors on SCL and SDA. The accuracy for the 35 −45 °C interval is below 0.2 °C and the conversion time for 12 data bits is 220 ms typical.

Personal Server was implemented by means of a PDA (HTC X7500) running Windows Mobile 5 as operating system. The software working on PDA was written by using C# from Microsoft Visual Studio 2008 and it is represented in Fig. 10.5 [43]. The software displays temporal waveform of physiologic signals, numerical

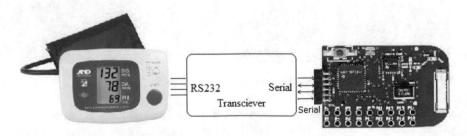

Fig. 10.4 The UA-767PC blood pressure monitor and the ez430-RF2500

values for monitored parameters and the status of the device (the battery voltage and distance of the device from the PDA). The distance is represented in percent of 100 computed based on RSSI (received signal strength indication measured on the power present in a received radio signal).

The USB interface is realized by using a serial to USB transceiver (FT232BL from Future Technology Devices International) and enables the eZ430-RF2500 to remotely send and receive data through USB connection using the MSP430 Application UART. It also contains a low dropout voltage regulator (TPS77301 from Texas Instruments) to provide 3.3 V to the eZ430-RF2500.

Wireless data transmission between the devices and PDA may be performed by using Bluetooth or WiFi standard, but they are more expensive, consume more power and are useful for applications that require high bandwidth. As a limitation Bluetooth allows only a limited number of nodes to communicate. The wireless protocol used to transfer data from device to PDA was SimpliciTi (from Texas Instruments). SimpliciTi is an open-source protocol for networks that typically contain battery operated devices that require long battery life and low data rate.

The software running on the PDA receives the real-time user raw data from wireless devices and performs a local signal processing in order to detect anomalies. When an anomaly is detected in the measured physiological parameters, the software application sends an alarm to the medical database server.

Table 10.1 presents the default physiological conditions that can cause alarms. These values are used by the alarm routine running on PDA. The healthcare specialist may modify these values according to the user's age or weight.

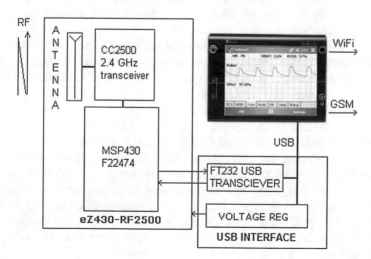

Fig. 10.5 The Personal Server (PDA)

Table 10.1 Default physiological conditions for alerts

Type of alarm	Physiological condition
Low oxygen saturation	$SpO_2 < 90\%$
Heart bradycardia	HR < 40 bpm
Heart tachycardia	HR > 120 bpm
Low respiratory rate	RR < 10 rpm
High respiratory rate	RR > 30 rpm
Blood pressure hypotension	Systolic BP < 90 mmHg or diastolic BP < 60 mmHg
Blood pressure hypertension	Systolic BP > 160 mmHg or diastolic BP > 100 mmHg
Low body temperature	BT < 35 °C
High body temperature	BT > 38 °C

10.4 ECG Signal Processing for Arrhythmia Detection

Atrial fibrillation (AF) is the most common chronic cardiac arrhythmia, resulted from abnormal electrical activity in the heart, affecting about 0.4 to 1.0% of the entire population. Its prevalence increases with advancing age and can reach 10% on the population older than 80. Moreover, the atrial fibrillation increases the mortality rate and risk of stroke. Several risk factors for atrial fibrillation include high blood pressure, myocardial infarction or congestive heart failure [21].

The normal sinus rhythm for most people is between 60 and 100 beats per minute (bpm). In atrial fibrillation episodes, electrical pulses from the atrium to the ventricles are irregular, resulting heart rates between 400 and 800 bpm. The irregularity of the RR intervals (RRI) can be used for the detection of atrial fibrillation episodes [22].

The process of detections the heart beats constitutes a significant part of the most ECG analysis systems. In applications were rhythm detection is performed, only the location of the R wave is required. In other applications it is necessary to find and recognize the features of the ECG signal, such as the P and T waves, or the ST segment, for the automated classification and diagnosis. Many algorithms for the extraction of the ECG features based on the digital filters have been reported in the literature, especially algorithms for the QRS complex recognition.

In order to compute the heart rate and RRI, the algorithm implemented on the software running on MSP430F2274 microcontroller detects the QRS complexes from ECG signal. The framework for QRS detection algorithm was derived from the Pan-Tompkins algorithm [39], and it is represented in Fig. 10.6. The Pan-Tompkins algorithm is the most widely used algorithm for the detection of QRS complexes from ECG signal, detects correctly more than 95% of the QRS complexes from the MIT/BIH Arrhythmia Database.

The ECG processing stage uses the raw signal to generate a windowed estimate of the energy in the QRS frequency band by using the following filters: low pass (5 Hz), high pass (11 Hz), absolute value of the derivative and averaging the absolute value over a moving average window. The combined high-pass, low-pass

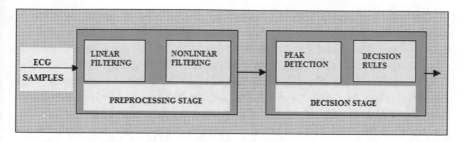

Fig. 10.6 Pan-Tompkins QRS detector

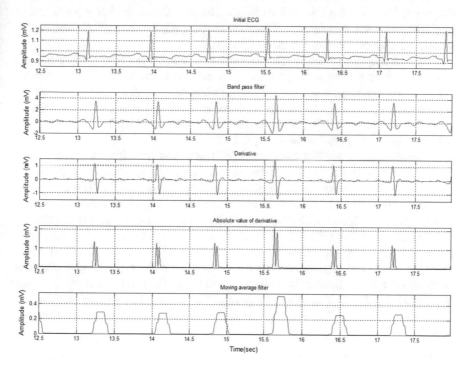

Fig. 10.7 QRS detector—preprocessing stage

and derivative filters produces a filter with the bandwidth that contains most of the energy in the QRS complex [53]. The averaging window was chosen to be the width of a typical QRS complex (80 ms). This window allows detections for the wider QRS complexes produced by Premature Ventricular Contractions. The results of the preprocessing stage are represented in the Fig. 10.7.

The peak detection stage uses two set of adaptive thresholds to detect QRS complexes. The decision rules used to classify the detected QRS complexes are based on information about RR interval, heart refractory period, or T wave.

The basic methodology used to detect atrial fibrillation is similar to [31] with some improvements. The cardiac rhythm was classified by use a RR versus dRR map in order to evaluate the dynamic of the heart for identification of atrial fibrillation. The chaotic characteristic is different from the Lorentz plot that uses for a discrete signal $x(k)$ the map $x(k)$ versus $x(k-1)$. The RR interval is computed by the interval between two successive RR and $dRR = RR(k) - RR(k-1)$ in the discrete time domain. Each record is divided in successive non-overlapping windows contained 32, 64 and 128 RR intervals. Because our application is implemented on a microcontroller and the computational effort is limited by this real-time approach we used in this stage only a 32 interval length window [30].

The plot of RR intervals versus dRR intervals (RdR) revealed a complex dynamic of the Normal (N) and Atrial Fibrillation cluster, as it is represented in Fig. 10.8. Also the dynamic of N+AF showed a complex variety of clusters that present a larger or a smaller scatter of the points.

The RdR map is divided in a rectangular grid with 25 ms resolution. Each cell can contain 0 or a number of points (NP) greater or equal to one. The nonempty cells (NC) are counted and the detection of atrial fibrillation is made when NC is greater than a predefined threshold. In order to avoid the situation that a point is counted twice or more times (Fig. 10.9), our algorithm uses a list of points with marked points when one of them is used.

In order to test the atrial fibrillation (AF) detection algorithm MIT/BIH atrial fibrillation database was used. The database includes 23 long-term ECG recordings of human subjects with atrial fibrillation (mostly paroxysmal). The individual recordings are each 10 h in duration, and contain two ECG signals each sampled at 250 samples per second with 12-bit resolution over a range of ±10 millivolts.

The statistic of results for the atrial fibrillation detection algorithm is presented in Table 10.2: true positive (TP)—AF is classified as AF; true negative (TN)—N is classified as N; false negative (FN)—AF is classified as N; false positive (FP)—N is classified as AF. Sensitivity (SE) and specificity (SP) are given by formulas TP/(TP+FN) and TN/(TN+FP), respectively. The predictive value of positive test (PV+) and the predictive value for negative test (PV−) are given by formulas

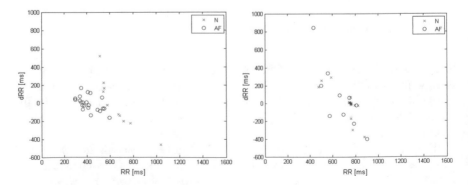

Fig. 10.8 Map plotting RR versus dRR

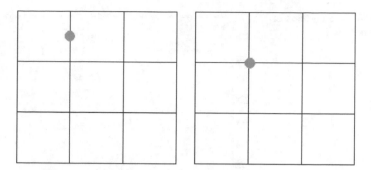

Fig. 10.9 Points being detected in two or more cells

Table 10.2 Detection results of atrial fibrillation using RdR maps

Subject	TP	TN	FN	FP	SE	SP	PV+	PV-	TH
04015_ch0	18	1236	2	133	0.900	0.903	0.119	0.998	22
04015_ch1	22	1452	1	148	0.957	0.908	0.129	0.999	23
04048_ch0	29	1125	2	111	0.935	0.910	0.207	0.998	17
04048_ch1	26	1117	1	117	0.963	0.905	0.182	0.999	14
04908_ch0	132	1854	10	194	0.930	0.905	0.405	0.995	17
04908_ch1	130	1668	9	162	0.935	0.911	0.445	0.995	20
04936_ch0	1465	539	129	95	0.919	0.850	0.939	0.807	22
04936_ch1	1139	408	115	51	0.908	0.889	0.957	0.780	18
05261_ch0	34	1312	2	101	0.944	0.929	0.252	0.998	21
06453_ch1	15	1269	2	296	0.882	0.811	0.048	0.998	18
07879_ch0	1238	532	18	26	0.986	0.953	0.979	0.967	19
07879_ch1	1233	529	14	28	0.989	0.950	0.978	0.974	18
07910_ch0	78	931	6	164	0.929	0.850	0.322	0.994	26
07910_ch1	84	965	6	221	0.933	0.814	0.275	0.994	27
08219_ch1	434	1217	41	198	0.914	0.860	0.687	0.967	23
08434_ch0	50	1125	1	53	0.980	0.955	0.485	0.999	24
08434_ch1	49	1155	1	58	0.980	0.952	0.458	0.999	23

TP/(TP+FP) and TN/(TN+FN), respectively. The predefined threshold TH (optimal threshold that maximize simultaneously both SE and SP) is predefined and it is selected in the first stage manually. We can see that for each subject, the optimal threshold can be different. We propose a simple algorithm that proceeds in an exhaustive manner by evaluating for each threshold from 2 to size of window all the pairs (sensitivity and specificity). The optimal value is selected for the smallest value of the difference Dop = SE–SP, as it is represented in the Fig. 10.10.

The result for data in database for all subjects is given in Table 10.2. We can see that the optimal threshold for entire database is 22 (Table 10.3), and in this case sensitivity = 90.3% and specificity = 91.2%.

Fig. 10.10 Determination of
optimal point (OP)

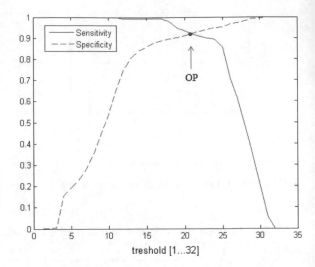

Table 10.3 Selection of
optimum threshold

TH	SE	SP	PV+	PV−
14	0.982	0.720	0.527	0.992
17	0.969	0.813	0.622	0.988
18	0.963	0.838	0.653	0.986
19	0.955	0.860	0.684	0.984
20	0.942	0.880	0.714	0.980
21	0.926	0.897	0.740	0.975
22	0.903	0.912	0.764	0.967
23	0.877	0.923	0.783	0.959
24	0.843	0.934	0.802	0.949
26	0.734	0.952	0.829	0.918
27	0.654	0.958	0.832	0.897

The flowchart of the software working on MSP430F2274 microcontroller from
the End Device is represented in Fig. 10.11. In this instance, the eZ430RF2500
module initializes onto the network, then, after a START command, wakes up to
read the ECG signals. In order to compute the heart rate, the ECG signals are
sampled at 200 Hz with 10 bits/sample. Also, the MSP430F2274 reads the battery
voltage, runs a heart rate detection algorithm, and communicates the results to
central monitoring station through Range Extender and Access Point. In order to
minimize the energy waste, since an important power consumer element is the radio
transceiver, the CC2500 entered into low power mode after each transmission
cycle.

The rhythm classification (Atrial Fibrillation detection), as it is implemented on
the MSP430F2274 microcontroller, is shown in Fig. 10.12.

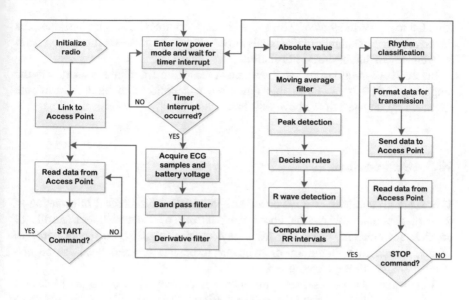

Fig. 10.11   Firmware running on the MSP430F2274 microcontroller

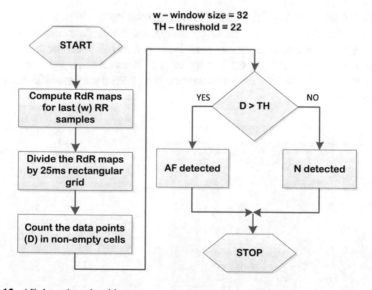

Fig. 10.12   AF detection algorithm

The prototype of the system described above has been implemented and tested. The accuracy of QRS detection algorithm was performed by computing the heart rate for test, by using the METRON 430 patient simulator, as it is shown in the Fig. 10.13. The simulator has following technical specifications: selectable heart rate from 30 to 300 bpm with $\pm 1\%$ accuracy and output ECG signal amplitude from 0.5 to 2 mV in steps of 0.5 mV, with ±2% accuracy. The simulator has also the possibility to generate ECG signals corresponding to the several common heart arrhythmias, including atrial fibrillation.

Figure 10.13 summarizes the measured values from simulated heart rate in the range of 30–300 bpm. From this figure we can notice that highest heart rate measured by the system is above 280 bpm.

10.5 Server Side

The whole system acts as a client/server application. The medical information in form of personal information about user, history of his/her illness, results of physical examination, laboratory tests, physiological signals and medical images (X-ray, computed tomography or magnetic resonance images) needs to be distributed among all registered users.

The medical database server application was developed by using the Microsoft Visual Studio 2008. The database is based on SQL Server 2008 and contains ten relational tables. The diagram of the database structure is represented in Fig. 10.14 [43].

The User table contains the user's basic personal health information, the physiologic signals are stored in the SignalData table, the results from laboratory tests or medical images are stored in the DataLab table, the alarms received from the users are stored in the table Alarm, the history of user illness and allergies are stored in the table PatientFile, and the user location based on GPS coordinates is stored in the table Location.

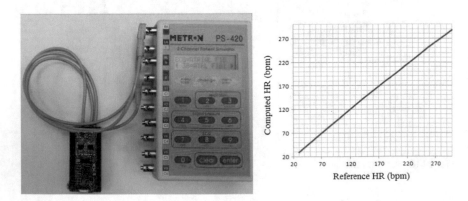

Fig. 10.13 Sensor Node test hardware and measured results for different simulated heart rates

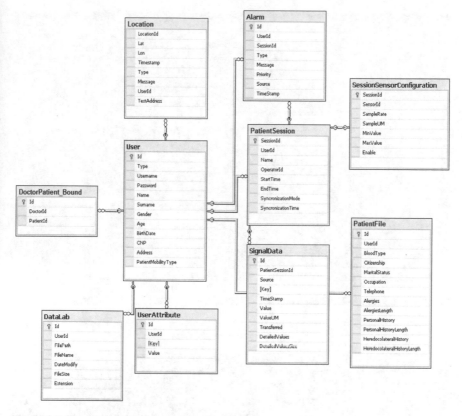

Fig. 10.14 The structure of the used relational database

10.6   Client Side

The client application (called MedApp), including the communication with the medical database server, was developed to run on windows based systems. While the server is running, the client can start a session from anywhere in the Internet by providing a proper log in and password on the MedApp. The display area has been divided into three parts: the graphics area, the user list area, and the buttons area. The graphics area is used to display temporal waveforms retrieved from medical database server.

When an alarm is received from the user, the client's software application generates an alert in the user interface. Figure 10.15 represents an ECG type HR alert in the client software application.

In order to track the location of users we integrated into our monitoring system a GPS sensor. Of course, modern smartphones have their own GPS modules. The ability to track the location of the users outdoor allows the ambulance center quickly locate a specific user who needs medical assistance immediately.

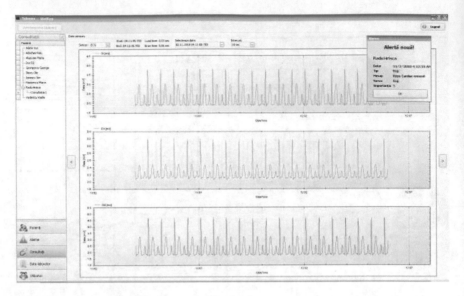

Fig. 10.15 MedApp—ECG signals and a HR alert

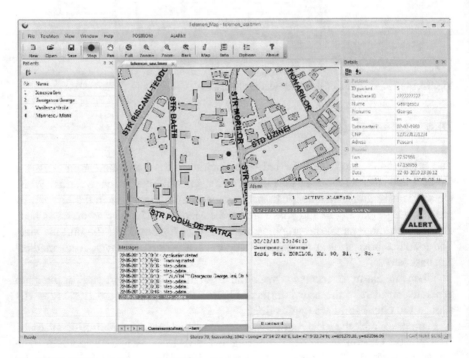

Fig. 10.16 TelemonMap user interface

The ambulance center has the possibility to locate the user with the aid of TelemonMap application (Fig. 10.16). This application uses a vector map with 16 layers, created with GIS NetSET Map Professional.

10.7 Conclusions

Such a regional telecenter as presented here performs as a support for a complex range of teleservices such as teleradiology, telepathology, teleconsulting in internal medicine and gerontology, telediagnosis, and telemonitoring of patient's medical status. It should also be a center for continuous training and e-learning tasks, both for medical personal and for patients. It meets certain requirements emphasized by various studies about the need for a new healthcare model that uses unobtrusive smart systems for vital signs and physical activity monitoring in many applications of mHealth technologies for pervasive health monitoring and pervasive healthcare. These new technologies can reduce the long-term monitoring cost of healthcare services and improve quality of life.

In this way, the new paradigms of ubiquitous computing, as well as pervasive computing, become reality and effective tools for modern medicine [55]. The first term means network connectivity everywhere, linking devices and systems, from the smallest to the largest ones. Pervasive computing, a similar term with ubiquitous computing, relies on the convergence of wireless technologies, advanced electronics and Internet. Both paradigm features allow continuous and (almost in) real time monitoring of human health in any environment, be it home, hospital, outdoors or the workplace. Pervasive computing addresses four key issues expressed by (i) smart spaces, (ii) invisibility, (iii) localized scalability and (iv) masking uneven conditioning. Also, the pervasive healthcare tries to change the healthcare delivery model, from doctor-centric to patient-centric, from acute reactive to continuous preventive, from sampling to monitoring [37]. This shifting process will maybe take much time to be a daily reality, and is represented by smart objects (medical sensors), characterized by the unobtrusiveness of sensing and computing in a pervasive system for health monitoring. These smart sensors deliver information to mobile platforms such as smartphone or tablet computers programmed to locally process the received data and to perform data synchronization with Web healthcare servers as Cloud computer systems. These computer resources expressed by networks servers, storage applications and Web services might be rapidly provisioned and released with minimal management effort or service provider interaction, by using computational intelligence and Semantic Web.

References

1. Adochiei, F., Rotariu, C., Ciobotariu, R., Costin, H.: A Wireless pulse oximetry system for patient, Advanced Topics in Electrical Engineering. Proceedings of the 7th International Symposium ATEE 2011, pp. 155−158 (2011)
2. Agrawal, D.P.: Personal/Body Area Networks and Healthcare Applications, Book chapter in Embedded sensor systems, pp. 353−390. Springer (2017). https://doi.org/10.1007/978-981-10-3038-3_16
3. Anliker, U., Ward, J.A., Lukowicz, P., et. al.: AMON: A wearable multiparameter medical monitoring and alert system. IEEE Trans. Inf. Technol. Biomed. **8**, 415–427 (2004). ISSN: 1089–7771
4. Augustyniak, P.: Remotely programmable architecture of a multi-purpose physiological recorder. Microprocess. Microsyst. **46**, 55–66 (2016). https://doi.org/10.1016/j.micpro.2016.07.007
5. Berna, S., Bostanci, E.: Opportunities, threats and future directions in big data for medical wearables. Proceedings of the International Conference on Big Data and Advanced Wireless Technologies (2016). https://doi.org/10.1145/3010089.3010100
6. BIOMED-TEL: Acquisition of biomedical signals and their teletransmission through mobile computing devices (in Romanian). http://tc.unitbv.ro/biomed/index_ro.html
7. Boron, W., Boulpaep, E.: Medical Physiology 3. Edn. Elsevier Publ. House (2016). ISBN:9780323427968
8. Bülbül, A., Küçük, S.: Pulse oximeter manufacturing & wireless telemetry for ventilation oxygen support. Int. J. App. Math. Electron. Comput. **4**, 211−215 (2016). ISSN: 2147-82282
9. Cardionet: Integrated system for continuous telesurveillance in e-health intelligent network of patients with cardiologic diseases (in Romanian). http://cardionet.utcluj.ro/Raport_tehnic_et2.pdf
10. Castanié, F., Lareng, L., et al.: Universal Remote Signal Acquisition For E-health (U-R-SAFE). HomeCare Concertation meeting June30−July 1, Toulouse, France. http://ursafe.tesa.prd.fr/ursafe/new/ConferencePapers/ConcertationMeeting2003_Ursafe (2003)
11. Chiuchisan, I., et al.: Health care system for monitoring older adults in a "Green" environment using organic photovoltaic devices. Env. Eng. Manag. J. **15**(12), 2595−2604 (2016)
12. Ciobotariu, R., Adochiei, F., Rotariu, C., Costin, H.: Wireless breathing system for long term telemonitoring of respiratory activity, Advanced Topics in Electrical Engineering. Proceedings of the 7th International Symposium ATEE 2011, pp. 635−638 (2011)
13. Ciucu, R.-I., et al.: A non-contact heart-rate monitoring system for long-term assessments of HRV, Advanced Topics in Electrical Engineering (ATEE), 2017 10th International Symposium on. IEEE (2017). https://doi.org/10.1109/ATEE.2017.7905060
14. CodeBlue: CodeBlue—Wireless Sensor Networks for Medical Care. http://fiji.eecs.harvard.edu/CodeBlue
15. Connelly, K., et al.: The future of pervasive health. IEEE Pervasive Comput. **16**(1), 16−20 (2017). https://doi.org/10.1109/MPRV.2017.17
16. Costin, H., Rotariu, C., et al.: MEDCARE—system for cardiologic telemonitoring through Internet. Med-Surg. J., Iași, Rom. **107**, No. 3, Supl. 1, 528−533 (2003)
17. Costin, H., Puscocim,S., Rotariu, C., Dionisie B., Cimpoesu, M.: A multimedia telemonitoring network for healthcare. Proceedings of XVII Int. Conference on Computer and Information Science and Engineering, ENFORMATIKA 2006, pp. 113−118. Cairo, Egypt (2006a). ISSN 1305–5313
18. Costin, H., Rotariu, C., Dionisie, B., Ciofea, R., Puscoci, S.: Telemonitoring system for complex telemedicine services. Proceedings of Int. Conference on Computers, Communications & Control, ICCCC 2006, pp. 150−155, June 1−3, Baile Felix Spa, Oradea (2006b)
19. Costin, H., Rotariu, C., et al.: Real time telemonitoring of Medical Vital Signs. In Long C., et al. (eds.) Recent Advances in Biomedical Electronics and Biomedical Informatics. Proceedings of the 2-nd Int. Conf. Biomedical Electronics and Biomedical Informatics-BEBI'09, Moskow, pp. 127−135, ISBN: 978-960-474-110-6 (2009)

20. Costin, H., Rotariu, C., Păsărică, A.: Identification of Psychological Stress by Analyzing Electrocardiographic Signal, Env. Eng. Manag. J. (EEMJ), **12**(6), 1255−1263 (2013a). print ISSN: 1582-9596, eISSN: 1843-3707
21. Costin, H., Rotariu, C., Păsărică, A.: A new method for paroxysmal atrial fibrillation automatic prediction. Buletinul Institutului Politehnic din Iasi (Bulletin of Polytechnic Institute of Iasi), Automatic Control and Computer Science Section, Vol. LIX (LXIII), Fasc.1, pp. 71−83 (2013b), ISSN 1220-2169
22. ElMoaqet, H., Almuwaqqat, Z., Saeed, M.: A new algorithm for predicting the progression from paroxysmal to persistent atrial fibrillation. Proceedings of the 9th International Conference on Bioinformatics and Biomedical Technology, pp. 102−106 (2017). https://doi. org/10.1145/3093293.3093311
23. Furmankiewicz, M., Sołtysik-Piorunkiewicz A., Ziuziański P.: Artificial Intelligence and Multi-agent software for e-health Knowledge Management System (in Polish). Informatyka Ekonomiczna **2**(32), 51−63 (2014)
24. Ghayvat, H., Mopadhyay, S., X G.: Sensing Technologies for Intelligent Environments: A Review, Intelligent Environmental Sensing, pp. 1–31. Springer International Publishing (2015). https://doi.org/10.1007/978-3-319-12892-4_1
25. Gogate, U., Bakal, J.W.: Smart healthcare monitoring system based on wireless sensor networks, Computing, Analytics and Security Trends (CAST). Int. Conf. IEEE (2016). https:// doi.org/10.1109/CAST.2016.7915037
26. Haulică, I.: Human physiology (in Romanian) 3. Eds. Medical Publ. House, Bucharest (2007). ISBN: 973-39-0597-4
27. HealthService24: Continuous mobile services for healthcare. Final report. http://www. healthservice24.com/Internet/external/healthservice24/images_/D1.5_HS24%20Final%20 Report.pdf (2006)
28. Humayun, A., et al.: Impact on the usage of wireless sensor networks in healthcare sector. JCSNS Int. J. Comput. Sci. Netw. Secur. **17**(4), 102–105 (2017)
29. Huzooree, G., Khedo, K.K., Joonas, N.: Pervasive mobile healthcare systems for chronic disease monitoring. Health Informat. J. (2017). https://doi.org/10.1177/1460458217704250
30. Li, Y., et al.: Probability density distribution of delta RR intervals: A novel method for the detection of atrial fibrillation. Australasian Physical & Engineering Sciences in Medicine, pp. 1−10 (2017). https://doi.org/10.1007/s13246-017-0554-2
31. Lian, J., Wang, L., Muessig, D.: A simple method to detect atrial fibrillation using RR intervals. Am. J. Cardiol. 15; **107**(10), 1494−1497 (2011)
32. Lin, B-S., Wong, A.M., Tseng, K.: Community-based ECG monitoring system for patients with cardiovascular diseases. J. Med. Syst. **40**, (2016). https://doi.org/10.1007/s10916-016-0442-4
33. Malan, D., Fulford-Jones, T., Welsh, M., Moulton, S.: CodeBlue: An Ad Hoc sensor network infrastructure for emergency medical care. MobiSys 2004 Workshop on Applications of Mobile Embedded Systems. http://www.eecs.harvard.edu/~mdw/papers/codeblue-bsn04.pdf (2004)
34. Milenkovic, A., Otto, C., Jovanov, E.: Wireless sensor networks for personal health monitoring: Issues and an implementation. Comput. Commun. Spec. issue: Wireless Sens. Netw.: Perform. Reliab. Secur. Beyond **29**, 2521–2533 (2006)
35. MobiHealth, Jones, V., Van Halteren, A., Bults, R., Konstantas, D.: University of Twente, MobiHealth: Mobile Healthcare, Center for Telematics and Information Technology—APS, Netherlads. https://doi.org/http://www.itu.int/itudoc/itu-t/workshop/e-health/wcon/s6con002. pdf
36. Moumtzoglou, A., (ed.): M-health Innovations for Patient-centered Care, Volume in the Advances in Healthcare Information Systems and Administration (AHISA) Book Series IGI Global (2016). ISSN:2328-1243
37. Mukhopadhyay, S.C., Postolache, O.A. (Eds.): Pervasive and mobile sensing and computing for healthcare. Technolo. Soc. Issues; Smart Sens. Meas. Instrum. Volume 2. Springer (2013)
38. Naddeo, S., et. al.: A real-time m-Health monitoring system: An integrated solution combining the use of several wearable sensors and mobile devices. Proceedings of the 10th

International Joint Conference on Biomedical Engineering Systems and Technologies (BIOSTEC 2017)—Vol. 5: HEALTHINF, pp. 545−552 (2017). ISBN:978-989-758-213-4

39. Pan, J., Tompkins, W.J.: A real-time QRS detection algorithm. IEEE Transact. Biomed. Eng. Vol. BME-32, 230−236 (1985)

40. Puscoci, S., Costin, H., Rotariu, C., et al.: TELMES—Regional medical telecentres. Proceedings of XVII Int. Conference on Computer and Information Science and Engineering, ENFORMATIKA 2006, pp. 243−246. Cairo, Egipt (2006). ISSN 1305-5313

41. Rotariu, C., Costin, H., et al.: E-health system for medical telesurveillance of chronic patients. Int. J. Comput. Commun. Control 5(5), 900−909 (2010). ISSN:1841-9836

42. Rotariu, C., Pasarica, Al., Costin, H., et. al.: Telemedicine system for remote blood pressure and heart rate monitoring. Proceedings of the 3rd International Conference on E-Health and Bioengineering—EHB 2011, 24–26 Nov. 2011, pp. 127−130 (2011a)

43. Rotariu, C., Costin, H., Andruseac, G., Ciobotariu, R., Adochiei, F.: An integrated system for wireless monitoring of chronic patients and elderly people. Proceedings of the 15th International Conference on System Theory, Control and Computing, Sinaia, Oct. 14−16, pp. 527−530 (2011b)

44. Rotariu, C., Arotaritei, D., Manta, V.: Wireless system for remote monitoring of atrial fibrillation. Proceedings of the 5th European DSP Education & Research Conference EDERC 2012, Amsterdam, The Netherlands, September 13−14, pp. 129−133 (2012a)

45. Rotariu, C., Manta, V.: Wireless system for remote monitoring of oxygen saturation and heart rate. Proceedings of the Federated Conference on Computer Science and Information Systems, FedCSIS 2012, Wrocław, Poland, September 9−12, pp. 215−218 (2012b)

46. Rotariu, C, Costin, H.: Remote respiration monitoring system for sleep apnea detection. Med-Surg. J. 117(1), 268−274 (2013a)

47. Rotariu, C., Manta, V., Ciobotariu, R.: Integrated system based on wireless sensors network for cardiac arrhythmia monitoring. Adv. Electr. Comput. Eng. 13(1), 95−100 (2013b)

48. Rotariu, C., Costin, H., Păsărică, A., Cristea, A., Dionisie, B.: Wireless skin temperature measurement system for circadian rhythm monitoring. Procedingd. of the 4th IEEE International Conference E-Health and Bioengineering—"EHB 2013", Iasi, Romania, 21−23 (2013c). ISBN 978-1-4799-2373-1

49. Rotariu, C., Costin, H., Ciobotariu, R., Păsărică, Al., Cristea, C.: Real-time system for continuous and remote monitoring of respiration during sleep using wireless sensors networks. IFMBE Proceedings, 44, 83−86, Proceedings of "MediTech 2014" conference, Cluj-Napoca, Romania, 5−7th June (2014)

50. Rotariu C., et al.: Continuous respiratory monitoring device for detection of sleep apnea episodes. Design and Technology in Electronic Packaging (SIITME), 2016 IEEE 22nd International Symposium for. IEEE, (2016). https://doi.org/10.1109/SIITME.2016.7777255

51. Rubel, P., Fayn, J., Atoui, H., Télisson, D.: Beyond EPI-MEDICS, 2nd OpenECG Workshop, Berlin, Germany (2004)

52. Rubel, P., Fayn, J., et al.: Toward personal eHealth in cardiology. Results EPI-MEDICS Telemedicine Proj. 38(4), 100−106 (2005)

53. Sokolova, A.A., et al.: Analysis of QRS detection algorithms barely sensitive to the QRS shape, Young Researchers in Electrical and Electronic Engineering (EIConRus), 2017 IEEE Conference of Russian pp. 738−740 (2017). https://doi.org/10.1109/EIConRus.2017.7910663

54. TELEASIS: Complex System on NGN support for home tele-assistance of elderly people (in Romanian). http://www.teleasis.ro

55. Triantafyllidis, A.K., et al.: A survey of mobile phone sensing, self-reporting, and social sharing for pervasive healthcare. IEEE J. Biomed. Health Inform. 21(1), 218–227 (2017). https://doi.org/10.1109/JBHI.2015.2483902

56. Varshney, U.: Pervasive healthcare computing. Springer LLC (2009). ISBN: 978-1-4419-0214-6

57. Vizza, P., et al.: A framework for the atrial fibrillation prediction in electrophysiological studies. Comput. Methods and Programs Biomed. 120(2), 55−76 (2015). https://doi.org/10.1016/j.cmpb.2015.04.001

58. Xu, H., Hua, K.: Secured ECG signal transmission for human emotional stress classification in wireless body area networks. EURASIP J. Inform. Secur. (2016). https://doi.org/10.1186/s13635-015-0024-x
59. Yang, X., Hui, C. (Eds.): Mobile telemedicine: A computing and Networking Perspective. CRC Press (2008)

Resource List

I. Books and Articles

60. Mukhopadhyay, S.C., Postolache, O.A. (Eds.): Pervasive and mobile sensing and computing for healthcare. Technological and Social Issues, Springer (2013). ISBN: 978-3-642-32537-3 (Print) 978-3-642-32538-0 (Online)
61. Chana, M., Campoa, E., Estèvea, D., Fourniolsa, J.-Y.: Smart homes—current features and future perspectives. Elsevier, Maturitas 64(2), 90–97 (2009)
62. Choi, J., Kim, D.K.: A remote compact sensor for the real-time monitoring of human heartbeat and respiration rate. IEEE Trans. Biomed. Circuits Syst. 3(3), 181–188 (2009)
63. Dilmaghani, R.S., Bobarshad, H., Ghavami, M., Choobkar, S., Wolfe, C.: Wireless sensor networks for monitoring physiological signals of multiple patients. IEEE Trans. Biomed. Circuits Syst. 5(4), 347–356 (2011)
64. Krishnamachari, B.: Networking Wireless Sensors. Cambridge University Press (2006)
65. Istepanian, R., Laxminarayan, S., Pattichis, C.S. (Eds.): M-Health: Emerging Mobile Health Systems. Springer (2005)
66. Lane, N.D., Miluzzo, E., Lu, H., Peebles, D., Choudhury, T., Campbell, A.T.: A survey of mobile phone sensing. IEEE Commun. Mag. 140–150 (2010)
67. Briggs, J.S., Adams, C., Fallahkhair, S., Iluyemi, A., Prytherch, D.: M-health review: Joining up healthcare in a wireless world. University of Portsmouth, UK, Technical Report. http://eprints.port.ac.uk (2012)
68. Axisa, F., Schmitt, P.M., Gehin, C., Delhomme, G., McAdams, E., Dittmar, A.: Flexible technologies and smart clothing for citizen medicine home healthcare, and disease prevention. IEEE Trans. Inform. Technolo. Biomed. 9(3), 325–336 (2005)
69. Kulkarni, P., Öztürk, Y.: Requirements and design spaces of mobile medical care. ACM SIGMOBILE Mobile Computing and Communications Review 11, 12–30 (2007)
70. Varshney, U.: Pervasive healthcare and wireless health monitoring. Mob. Netw. Appl. 12, 113–127 (2007)
71. Akkaya, K., Younis, M.: A survey on routing protocols for wireless sensor networks. Elsevier Ad Hoc Netw. J. 3(3), 325–349 (2005)
72. Falk, T., Baldus, H., Espina, J., Klabunde, K.: Plug 'n play simplicity for wireless medical body sensors. Mob. New. Appl. 12, 143–153 (2007)
73. Neves, P., Stachyra, M., Rodrigues, J.: Application of wireless sensor networks to healthcare promotion. J. Commun. Softw. Syst. 4(3), 181–190 (2008). ISSN: 1845-6421
74. Shnayder, V., Chen, B.-r., Lorincz, K., Fulfor-Jones, T., Welsh, M.: "Sensor Networks for Medical Care." Harvard University (2005)
75. Tang, C.: Comprehensive energy efficient algorithm for WSN, INT. J. Comput. Commun. Control 9(2):209–216 (2014). ISSN 1841-9836
76. Chang, K.-M., Liu, S.-H.: Wireless portable electrocardiogram and a tri-Axis accelerometer implementation and application on sleep activity monitoring. TELEMEDICINE and e-HEALTH, MARY ANN LIEBERT, Inc. 17(3), 177–184 (2011)
77. Stiel, I.G., Spaite, D.W., Field, B., et al.: OPALS Study group. Advanced life support for out-of hospital respiratory distress. New England J. Med. 356, 2156–2164 (2007)

78. Wu, W.H., Bui, A.A.T., Batalin, M.A., Au, L.K., Binney, J.D., Kaiser, W.J.: MEDIC: Medical embedded device for individualized care. Artif. Intell. Med. Elsevier **42**, 137–152 (2008)

II. Web Resources

79. IEEE 802.15 WPAN Task Group 6 (TG6)—Body Area Networks, URL: http://www.ieee802. org/15/pub/TG6.html
80. The promise of Wireless Sensor Networks for Medicine, URL: http://www.intel.com/ research/exploratory/wireless_promise.htm
81. LifeShirt from VivoMetrics http://www.vivometrics.com/research/clinical_trials/about_the_ system/what_is_the_lifeShirt_system.php
82. AlarmNet—Assisted-Living and Residential Monitoring Network, URL: http://www.cs. virginia.edu/wsn/medical/index.html
83. Economist Intelligent Unit "The future of healthcare in Europe", The Economist, on-line http://www.eufutureofhealthcare.com/sites/default/files/EIUJanssen%20Healthcare_Web%20 version.pdf
84. National Center for Health Statistics, URL: http://www.cdc.gov/nchs/Default.htm
85. Mobihealth project, URL: http://www.mobihealth.org
86. THALEA project, URL: www.thalea-pcp.eu
87. Telecare Services Association, http://www.telecare.org.uk/about-us
88. VitalJacket form Biodevices, SA (Portugal), URL: http://www.vitaljacket.com/
89. "Telemedicine and e-Health" journal, http://www.liebertpub.com/TMJhttp://www.liebertpub. com/TMJ
90. United States Department of Health and Human Services. (2007) Personalized Health Care: opportunities, pathways, resources. on line at http://www.liebertpub.com/TMJhttp://www. hhs.gov/myhealthcare/news/phc-report.pdf

Chapter 11
Preprocessing in High Dimensional Datasets

Amparo Alonso-Betanzos, Verónica Bolón-Canedo,
Carlos Eiras-Franco, Laura Morán-Fernández and Borja Seijo-Pardo

Abstract In the last few years, we have witnessed the advent of Big Data and, more specifically, Big Dimensionality, which refers to the unprecedented number of features that are rendering existing machine learning inadequate. To be able to deal with these high-dimensional spaces, a common solution is to use data preprocessing techniques which might help to reduce the dimensionality of the problem. Feature selection is one of the most popular dimensionality reduction techniques. It can be defined as the process of detecting the relevant features and discarding the irrelevant and redundant ones. Moreover, discretization can help to reduce the size and complexity of a problem in Big Data settings, by diminishing data from a large domain of numeric values to a subset of categorical values. This chapter describes in detail these preprocessing techniques as well as providing examples of new implementations developed to deal with Big Data.

Keywords Preprocessing · Big data · Big dimensionality · Discretization · Feature selection

A. Alonso-Betanzos (✉) · V. Bolón-Canedo · C. Eiras-Franco · L. Morán-Fernández ·
B. Seijo-Pardo
Departamento de Computación, Universidade Da Coruña, Campus de Elviña S/N, 15071 A
Coruña, Spain
e-mail: ciamparo@udc.es

V. Bolón-Canedo
e-mail: vbolon@udc.es

C. Eiras-Franco
e-mail: carlos.eiras.franco@udc.es

L. Morán-Fernández
e-mail: laura.moranf@udc.es

B. Seijo-Pardo
e-mail: borja.seijo@udc.es

© Springer International Publishing AG 2018
D.E. Holmes and L.C. Jain (eds.), *Advances in Biomedical Informatics*,
Intelligent Systems Reference Library 137,
https://doi.org/10.1007/978-3-319-67513-8_11

247

11.1   Introduction

In the last few years, we have witnessed the advent of the so-called Big Data, a term which was coined in 1997 referring to the area of scientific visualization, in which the datasets are usually very large. Giving a proper definition of this term is not easy, but broadly speaking it can be defined as data that is too big to be processed comfortably on a single machine, either because of processor, memory or disk bottleneck. Big Data is also defined by its properties, which are commonly known as the 7 V's: volume, velocity, variety, veracity, value, variability and visualization.

Because the volume, velocity, variety and complexity of datasets are continuously growing, machine learning techniques have become indispensable in order to extract useful information from huge amounts of otherwise meaningless data. Analogous to Big Data, the term Big Dimensionality has been coined to refer to the unprecedented number of features arriving at levels that are rendering existing machine learning methods inadequate [1]. As an example, the maximum dimensionality of datasets posted at the widely-used UCI Machine Learning Repository [2] was only about 100 in the 1980s, while nowadays it is more than 3 millions. As another example, in the popular LIBSVM Database [3], one can find datasets with more than 29 million features.

This ultrahigh dimensionality requires massive memory, which implies a high computational cost for training. Moreover, having such a large number of features deteriorates the generalization ability of learning algorithms, due to the so-called "curse of dimensionality". This term was coined by Bellman [4] to express the difficulty of optimization by exhaustive enumeration on product spaces. It is a phenomenon that arises when dealing with high-dimensional spaces, which does not occur in low-dimensional settings. A common solution to tackle the "curse of dimensionality" is to reduce the dimensionality of the problem, a task that can be achieved by means of data preprocessing.

Data preprocessing is a fundamental step in real applications of machine learning and data mining [5–7]. As methods for data gathering do not usually include a strict quality control, missing data, impossible data, out-of-range values, noisy data, etc., appear rather frequently. For that reason, as quality data is almost imperative for assuring quality results for classification, prediction, etc. tasks, preprocessing data becomes a must-do process in most situations, accounting for a considerable amount of time. Data preprocessing aims at eliminating as much irrelevant, redundant, unreliable and noisy data as possible. After data preprocessing quality is assured in several dimensions, such as accesibility, accuracy, consistency, completeness, timeliness, believability, interpretability, accesibility and value added. Several common preprocessing operations are (see Fig. 11.1):

- Data cleansing, that encompasses filling-in missing values, smoothing or removing noisy data, identifying and removing outliers, resolving inconsistencies, etc.
- Data integration, when multiple databases or files need to be combined.
- Data transformation, performing operations that normalize data, or aggregate it.

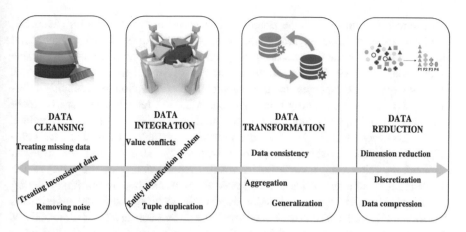

Fig. 11.1 Preprocessing operations

- Data reduction, that includes the selection and extraction of features and/or instances from a data set. The aim is to find a reduced representation of data in volume but producing the same or similar (even better) analytical results. The main data reduction strategies are dimensionality reduction, aggregation and clustering features (both reduce the number of input features), and sampling (that reduces the number of samples of the dataset).
- Data discretization, that might be considered a type of data reduction, as it divides the range of a continuous attribute into intervals, thus reducing the amount of data, as interval labels can then be used to replace actual data values. One of the reasons for using this process might be that some data mining algorithms can only deal with categorical attributes as input.

Although all operations are equally important for obtaining an operative high-quality dataset, data reduction, including data discretization, is specially important for high dimensional datasets. This is because complex data analysis on datasets with a huge quantity of samples and/or features may consume high amounts of computational resources, mainly temporal. As mentioned above, data reduction may refer to reducing either the volume or the dimensions (number of attributes or features) of the dataset, while still yielding useful knowledge after the data mining process. While several strategies for data reduction might be used, as data cube aggregation, data compression, numerosity reduction [6, 7], etc. we will center this chapter on two operations: (1) discretization and (2) a specific type of methods of dimensionality reduction, named feature selection [8]. The reason for this is that feature selection facilitates the understanding of the patterns extracted afterwards, as the original features are not transformed (as it happens in other dimensionality reduction techniques such as Principal Component Analysis—PCA —or Wavelets transforms [7]), but the relevant features are maintained while eliminating irrelevant and redundant ones. On the other hand, as mentioned above, discretization is a necessary step for several feature selection methods.

Feature selection (FS) is the process of selecting relevant variables (features), eliminating the irrelevant or redundant ones, with the aim of obtaining better and simpler models. As the process does not transform the original features, it obtains models that might be easier to interpret for researchers of the application fields. Other interesting advantages are the enhancing generalization by reducing over-fitting, and the shorter training times needed. Although feature selection is widely used in many fields, it is essential in those domains in which the number of features is comparatively much higher than the number of samples, as in this case computational models are more prone to overfitting [9, 10]. Among the different applications, microarray data is an interesting representative of the kind, as the problem and its characteristics have inspired many researchers in machine learning [9, 11–15]. Microarray gene expression data are the result of biological experiments that monitor (using microarray technology) the expression levels of thousands of genes simultaneously under certain conditions. A microarray is a glass slide on to which DNA molecules are fixed orderly at specific locations (features). Microarrays might have thousands of features and each of the latter might contain some million copies of identical DNA molecules that correspond uniquely to a gene. The datasets contain a small number of samples, usually under 100 patients, but thousands of features that are normally used to separate healthy from unhealthy subjects, or to classify different types of tumors. Besides this problem of high dimensionality, other characteristics of microarray datasets, such as class imbalance (one class, or several in multiclass problems, which is called the majority one has considerably much more samples than other named the minority classes, causing bias in the classifiers), overlapping among classes (making it difficult to separate them with high accuracy), dataset shift (feature distribution is different on the test data regarding the training dataset), and the existence of outliers (samples that are incorrectly labeled or identified mainly due to different possible contamination processes), make the study of these datasets a challenge for machine learning researchers. In this scenario, feature selection takes a main role in the analysis, in order to remove redundant and irrelevant features, avoid overfitting and facilitate interpretability of the results to the expert biologists that aim to relate gene expressions to diseases. Although less frequent in the literature, discretization is also needed for microarray data. Perhaps one of the reasons for the scarcity of research works in this area is that in many tools some filters, one of the most employed feature selection techniques [8, 9] use a default discretization prepro-cessing, transparent for the user. Bolón-Canedo et al. [16] added a previous dis-cretization step before a feature selection filter, demonstrating an important improvement of the classification results. The rationale under the approach is twofold: on the one hand, being able to select the discretizer to be employed improves performance of the filter and, on the other hand, there used to be a large number of genes with very unbalanced values (see Fig. 11.2 as an example of an unbalanced attribute in a microarray dataset).

In this chapter we will describe in more detail several feature selection and discretization techniques. Section 11.2 is centered on discretization techniques

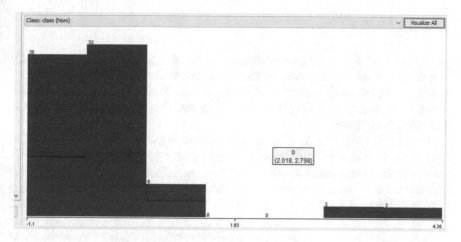

Fig. 11.2 Unbalanced attribute in a microarray dataset (Leukemia)

while Sect. 11.3 is focused on feature selection. Finally, Sect. 11.4 comments on new and future research directions.

11.2 Discretization

As mentioned before, discretization consists in transforming the continuous variables of a dataset into nominal (discrete) variables. After performing this process, data is simplified since the number of possible values for each variable is reduced to a few discrete values, so discretization is oftentimes regarded as a data reduction mechanism [17].

This can be a mandatory step if the model used is not able to handle continuous variables. This is the case for a number of classification models such as, for instance, Naïve Bayes [18], and also for some feature selection algorithms such as InfoGain [19]. In other cases, even though the model or algorithm is able to work with continuous variables, both speed and accuracy are greatly improved when data is discretized before training. Moreover, those algorithms that employ decision trees achieve shorter and more accurate models if discretization is used [20]. However, discretization implies also a loss of information and it is the job of the data scientist to make this loss as irrelevant as possible for the problem at hand so that the resulting simplification has a positive effect in the machine learning system.

In general, discretization constitutes an important preprocessing step that can greatly influence the performance of a machine learning system, and thus numerous discretization methods have been developed, although very few are available for Big Data scenarios.

11.2.1 Methods

Over a hundred discretization algorithms are available to data scientists. Selecting which one to employ for each problem is a crucial decision that can have great impact on the overall performance of the model. Taxonomies such as the one proposed by Ramírez et al. [17] are a valuable tool that can be helpful in order to simplify this decision.

The goal of discretization is to obtain a nominal value for each possible value of a variable. This could be achieved in a variety of ways: for instance, one could transform any real number to one of two nominal values by checking whether the absolute value of the integer part of such number is a pair or an odd number. In this case, -3.71 would be transformed to "ODD", 182.4 to "PAIR", 0.004 to "PAIR" and so on. However, it is in the data scientist interest that the transformed values maintain their original significance, so preserving the ordering relations between the values is often desired. Consequently, the usual discretization procedure consists in splitting the variable range in a set of disjoint intervals and assigning to each value the ordinal of the interval it lies in. In general terms, a discretization process consists of four steps:

1. Sorting the continuous values. Performed only once at the beginning of the process, prepares the input space for exploration.
2. Calculating the cut points, defining the beginning and end of each interval.
3. Splitting or merging intervals. Some algorithms have a target number of output values, while others select the most significant intervals, regardless of their number. This step examines the current intervals and increments or reduces their number according to some criteria.
4. Stopping according to some criterion or returning to step 2 otherwise.

The complexity of each of these steps varies depending on the method, which results in great disparity in the computational complexity of the algorithms.

Ideally, the obtained nominal values should have a close association with the class in a classification problem; this is the goal of supervised discretization algorithms, which use information regarding the class of each item to decide the discretization intervals. Below we introduce three common discretization methods in order to provide the reader with a notion of the usual approaches for discretization. To illustrate each method, the resulting intervals and the number of elements inside each one is represented. The input variable consists of a mixture of gaussians, one thousand elements with average $\mu = 4$ and standard variation $\sigma = 3$ and another thousand elements with $\mu = 8$ and $\sigma = 3$. To illustrate supervised methods a different class was assigned to each gaussian. The resulting two classes are represented with two different shades in the figures.

- Equal width: This simple unsupervised discretization algorithm calculates the range of the variable and then divides it in equal parts. The resulting intervals will generally be unbalanced, with many items ending in a few of them while some are much less populated. The split/merge step is disregarded in this simple method (Fig. 11.3).

- Equal frequency: The aim of this algorithm is to obtain intervals that contain a constant number of items (Fig. 11.4).
  Obtaining a fixed number of intervals is the goal of the basic version of this method. Nevertheless, this is suboptimal for some classification algorithms, so a variation named Proportional k-Interval Discretization [21] can be used instead. This algorithm adjusts the number of intervals according to the number of samples.
- Minimum Descriptive Length (MDL): This popular algorithm uses information entropy minimization as an heuristic to calculate the most suitable cut points [22] (Fig. 11.5).

11.2.2 Big Data Approaches

The data simplification and performance improvement associated with discretization turns it into a very appealing preprocess step when dealing with Big Data. Nonetheless, classical data reduction methods were not designed to handle huge amounts of data which makes its use difficult or impossible. To solve this, new implementations of the most popular methods which take advantage of distributed computation frameworks have become available. Although when dealing with Big Data the data scientist should avoid algorithms with computational complexity larger than $O(n)$ to ensure scalability, leveraging distributed computing can open

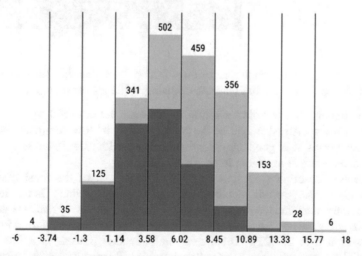

Fig. 11.3 Equal width discretization. *Colors* represent the two classes. *Bars* indicate the number of elements in each interval

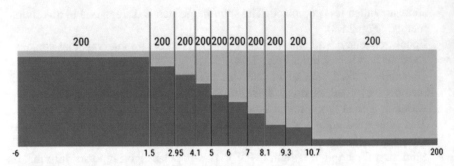

Fig. 11.4 Equal frequency discretization. *Colors* represent the two classes. *Bars* indicate the number of elements in each interval

Fig. 11.5 Minimum descriptive length discretization. *Colors* represent the two classes. *Bars* indicate the number of elements in each interval

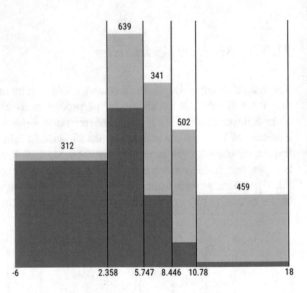

the door to using more complex algorithms in very large datasets. Listed below are some of the options available when discretizing variables for a large dataset.

- Proportional k-Interval discretization. This simple yet powerful equal frequency discretization method is one of the most popular. Its low computational complexity makes it a good choice when dealing with Big Data. A distributed implementation is available in Spark MLLib [23].
- Minimum Descriptive Length (MDL). MDL is one of the most often used supervised discretization methods, as proven by the fact that it is the default in the popular machine learning suite Weka [24]. Data scientists that lack access to a computing cluster can leverage the parallel Weka implementation proposed by Eiras-Franco et al. [25], which uses multithreaded processors to perform in parallel the computations required to discretize different variables. Also, a distributed implementation is available for Apache Spark [17, 26] which leverages

a computer cluster to speed up the sorting and cut point evaluation steps involved in this method, enabling it to deal with large datasets.

Despite the existence of these ready-to-use options, often the selection of the discretization algorithm depends heavily on the nature of the problem at hand. Sometimes a complex discretization algorithm run on a small subsample and then applied to the whole dataset will offer the best results, while on other occasions a fast and simple discretizer on the whole dataset will be a better choice. As with many other design decisions, it is the responsibility of the data scientist to perform cross-validation tests to assess the best alternative for their problem.

11.3 Feature Selection

Since the advent of Big Data, machine learning methods need, more than ever, to reduce the unprecedented scale of data, and thus feature selection has become almost essential. Typically, feature selection methods are classified into filters, wrappers, and embedded methods, according to their relationship with the induction algorithm (see Fig. 11.6). Filters are the simplest models, and they rely on the general characteristics of the training data and are independent of the induction algorithm. Typically, they are based on statistical measures, correlation, mutual information, etc. On the contrary, wrappers make use of the prediction abilities of a given induction algorithm to determine which subset of features has the highest predictive power. Finally, embedded methods search for an optimal subset of features through the classifier construction. In other words, ensemble methods learn which features best contribute to the accuracy of the model while the model is being created. Because of the interaction with the learning algorithm, wrappers and embedded methods usually give more accurate subsets of features, but at the cost of being computationally expensive (especially wrappers). Filters, in turn, are advantageous for their low computational cost and good generalization ability. The three approaches are extensively used, although the filter model is more adequate for Big Data settings, particularly if the problem at hand has a large number of input features.

Feature selection methods can also be classified according to individual evaluation and subset evaluation approaches [9]; the former —also known as feature ranking— assesses individual features by assigning them weights according to relevance, whereas the latter produces candidate feature subsets based on a specific search strategy which are subsequently evaluated by some measure.

Because of its capacity to improve the performance of learning algorithms, feature selection has become very popular in the field of machine learning, in processes such as clustering [27, 28], regression [29, 30], and classification [31, 32] (both supervised and unsupervised).

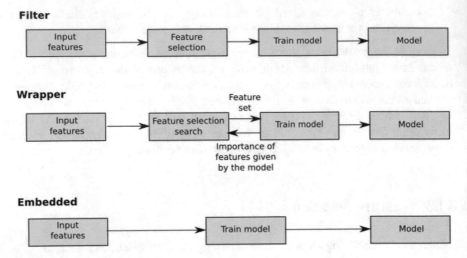

Fig. 11.6 Feature selection approaches

Table 11.1 Classical feature selection methods

	Type	References
Chi-Squared	Filter	Liu and Setiono [73]
F-score (Fisher score)	Filter	Duda et al. [74]
Information Gain	Filter	Quinlan [19]
ReliefF	Filter	Kononenko [65]
mRMR	Filter	Peng et al. [35]
SVM-RFE	Embedded	Guyon et al. [12]
CFS	Filter	Hall [75]
FCBF	Filter	Yu and Liu [76]
INTERACT	Filter	Zhao and Liu [77]
Consistency	Filter	Dash and Liu [78]

Among the large number of feature selection algorithms available, some of them are widely-used by researchers. Table 11.1 briefly displays the most popular feature selection methods.

Curiously enough, although feature selection is almost a must for machine learning algorithms being able to manage high dimensional datasets, most available methods were not developed taking this scenario into account, and unfortunately their computational complexities prevent their use in these cases. In the next sub-sections different approaches aimed at specific strategies used to cope with the requirements of large numbers of both samples and features are devised. First, Graphical Processing Units (GPU) can be used to implement faster versions of well-known algorithms. Another possibility is to distribute the datasets in several partitions, instead of maintaining a unique dataset, and thus execute a feature selection method in each partition, using afterwards some strategy to merge the results of each partition and obtaining a final overall result. Ensemble approaches come close to this same idea when trying to cope with high dimensions, allowing for employing different strategies, such as using the same feature selection method over different partitions of the dataset. In general, all these distributed approaches aim at obtaining a result in a competitive time, while trying to maintain performance in similar levels as the original performance if data were in a unique node. A similar philosophy is followed by parallelization, that takes advantage of distributing calculations among the different computers of a cluster, and thus assuring equal performance as in an original centralized algorithm, but with reduced temporal requirements.

11.3.1   GPU-Based Approaches

GPUs appeared for the first time in mid 1999, in an attempt to offload a significant amount of graphics processing from the CPU, until then the only responsible for both standard calculations and graphic operations. The success of the units was impressive, and they were immediately adopted by both hardware and software developers. GPUs are massively parallel floating point processors attached to dedicated high speed memory, relying on multithreading to provide throughput-oriented performance at a considerably lower cost than traditional parallel processing computers. Therefore GPUs are well suited for data-parallel applications, resulting in large improvements in running time. As one of feature selection problems is that most algorithms are unfeasible in large dimensional datasets due to memory and time limitations, GPU design and/or implementations of feature selection algorithms can revert the situation, making it possible to reduce the dimension of the problem, while maintaining or improving accuracy of the results. Despite its indisputable interest, there are only a few works on the subject yet. In [33] the authors implement a feature selection algorithm that can be employed for outlier detection, and which aim is to pick those features that best capture the behavior of normal data while making outliers more distinct from normal data. For doing this, the algorithm takes advantage of the parallelism that exists during the k-NN search and the forward feature search technique employed. Thus, for each step of the feature search, it is possible to calculate the criterion function for all of the possible feature subsets concurrently on a GPU. This, combined with the parallelism of the k-NN search, enables the algorithm's performance to scale efficiently in terms of both the number of features and the number of data points. An implementation of a genetic algorithm that has been parallelized using the classical island approach, but also considering GPUs to speed up the computation of the fitness function is presented in [34]. An extension of the state-of-the-art mRMR(minumum Redundancy Maximum Relevance) algorithm [35], is available at [36]. The algorithm is named fast-mRMR, and beside its implementation in C++ suitable for small-medium datasets, two other versions aimed at solving the lack of efficiency and scalability of the algorithm in big datasets are accessible, a distributed version using Apache Spark [37], and a GPU-CUDA (Compute Unified Device Architecture) [38] parallel version. The optimizations for both platforms implied important changes derived from the characteristics of both, and are described in detail in [39]. Fast-mRMR was tested using both synthetic and real datasets, and as a conclusion the authors demonstrated the advantage in using the GPU implementation when the number of samples of the dataset is high, more than 100.000 samples, while the Spark version is preferable when the number of features is high, more than 2.000.

11.3.2 Distributed Feature Selection

Traditionally, feature selection methods have been designed to run in a centralized computing environment. However, with the coming age of network technologies, many distributed approaches have been developed during the last years. The main reason is that, when dealing with high dimensional datasets, most existing feature selection methods do not scale well—memory demands and impracticable runtimes —, and their efficiency may significantly deteriorate to the point of becoming inapplicable. Besides, data is often distributed across geographical and organizational boundaries, and it is not legal or affordable to gather it in a single location. For example, data concerning financial information might be owned by separate organizations which are not willing to share confidential information with each other, but may be interested in enhancing their own models by exchanging useful information.

Although distributed feature selection is a fairly new area, it has been receiving a growing amount of attention since its inception. Das et al. [40] addressed the problem of feature selection in a large peer-to-peer environment, developing a local distributed privacy preserving algorithm. Banerjee et al. [41] proposed a secure distributed protocol that allowed feature selection for multiple parties without revealing their own data. Tan et al. [42] presented a new adaptive feature scaling framework for ultrahigh dimensional feature selection on big datasets. The proposed method demonstrated competitive performance over existing feature selection methods in terms of training efficiency and generalization performance. However, the algorithm was designed to be executed in a single machine. Other authors had considered parallel computing environments to re-implement feature selection algorithms. Peralta et al. [43] presented a feature selection algorithm based on evolutionary computation that used the MapReduce paradigm to obtain subsets of features for large datasets.

Bolón-Canedo et al. [44, 45] proposed a distributed method that could be used to partition the data vertically—by features—or horizontally—by samples—(see Fig. 11.7), depending on the characteristics of a particular problem. The general idea consisted of dividing the data into several nodes and then applying a filter at each partition performing several rounds to obtain a stable set of features. Later, a merging procedure is carried out to combine the results into a single subset of relevant features. Particularly, the partial feature subsets were combined based on improvements in classification accuracy. Experiments on six benchmarking UCI datasets showed that the proposed methodology was able to match and in some cases even outperform the classification accuracy obtained by standard algorithms applied to the non-partitioned datasets. In Fig. 11.8 we can see the results obtained with the horizontal partitioning by a C4.5 classifier after applying CFS feature selection method, as an example.

Figure 11.8a shows that classification accuracy is similar for both approaches: for some datasets the distributed version (CFS-D) even slightly outperforms the standard centralized one (CFS-C). If we focus on the computational time (Fig. 11.8b), we can see that the reduction in large datasets such as Mnist (which has 717 features and 40,000 training samples) is more than notable. Notice that, in the case of the distributed approach, we are taking as computational time the maximum time among those obtained in the different nodes, plus the time necessary to merge the partial results.

For the sake of a better visualization, in Fig. 11.8c we removed the results for Mnist. Now, we can see that for smaller datasets (such as Spambase, Ozone, Madelon or Connect4), the reduction in time is not so remarkable or even it does not exist. This is because although the computational time to perform feature selection in each node is shorter than in the centralized node, in some cases the

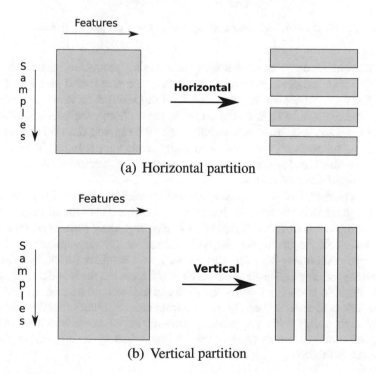

(a) Horizontal partition

(b) Vertical partition

Fig. 11.7 Different approaches to divide data into available nodes

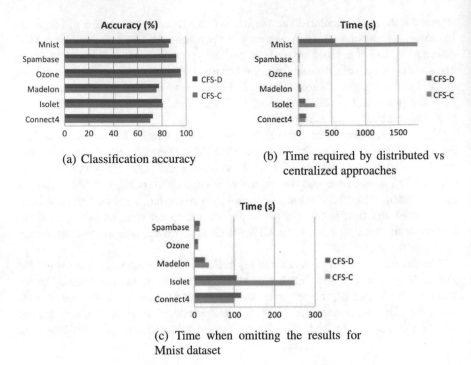

(a) Classification accuracy

(b) Time required by distributed vs centralized approaches

(c) Time when omitting the results for Mnist dataset

Fig. 11.8 Results obtained with horizontal partitioned, CFS filter and C4.5 classifier

computational burden that comes with the merging procedure makes the use of distributed approaches inappropriate. However, if the user would prefer to save this merging time, it is possible to establish a fixed threshold in an attempt to avoid this specific calculation. In light of these results, the authors concluded that their distributed proposal performed successfully, since the running time was considerably reduced and the accuracy did not drop to inadmissible values. In fact, their approach was able to match and in some cases even improve the standard algorithms applied to the non-partitioned datasets.

The same methodology was also applied to microarray data [46], which is a challenging scenario for machine learning researchers. This type of data is used to collect information from tissue and cell samples regarding gene expression differences that could be useful for diagnosis disease or for distinguishing a specific tumor type. Although there are usually very few samples (often less than 100 patients) for training and testing, the number of features in the raw data ranges from 6000 to 60,000, since it measures the gene expression en masse. A typical classification task is to separate healthy patients from cancer patients based on their gene expression "profile" (binary approach). There are also datasets where the goal is to distinguish among different types of tumors (multiclass approach), making the task even more complicated.

Several studies have shown that most genes measured in a DNA microarray experiment are not relevant for an accurate classification among different classes of the problem [9], so feature selection plays a crucial role in DNA microarray analysis. Although the experimental results for eight microrray datasets showed that the runtime was considerably shortened whereas performance was maintained or even improved compared to centralized algorithms [46], the merging procedure was again dependent on the classifier used and in some cases it introduced an important computational burden. Thus, in order to overcome this drawback, a framework for distributing the feature selection process [47, 48] was proposed, which performed a merging procedure to update the final feature subset according to the theoretical complexity of these features, by using some data complexity measures [49] instead of the classification error. Hence, the new framework was not only independent of the classifier chosen, but also reduced drastically the computational time needed for the algorithm. Trying to check the behavior of the approach on high input dimensionality, the novel procedure was tested in five microarray datasets (see Table 11.2 [50]. The microarray data is partitioned vertically, by assigning groups of t features to each subset, where the number of features t in each subset is half the number of the samples, to avoid over-fitting.

Figure 11.9a displays the number of features selected by the CFS filter for the centralized approach (C) and distributed approaches (D-F1, D-F2 and D-N2) based on the data complexity measures proposed by Ho and Basu [49]: Fisher's multiple discriminant ratio (F1), length of the overlapping region (F2) and ratio of average intra-/interclass nearest neighbor distance (N2), respectively.

As can be seen, the number of features selected by the distributed methods is larger than that selected by the centralized approach. This is caused by the fact that, with the vertical partition, the features were distributed across the packets and it was more difficult to detect redundancy between different partitions. Even so, the distributed approaches were using—in the worst case—5.5% (Breast), 8.9% (Gli85), 6.4% (CLL-SUB-111), 12.2% (Lung-Cancer) and 12.1% (11-Tumors) of the total features. In terms of classification accuracy, Fig. 11.9b shows the average classification accuracy over the four classifiers used (C4.5, NB, kNN and SVM) for each microarray dataset. As can be seen, in general there were no differences in terms of accuracy. Except for CLL-SUB-111, the best classification performance was reported by one of the distributed approaches.

Finally, Table 11.3 shows the runtime required by the CFS filter. Again, the execution time was drastically shortened by applying the distributed approach. Thus, in light of the above, the authors suggest using the distributed approach and,

Table 11.2 Number of classes, features and samples of the five microarray datasets

Dataset	Classes	Features	Samples
Breast	2	24,481	78
Gli85	2	22,283	85
CLL-SUB-111	3	111,340	111
Lung-Cancer	5	12,600	203
11-Tumors	11	12,533	174

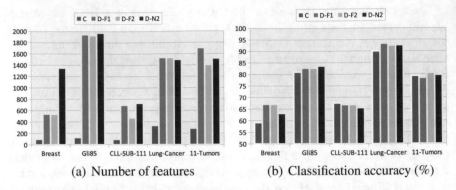

(a) Number of features   (b) Classification accuracy (%)

Fig. 11.9 Comparing the centralized and distributed approaches using vertical partition in terms of number of selected features and classification accuracy for the CFS filter

Table 11.3 Runtime (seconds) for the CFS filter on the five microarray datasets

	Centralized	Distributed		
		D-F1	D-F2	D-N2
Breast	7669.2	3.6	1.3	2.8
Gli85	7652.1	4.4	2.1	3.6
CLL-SUB-111	1335.0	3.4	1.1	2.6
Lung-Cancer	9434.4	3.4	1.1	2.6
11-Tumors	7959.8	3.4	1.1	2.6

more specifically, D-F2, due to its important savings in runtime and storage requirements.

11.3.2.1 Ensemble Feature Selection

Typically, ensemble learning has been applied to classification, where the most popular methods are *bagging* [51] and *boosting* [52]. In recent works it is proposed to improve the robustness and/or performance of a feature selection algorithm by using multiple feature selection evaluation criteria. For example, in the work [53] five different filters were employed—selecting five different subsets of features to train and to test a specific classifier—and the outputs are combined by simple voting. In the study [54], the adequacy of using an ensemble of filters rather than a single filter was demonstrated on synthetic and real data. There are several ways to design an ensemble [55], but an important problem in the context of this chapter is the computational time taken with regard to individual methods. One possible way of dealing with this is to parallelize the training task. *N* models are generated using the same method, all with different training data (See Fig. 11.10). Therefore, this ensemble approach consists in distributing the training data among a set of nodes. The same feature selection method is then executed on each node.

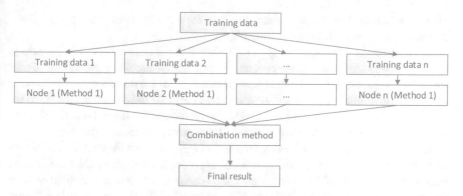

Fig. 11.10 Diagram of the homogeneous distributed ensemble

Following this scheme in the work reported in the article [56], the methods that conform the ensemble were ranker methods. The outputs of the components of the ensemble have to be joined by a combination method in order to produce a unique final output. The combination method, also known as "aggregator", is responsible for conducting the fusion of several rankings. A single final ranking is obtained as output of the combination methods that join all the input rankings. To perform the combination of the different input rankings several different measures are used, ranging from simple calculation measures —minimum, maximum, mean, etc.—to more sophisticated ones like *SVM-Rank* [57]. Since the ranker methods sort all the features, it is necessary to set a threshold in order to obtain a practical subset of features. Most works in the literature use several thresholds that retain different percentages of features [58, 59]. As optimal thresholds are dependent on the particular dataset being studied, several attempts have been made to derive a general automatic threshold [60, 61], thus in the ensemble being described several different threshold values to delimit data dimensionality have been employed, some with fixed retaining percentages (10, 25 and 50%), and other dataset-based [48, 62–64].

Some specific experimental results are shown below for the ensemble approach described, analyzing the average time gains and the test error stability between individual and ensemble approaches. The experiments were performed on seven different datasets—including microarray DNA datasets—using five individual feature selection methods to conform the final ensemble—three filter methods *Information Gain* [19], *mRMR* [35] and *ReliefF* [65] and two embedded methods *SVM-RFE* [12] and *FS-P* [61]—with seven combination methods and five cutoff threshold values (two automatic and three fixed values).

Figure 11.11 show the average time gains for the homogeneous distributed ensemble versus the individual approaches. As can be observed, the homogeneous ensembles compared to the individual approaches improved times by a factor of 100 on average in the best case. The feature selection method which average training times most improved in the distribution process was the embedded *SVM-RFE*. The fact that the *InfoGain* filter yielded the poorest improvement is not surprising, since it is an univariate and fast method, so even attempts at paral-lelization produced no improvement [66].

In analyzing performance, Fig. 11.12 shows the number of cases for which results obtained by the homogeneous distributed ensemble approach were not significantly different than the individual method result (in other words, the number of times that ensemble accuracy were comparable with individual accuracy). This comparison has been made for the *SVM-RFE* method since it has been the method for which the distribution process obtained the greatest average speedup.

As can be observed, the homogeneous *SVM-RFE* ensemble approach obtained very similar results to *SVM-RFE* individual approach regardless of the combination method used. A total of 35 experiments have been performed where it could be seen that the best combination method was *Min*—with 32 of 35 not significantly different results than individual approach—and the worst were *Mean* and *Stuart*—with 28 of 35 not significantly different results than individual approach. Although the homogeneous distributed ensemble approach did not achieve the same classification performance as the individual approach in some specific cases, this is maintained at a reasonable levels. In addition, with the homogeneous distributed ensemble it is possible to improve times by a factor of 100 on average.

The most important advantage of the homogeneous distributed ensemble approach was clearly the great reduction in training times while classification performance held at reasonable levels. This outcome reflects the notion of divide-and-conquer since, in some cases, the result obtained by a feature selection method may be more accurate when the focus is on a local region of the data.

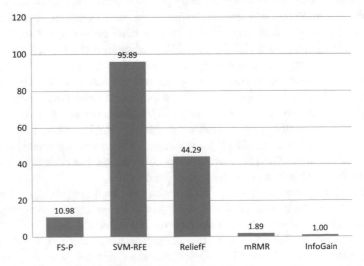

Fig. 11.11 Average speedup for homogeneous distributed ensembles versus individual approaches

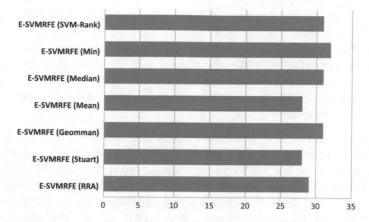

Fig. 11.12 Number of cases when results obtained by SVM-RFE homogeneous distributed ensemble approach were not significantly different than individual result

11.3.3 Parallelization

Another way to speed up feature selection is to distribute calculations across several computing units, either in a single machine or in a computer cluster. Unlike in distributed feature selection, the result obtained is the same as it would be if the computation was run on a single processor. Parallelization is a good option when data is centralized in a location and can be accessed by numerous computing units. In this case, instead of using a single processor to perform all the computations a parallel version of the algorithm can be used to obtain the same results in a shorter time.

Some feature selection algorithms lend themselves well to parallel implementation, usually because they perform a high number of independent computations that are then aggregated to obtain a global result [25]. The parallel version of the algorithm assigns a disjoint subset of these calculations to each computing unit, which effectively breaks the link between computational complexity and time complexity that is, in some cases a larger dataset can be processed in the same time by just using more executors to perform the computation. This transformation allows the use of elaborate feature selection algorithms with high computational complexity to very large datasets if enough computational units are available. Nonetheless, there are a few caveats to this apparently endless scalability. Each computational unit need to have access to the dataset (or at least a fraction of it), which requires that either (1) data is stored in a single location and accessed by the executors or (2) data is distributed to each executor. Both mechanisms suffer when the amount of computational units grows, which limits scalability. Moreover, some algorithms can not be completely parallelized and have sequential parts that maintain the original time complexity, which reduces the overall speedup. Finally, the parallel version of the algorithms can introduce additional code that takes time on itself, incrementing the execution time.

The influence of these factors varies depending on the implementation, so it is up to the data scientist to know the limitations of each algorithm to determine if it is suitable for their particular problem. Listed below are some of the available options in various platforms which will hopefully help the reader decide the most suitable one for their situation.

- **Single machine**: The Weka suite offers data scientist implementations of numerous machine learning algorithms which has contributed to making it a very popular tool that has been downloaded over six million times [24]. Among its features there are several feature selection algorithms, although only sequential versions are provided. In [25], parallel implementations of four of the most popular algorithms, most with quadratic complexities originally, are described and made available.
- **Multiple machines**: Spark implementations of the aforementioned four methods are also provided in the same work [25]. Additionally, a variety of feature selection methods based on information theory are available in the same platform as a package [67] (Fig. 11.13).

11.4   New and Future Research Directions

There are several research lines open in this area. First, there is a clear need for the development of a taxonomy of discretization methods, making it easier for data scientists to choose the most adequate to their problems, while new schemes and

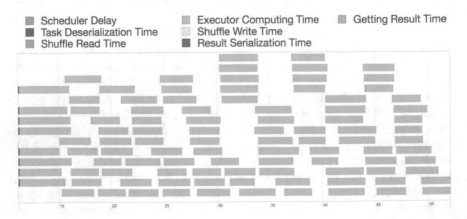

Fig. 11.13 Parallel execution of one operation involved in the Relief-F algorithm across eight processors as depicted by the log of Apache Spark. X axis depicts time. Note how up to eight tasks are performed at the same time

implementations of discretizers should be made available to be able to cope with the high dimensional problems of nowadays. On the feature selection area, there are also open questions. One interesting line of research is the use of data complexity measures. Data complexity analysis is a relatively recent proposal aimed at representing intrinsic data characteristics that are considered complex to classification tasks—such as the amount of overlap between the classes, the shape of the decision boundary or the proximity among the classes—and at identifying correlations with classification performance. The most commonly employed data complexity measures are those proposed by Ho and Basu [49]. More related to our particular interest inhere, there exist proposals which do not use the data complexity measures with respect to a classifier directly, but taking into account a preprocessing technique. Dong and Kothari [68] developed a feature selection method based on the F1 definition. In this context, other works show that, by applying a proper feature selection method, one can reduce the influence of the particular characteristics of microarray data in the error rates of the classifiers induced [69, 70]. Considering class imbalance problems, Luengo et al. [71] analyzed the usefulness of the data complexity measures in order to evaluate the behavior of undersampling and oversampling methods. Morán-Fernández et al. [72] computed several data complexity measures over imbalanced microarray datasets in order to support the application or not of an oversampling method before and after applying feature gene selection. Furthermore, as explained previously, others researchers proposed using data complexity measures to establish threshold values in distributed feature selection [50] and ensemble feature selection [56].

Throughout this chapter, we have highlighted the necessity of preprocessing techniques to deal with high-dimensional settings, as well as presented novel methods and implementations. However, although researchers are intensively working in this area, there are still some challenges that need to be confronted, as pointed out by Bolón-Canedo et al. [58]. For example, many feature selection algorithms are based in computing pairwise correlations, which implies that when dealing with a million features the computer would need to handle a trillion correlations. Another problem is real-time processing, as nowadays we are surrounded by social media networks and portable devices which require sophisticated methods capable of dealing with vast amount of data in real time. And, finally, visualization of the results is an issue that must be carefully considered. The challenge is to enable user-friendly visualization of results so as to enhance interpretability.

In conclusion, data preprocessing techniques are more in need than ever, and their efficiency has been broadly demonstrated in a variety of applications that require dealing with huge amounts of data. However, it seems clear that this high number of features will only continue to increase, so researchers must be prepared to the new challenges that are to come.

References

1. Zhai, Y., Ong, Y., Tsang, I.: The emerging "Big Dimensionality?". IEEE Comput. Intell. Mag. **9**(3), 14–26 (2014)
2. Bache, K., Lichman, M.: UCI machine learning repository. http://archive.ics.uci.edu/ml. Accessed November 2016
3. Chang, C.C., Lin, C.J.: LIBSVM: a library for support vector machines. ACM Trans. Intell. Syst. Technol. (TIST) **2**(3), 27 (2011)
4. Bellman, R.: Dynamic Programming. Princeton UP, Princeton (1957)
5. García, S., Luengo, J., Herrera, F.: Data Preprocessing in Data Mining. Springer, New York (2015)
6. Han, J., Pei, J., Kamber, M.: Data Mining: Concepts and Techniques. Elsevier, MA, USA (2011)
7. Witten, H., Frank, E., Hall, M.A.: Data Mining: Practical Machine Learning Tools and Techniques. Elsevier, Philadelphia (2011)
8. Guyon, I., Elisseeff, A.: An introduction to variable and feature selection. J. Mach. Learn. Res. **3**, 1157–1182 (2003)
9. Bolón-Canedo, V., Sánchez-Maroño, N., Alonso-Betanzos, A.: Feature Selection for High-Dimensional Data. Springer-Verlag, Berlin (2015)
10. Zhai, Y., Ong, Y.S., Tsang, I.W.: The emerging "Big Dimensionality". IEEE Comput. Intell. Mag. **9**, 16–26 (2014)
11. Ding, C., Peng, H.: Minimum redundancy feature selection from microarray gene expression data. J. Bioinform. Comput. Biol. **3**, 185–205 (2005)
12. Guyon, I., Weston, J., Barnhill, S., Vapnik, V.: Gene selection for cancer classification using Support vector Machines. Mach. Learn. **46**, 389–422 (2002)
13. Shah, M., Marchand, M., Corbeil, J.: Feature selection with conjunctions of decision stumps and learning from microarray data. IEEE Trans. Pattern Anal. Mach. Intell. **34**, 174–186 (2012)
14. Ramírez-Gallego, S., Lastra, I., Martínez-Rego, D., Bolón-Canedo, D., Benítez, J.M., Herrera, F., Alonso-Betanzos, A.: Fast-mRMR: Fast minimum redundancy maximum relevance algorithm for high-dimensional big data. Int. J. Intell. Syst. **0**, 1–19 (2016)
15. Bolón-Canedo, V., Sánchez-Maroño, N., Alonso-Betanzos, A., Benítez, J.M., Herrera, F.: A review of microarray datasets and applied feature selection methods. Inf. Sci. **282**, 111–135 (2014)
16. Bolón-Canedo, V., Sánchez-Maroño, N., Alonso-Betanzos, A.: On the effectiveness of discretization on gene selection of microarray data. In: Proceedings International Joint Conference on Neural Networks (IJCNN) 2010, pp. 167–174 (2010)
17. Ramírez-Gallego, S., García, S., Mouriño-Talín, H.: Martínez-Rego, D., Bolón-Canedo, V., Alonso-Betanzos, A., Benítez, J.M., Herrera, F.: Data discretization: taxonomy and big data challenge. Wiley Interdisciplinary Reviews. Data Min. Knowl. Disc. **6**, 5–21 (2016)
18. Yang, Y., Webb, G.I.: Discretization for naive-Bayes learning: managing discretization bias and variance. Mach. Learn. **74**(1), 39–74 (2009)
19. Quinlan, J.R.: Induction of decision trees. Mach. Learn. **1**(1), 81–106 (1986)
20. Hu, H.W., Chen, Y.L., Tang, K.: A dynamic discretization approach for constructing decision trees with a continuous label. IEEE Trans. Knowl. Data Eng. **21**(11), 1505–1514 (2009)
21. Yang, Y., Webb, G.I.: Proportional k-interval discretization for naive-Bayes classifiers. In: European Conference on Machine Learning, pp. 564–575, Springer, Berlin (2001)
22. Fayyad, U., Irani, K.B.: Multi-interval discretization of continuous-valued attributes for classification learning. In: Proc. IJCAI-93, pp. 1022–1027 (1993)
23. Machine Learning Library (MLlib) for Spark, Mllib.: [Online]. Available: http://spark.apache. org/docs/latest/mllib-guide.html (2015)
24. Hall, M., Frank, E., Holmes, G., Pfahringer, B., Reutemann, P., Witten, I.H.: The WEKA data mining software: an update. ACM SIGKDD Explor. Newsl. **11**(1), 10–18 (2009)

25. Eiras-Franco, C., Bolón-Canedo, V., Ramos, S., González-Domínguez, J., Alonso-Betanzos, A., Touriño, J.: Multithreaded and Spark parallelization of feature selection filters. J. Comput. Sci. (2016). https://doi.org/10.1016/j.jocs.2016.07.002
26. Ramírez-Gallego, S., García, S., Mouriño-Talín, H., Martínez-Rego, D., Bolón-Canedo, V., Alonso-Betanzos, A., Benítez, J.M., Herrera, F.: Distributed entropy minimization discretizer for big data analysis under Apache Spark. In: Proc. 9th IEEE International Conference on Big Data Science and Engineering (IEEE BigDataSE-15) Trustcom/BigDataSE/ISPA vol. 2, pp. 33–40, IEEE (2015)
27. Boutsidis, C., Drineas, P., Mahoney, M.W.: Unsupervised feature selection for the k-means clustering problem. Adv. Neural Inf. Process. Syst. **22**, 153–161 (2009). https://papers.nips.cc/book/advances-in-neural-information-processing-systems-22-2009
28. Roth, V., Lange, T.: Feature selection in clustering problems. Adv. Neural Inf. Process. Syst. **16** (2003). https://papers.nips.cc/book/advances-inneural-information-processing-systems-16-2003
29. Leardi, R., Lupiáñez González, A.: Genetic algorithms applied to feature selection in PLS regression: how and when to use them. Chemometr. Intell. Lab. Syst. **41**(2), 195–207 (1998)
30. Paul, D., Bair, E., Hastie, T., Tibshirani, R.: "Preconditioning" for feature selection and regression in high-dimensional problems. Ann. Stat. **36**(4), 1595–1618, 2008
31. Dash, M., Liu, H.: Feature selection for classification. Intell. data anal. **1**(3), 131–156 (1997)
32. Pal, M., Foody, G.M.: Feature selection for classification of hyperspectral data by SVM. IEEE Trans. Geosci. Remote Sens. **48**(5), 2297–2307 (2010)
33. Azmadian, F., Yilmazer, A., Dy, J.G., Aslam, J.A., Kaeli, D.R.: Accelerated feature selection for outlier detection using the local kernel density ratio. In: Proceedings IEEE 12th International Conference on Data Mining, pp. 51–60 (2012)
34. Guillén, A., García Arenas, M.I., van Heeswijk, M., Sovilj, D., Lendasse, A., Herrera, L.J., Pomares, H., Rojas, I.: Fast feature selection in a GPU cluster using the delta test. Entropy **16**, 854–869 (2014)
35. Peng, H., Long, F., Ding, C.: Feature selection based on mutual information criteria of max-dependency, max-relevance, and min-redundancy. IEEE Trans. Pattern Anal. Mach. Intell. **27**, 1126–1238 (2005)
36. Fast-mRMR package. https://github.com/sramirez/fast-mRMR. Accessed December 2016
37. Apache Spark: Lightning-fast cluster computing. http://shop.oreilly.com/product/063692002 8512.do (2015). Accessed December 2016
38. NVIDIA accelerated computing, CUDA platforms. https://developer.nvidia.com/additional-resources. Accessed December 2016
39. Ramírez-Gallego, S., Lastra, I., Martínez-Rego, D., Bolón-Canedo, V., Benítez, J.M., Herrera, F., Alonso-Betanzos, A.: Fast-mRMR: Fast minimum redundancy maximum relevance algorithm for high-dimensional big data. Int. J. Intell. Syst. **0**, 1–19 (2016)
40. Das, K., Bhaduri, K., Kargupta, H.: A local asynchronous distributed privacy preserving feature selection algorithm for large peer-to-peer networks. Knowl. Inf. Syst. **24**(3), 341–367 (2010)
41. Banerjee, M., Chakravarty, S.: Privacy preserving feature selection for distributed data using virtual dimension. In: Proceedings of the 20th ACM international conference on Information and knowledge management, pp. 2281–2284 (2011)
42. Tan, M., Tsang, I.W., Wang, L.: Towards ultrahigh dimensional feature selection for big data. J. Mach. Learn. Res., **15**(1), 1371–1429 (2014)
43. Peralta, D., del Río, S., Ramírez-Gallego, S., Triguero, I., Benítez, J.M., Herrera, F.: Evolutionary feature selection for big data classification: a mapreduce approach. Math. Probl. Eng. **2015** (2015). http://dx.doi.org/10.1155/2015/246139
44. Bolón-Canedo, V., Sánchez-Maroño, N., Cerviño-Rabuñal, J.: Toward parallel feature selection from vertically partitioned data. In: Proceedings of European Symposium on Artificial Neural Networks (ESANN), pp. 395–400 (2014)
45. Bolón-Canedo, V., Sánchez-Maroño, N. and Cerviño-Rabuñal, J.: Scaling up feature selection: a distributed filter approach. In: Proceedings of Conference of the Spanish Association for Artificial Intelligence (CAEPIA), pp. 121–130 (2013)

46. Bolón-Canedo, V., Sánchez-Maroño, N., Alonso-Betanzos, A.: Distributed feature selection: an application to microarray data classification. Appl. Soft Comput. **30**, 136–150 (2015)
47. Bolón-Canedo, V., Sánchez-Maroño, N., Alonso-Betanzos, A.: A distributed feature selection approach based on a complexity measure. International Work-Conference on Artificial Neural Networks, 15–28 (2015)
48. Morán-Fernández, L., Bolón-Canedo, V., Alonso-Betanzos, A.: A time efficient approach for distributed feature selection partitioning by features. Lecture Notes in Artificial Intelligence. LNAI-9422, 16th Conference of the Spanish Association for Artificial Intelligence, pp. 245–254 (2015)
49. Ho, T.K., Basu, M.: Complexity measures of supervised classification problems. IEEE Trans. Pattern Anal. Mach. Intell. **24**(3), 289–300 (2002)
50. Morán-Fernández, L., Bolón-Canedo, V., Alonso-Betanzos, A.: Centralized vs. distributed feature selection methods based on data complexity measures. Knowledge-Based Syst. **105**, 48–59 (2016)
51. Breiman, L.: Bagging predictors. Mach. Learn. **24**(2), 123–140 (1996)
52. Schapire, R.E.: The strength of weak learnability. Mach. Learn. **5**(2), 197–227 (1990)
53. Bolón-Canedo, V., Sánchez-Maroño, N., Alonso-Betanzos, A.: An ensemble of filters and classifiers for microarray data classification. Pattern Recogn. **45**(1), 531–539 (2012)
54. Bolón-Canedo, V., Sánchez-Maroño, N., Alonso-Betanzos, A.: Data classification using an ensemble of filters. Neurocomputing **135**, 13–20 (2014)
55. Bramer, M.: Principles of Data Mining, 2nd edn.. Springer, London (2013)
56. Seijo-Pardo, B., Porto-Díaz, I., Bolón-Canedo, V., Alonso-Betanzos, A.: Ensemble feature selection: homogeneous and heterogeneous approaches. Knowledge-Based Syst. **118**, 124–139 (2017)
57. Joachims, T.: Optimizing search engines using clickthrough data. In: Proceedings of the eighth ACM SIGKDD international conference on Knowledge discovery and data mining, ACM (2002)
58. Bolón-Canedo, V., Sánchez-Maroño, N., Alonso-Betanzos, A.: Recent advances and emerging challenges of feature selection in the context of big data. Knowl.-Based Syst. **86**, 33–45 (2015)
59. Bolón-Canedo, V., Sánchez-Maroño, N., Alonso-Betanzos, A.: A review of feature selection methods on synthetic data. Knowl. Inf. Syst. **34**(3), 483–519 (2013)
60. Khoshgoftaar, T. M., Golawala, M. and Van Hulse, J.: An empirical study of learning from imbalanced data using random forest. 19th IEEE International Conference on Tools with Artificial Intelligence (ICTAI 2007), **2**, IEEE (2007)
61. Mejía-Lavalle, M., Sucar, E. and Arroyo, G.: Feature selection with a perceptron neural net. In: Proceedings of the international workshop on feature selection for data mining (2006)
62. Seijo-Pardo, B., Bolón-Canedo, V. and Alonso-Betanzos, A.: Using a feature selection ensemble on DNA microarray datasets. In: Proceeding of 24th European Symposium on Artificial Neural Networks, pp. 277–282 (2016)
63. Seijo-Pardo, B., Bolón-Canedo, V. and Alonso-Betanzos, A.: Using data complexity measures for thresholding in feature selection rankers. Lecture Notes in Artificial Intelligence. LNAI-9868 Advances in Artificial Intelligence. 17th Conference of the Spanish Association for Artificial Intelligence, CAEPIA 2016, pp. 121–131 (2016)
64. Wang, H. and Khoshgoftaar, T. M. and Napolitano, A.: A comparative study of ensemble feature selection techniques for software defect prediction, Machine Learning and Applications (ICMLA), 2010 Ninth International Conference on. IEEE (2010)
65. Kononenko, I.: Estimating attributes: analysis and extensions of RELIEF. Mach. Learn.: ECML-94 **784**, 171–182 (1994)
66. Eiras-Franco, C., Bolón-Canedo, V., Ramos, S., González-Domínguez, J., Alonso-Betanzos, A. and Touriño, J.: Paralelización de algoritmos de selección de características en la plataforma Weka, CAEPIA 2015 (Workshop BigDADE), pp. 949–958 (2015)
67. Ramírez-Gallego, S., Mouriño-Talín, H., Martínez-Rego, D., Bolón-Canedo, V., Benítez, J. M., Alonso-Betanzos, A., Herrera, F.: An information theoretic feature selection framework

for big data under Apache Spark. arXiv preprint arXiv:1610.04154 (2016). Available: IEEE Trans Syst Man Cybern: Syst PP(99). doi:10.1109/TSMC.2017.2670926

68. Dong, M., Kothari, R.: Feature subset selection using a new definition of classifiability. Pattern Recogn. Lett. **24**(9), 1215–1225 (2003)
69. Lorena, A.C., Costa, I.G., Spolaôr, N., De Souto, M.C.P.: Analysis of complexity indices for classification problems: cancer gene expression data. Neurocomputing **75**(1), 33–42 (2012)
70. Morán-Fernández, L., Bolón-Canedo, V., Alonso-Betanzos, A.: Can classification performance be predicted by complexity measures? A study using microarray data, Knowledge and Information Systems, pp. 1–24 (2016)
71. Luengo, J., Fernández, A., García, S., Herrera, F.: Addressing data complexity for imbalanced data sets: analysis of SMOTE-based oversampling and evolutionary undersampling. Soft. Comput. **15**(10), 1909–1936 (2011)
72. Morán-Fernández, L., Bolón-Canedo, V., Alonso-Betanzos, A.: Data complexity measures for analyzing the effect of SMOTE over microarrays. European Symposium on Artificial Neural Networks, Computational Intelligence and Machine Learning (2016)
73. Liu, H., Setiono, R.: Chi2: Feature selection and discretization of numeric attributes. In: Proceedings of the Seventh International Conference on Tools with Artificial Intelligence, pp. 388–391 (1995)
74. Duda, R.O., Hart, P.E., Stork, D.G.: Pattern Classification. Wiley, New York (1999)
75. Hall, M.A.: Correlation-based feature selection for machine learning. PhD Thesis, The University of Waikato (1999)
76. Yu, L., Liu, H.: Feature selection for high-dimensional data: a fast correlation-based filter solution. Machine Learning-International Workshop then Conference-, 856–863 (2003)
77. Zhao, Z., Liu, H.: Searching for interacting features. IJCAI, **7**, 1156–1161 (2007)
78. Dash, M., Liu, H.: Consistency-based search in feature selection. Artif. Intell. **151**(1), 155–176 (2003)

Chapter 12
Gerotranscendental Change in the Elderly: An Analysis of 'Faith' and 'Leaving from Clinging to Self'

Takayuki Kawaura and Yasuyuki Sugatani

Abstract Recently, the number of elderly persons (65 years or older) has in-creased in Japan due to the aging of the general population. Recent changes in family structure have greatly increased the percentage of single-elderly households. Based on these social circumstances, participants were classified into living-alone (single-elderly households) and living-together (elderly households with a spouse only) groups to determine group differences in this study. Participants were 56 persons living in fee-based nursing homes for the elderly located in five pre-fectures in Japan. A questionnaire was developed based on Gerontranscendence: A Developmental Theory of Positive Aging by L. Tornstam to ask the elderly about eight gerontranscendence-related themes.

Of these eight themes, the responses regarding gerotranscendental changes, such as religiosity and separation from clinging to self, were examined and analyzed using quantification theory II. The results of the analyses confirmed a gerotran-scendental tendency of "superficial obsessiveness," both in the liv-ing-alone and living-together groups. According to residence status, a gero-transcendental ten-dency of "feeling being made to live" was confirmed in the living-alone group, and a gerotranscendental tendency for "feelings of famili-arity toward a supernatural being, such as God" was confirmed in the living-together group. In addition, there were items to which responses suggested a tendency opposite to gerontranscen-dence, which could not be clarified in this study. However, reasons explaining this tendency may be determined by an in-depth analysis of the eight themes comprising the entire questionnaire used in this study.

T. Kawaura (✉)
Department of Mathematics, Kansai Medical University, 2-5-1 Shinmachi,
Hirakata, Osaka 573-1010, Japan
e-mail: kawaurat@hirakata.kmu.ac.jp

Y. Sugatani
Center for Medical Education, Kansai Medical University, 2-5-1 Shinmachi,
Hirakata, Osaka 573-1010, Japan
e-mail: sugatani@hirakata.kmu.ac.jp

© Springer International Publishing AG 2018
D.E. Holmes and L.C. Jain (eds.), *Advances in Biomedical Informatics*,
Intelligent Systems Reference Library 137,
https://doi.org/10.1007/978-3-319-67513-8_12

Keywords Gerotranscendence · The elderly · Psychological characteristics · Habitation · Quantification theory type II

12.1 Introduction

According to the Statistics Bureau of the Ministry of Internal Affairs and Communications, there are 34,436,000 Japanese elderly aged 65 years or older, accounting for 27.1% of the total population (population aging rate) as of July 1, 2016 (definitive value) [1]. Although the population aged 65 or older did not reach 5% of the total population in 1950, it reached 7% in 1970 and doubled to 14% or more in 1994, beginning a period characterized by an "aging society." Currently, the percentage has exceeded 27% and one in four people is elderly. One in ten people is aged 75 years or older (accounting for 13.2% of the total population), which is termed a "full-scale aged society." [2]

According to residence status, there are 23,724,000 households with an elderly member aged 65 years or older, accounting for 47.1% of all households (50,361,000 households) as of June 4, 2015, as shown in the Comprehensive Survey of Living Conditions [3] performed by the Ministry of Health, Labour and Welfare. An upward trend has been shown in the percentage of households with an elderly member aged 65 years or older over time (27.3% in 1989, 31.1% in 1995, 35.8% in 2001, 40.1% in 2007, 44.7% in 2013).

The most common family structure for this age group was "living with children" (13,526,000 persons accounting for 39.0% of the elderly aged 65 years or more), followed by "households with a spouse only" (at least one spouse is 65 years old or older; 13,467,000 persons accounting for 38.9% of those) and "single-households" (6243,000 persons who account for 18.0% of this population). An upward trend has also been shown in the percentage of single-households over time (11.2% in 1989, 12.6% in 1995, 13.8% in 2001, 15.7% in 2007, 17.7% in 2013).

Previous studies and surveys have revealed that stress tends to increase due to disease or a sense of isolation in single elderly, which leads to new disease [4, 5].

In this study, we report the results of a survey that was performed with the elderly, using a questionnaire that was developed based on *Gerontranscendence: A Developmental Theory of Positive Aging* by L. Tornstam [6–8]. Using quantification theory II, the data were analyzed and compared by classifying the participants according to their residence status: single-elderly households (the living-alone group) and elderly households with a spouse only (the living-together group). Based on the analysis, characteristics of each group were determined to explore support measures for the elderly.

[Question D: Did any change in "state of mind" and "way of thinking" occur as a result of getting old?]

D-1: I have become less likely to focus on superficial things.
D-2: I have become uninterested in things and money.
D-3: I feel happiness even in small matters.
D-4: Increasingly I feel that I am made to live.
D-5: I have deeper faith.
D-6: I feel something familiar about a supernatural being, such as God.
D-7: I feel that a sense of fear of death has disappeared.
D-8: I strongly want to live out my life.

12.2   Data

The data were obtained from a total of 56 people who lived in fee-based nursing homes for the elderly located in five prefectures (Hyogo, Chiba, Shizuoka, Kanagawa, and Ehime) in Japan between May 2010 and March 2011.

A questionnaire was developed based on *Gerontranscendence: A Developmental Theory of Positive Aging* by L. Tornstam, comprising 35 questions pertaining to the following five categories: "A. Change in my feelings toward myself," "B. Change in how I associate with other people;" "C. Change in my feelings toward my family," "D. Transcendental changes such as faith and separation from clinging to self," and "E. Negative changes resulting from old age."

Items measuring general characteristics were also included in the questionnaire and the following seven characteristics were used in the analysis for this study:

Age group (early elderly (65–74), late-stage elderly (75–84), super elderly (85 +))
   Gender (male, female)
   Children (presence or absence)
   Awareness of health (cheerful, average, anxious)
   Medical examination (whether they consult a doctor)
   Lifestyle (affluent, average, poor)
   Care and support (whether they receive care services)

In this study, participants were classified into three age groups: early elderly (aged 65–74 years old), late-stage elderly (aged 75–84 years old), and super elderly (aged 85 years old or greater). In addition, care and support responses indicate whether the participants receive support such as physical care or housework.

12.3   Analysis

The data obtained from 56 participants were analyzed by classifying them into the living-alone group (37 participants) and the living-together group (19 participants). Using quantification theory II [9–12], the impact of residence status (single-elderly households or elderly households with a spouse only) on the responses to each question was analyzed: the 35 questionnaire items were treated as external criteria (objective variable) and the seven general characteristics were treated as explanatory variables.

This study focused on the eight responses for question D: Transcendental changes such as faith and separation from clinging to self (see Table 12.1). Based on the category scores calculated using quantification theory II, we determined whether there was a transcendental tendency with age group as an explanatory variable and examined the influence of the other explanatory variables.

12.4   Results

Table 12.1 shows the results of the association between residence status and the general characteristics: the p value indicates a p value for the chi square test; Cramer's V is the Cramer's coefficient of association calculated based on the value obtained from chi square test.

Table 12.1 An association was found between residence status and three of the general characteristics

Item	Category	χ^2 value	P value	Cramer's V
Age group	Early elderly Late-stage elderly Super elderly	7.139	0.028*	0.3570
Gender	Male Female	4.487	0.034*	0.2831
Children	Presence Absence	0.430	0.512	0.0876
Health	Cheerful Average Anxious	2.261	0.323	0.2009
Medical examination	Consult a doctor Do not consult a doctor	0.002	0.967	0.0055
Lifestyle	Affluent Average Poor	4.747	$0.093^\dagger$	0.2912
Care, support	Receive services Do not receive services	0.239	0.625	0.0653

*:p < 0.05, †:p < 0.10

$$\text{Cramer's V} = \sqrt{\frac{\chi^2}{N(\min[k, l] - 1)}}$$

N Number of the data;
k Number of the lines in the contingency table;
l Number of the columns in the contingency table

Cramer's coefficient of association is a value from 0 to 1 indicating the association among categorical data, where Cramer's V of 0.25 or more and less than 0.50 indicates an association and a value of 0.50 or more indicates a very strong association.

As shown in Table 12.1, an association was found between residence status and three of the general characteristics: age group, gender, and lifestyle.

The responses to the eight items for question D: Transcendental changes such as faith and separation from clinging to self were examined below.

12.4.1 Superficial Obsessiveness

To further elucidate cases where participants became less likely to focus on superficial things or did not become less likely to focus on superficial things, the associations between the general characteristic categories, including age group, were examined according to the residence status, with responses to item D-1: I have become less likely to focus on superficial things (yes or no) as the external criterion (target variable).

The range denotes differences between the maximum and minimum category scores for the explanatory variable; explanatory variables with higher range values indicate greater contributions to the target variable. The partial correlation coefficient indicates a correlation between the target variable and the explanatory variable and expresses the impact of the explanatory variable on the target variable. Negative values for category scores indicate a tendency to become less likely to focus on superficial things, while positive values indicate a tendency to not become less likely to focus on superficial things (Table 12.2).

Table 12.2 Average score, correlation ratio, and predictive value in item D-1

	D-1	Average score	Correlation ratio	Predictive value (%)
Living alone	Yes	−0.3367	0.3528	78.38
	No	1.0476		
Living together	Yes	−0.5309	0.7892	94.74
	No	1.4865		

With age group as an explanatory variable, a tendency to become less likely to focus on superficial things was exhibited by the super elderly in the living-alone group, and by the late-stage elderly and the super elderly in the living-together group. A tendency to not become less likely to focus on superficial things was exhibited by the early elderly and late-stage elderly in the living-alone group, and by the early elderly in the living-together group. n terms of aging, regardless of the residence status, there was a relationship between the age group categories and participants' tendency to not become less likely to focus on superficial things. Therefore, for superficial obsessiveness, a transcendental tendency was demonstrated.

As shown in Tables 12.3 and 12.4, when responses to item D-1 were used as external criterion, the explanatory variables with a higher range were age group in the living-alone group (1.9478) and lifestyle in the living-together group (5.2405). The explanatory variables with a higher partial correlation coefficient were age group in the living-alone group (0.4173), similar to the range results, and health in the living-together group (0.8543). The analysis using Cramer's coefficient of association demonstrated associations between responses to item D-1 and several general characteristics: age group (0.2912) and health (0.2505) in the living-alone group, and age group (0.2624), health (0.4966), and medical examination (0.2588) in the living-together group.

Table 12.3 Association and impact of each factor (living-alone group) in item D-1

Item	Category	χ^2 value	P value	Cramer's V	Category score	−	+	Range	Partial correlation coefficient
Age group	Early elderly	3.1372	0.2083	0.2912	1.2803			1.9478	0.4173
	Late-stage elderly				0.1456				
	Super elderly				−0.6675				
Gender	Male	0.2281	0.6329	0.0785	−0.0446			0.0532	0.0123
	Female				0.0086				
Children	Presence	1.6556	0.1982	0.2115	0.3617			0.8923	0.2923
	Absence				−0.5306				
Health	Cheerful	2.3223	0.3131	0.2505	0.2932			1.5936	0.2814
	Average				−0.3448				
	Anxious				1.2488				
Medical examination	Consult a doctor	2.3018	0.1292	0.2494	0.1717			1.0589	0.2592
	Do not consult a doctor				−0.8872				
Lifestyle	Affluent	1.4244	0.4906	0.1962	−0.2176			1.7079	0.3085
	Average				0.4758				
	Poor				−1.2321				
Care, support	Receive services	0.2398	0.6243	0.0805	0.1413			0.1936	0.0546
	Do not receive services				−0.0523				

Table 12.4 Association and impact of each factor (living-together group) in D-1

Item	Category	χ^2 value	P value	Cramer's V	Category score	−	+	Range	Partial correlation coefficient
Age group	Early elderly				0.2424				
	Late-stage elderly	1.3081	0.5199	0.2624	−0.1639			0.4748	0.2634
	Super elderly				−0.2323				
Gender	Male				−0.0132				
	Female	0.0123	0.9116	0.0255	0.0096			0.0228	0.0153
Children	Presence				0.0274				
	Absence	0.4211	0.5164	0.1489	−0.0593			0.0867	0.0714
Health	Cheerful				1.1710				
	Average	4.6850	0.0961	0.4966	−0.9699			2.5746	0.8543
	Anxious				1.6047				
Medical examination	Consult a doctor				0.4224				
	Do not consult a doctor	1.2723	0.2593	0.2588	−2.2527			2.6751	0.8195
Lifestyle	Affluent				2.1359				
	Average	0.3974	0.8198	0.1446	−0.3885			5.2405	0.8021
	Poor				−3.1046				
Care, support	Receive services				−2.0794				
	Do not receive services	0.0045	0.9464	0.0154	0.5545			2.6340	0.7511

12.4.2   Changes in Interest in Things and Money

To elucidate cases where participants became uninterested in things and money or did not become uninterested in things and money, the associations between the general characteristics, including age group, were examined according to residence status, with responses to item D-2: I have become uninterested in things and money (yes or no) as the external criterion (target variable).

The results of the analysis are shown in Tables 12.5, 12.6, and 12.7. Negative values for the category scores indicate a tendency to become uninterested in things

Table 12.5 Average score, correlation ratio, and predictive value in item D-2

	D-2	Average score	Correlation ratio	Predictive value (%)
Living alone	Yes	0.4807	0.1964	67.57
	No	−0.4086		
Living together	Yes	1.9029	0.6790	94.74
	No	−0.3568		

Table 12.6 Association and impact of each factor (living-alone group) in item D-2

Item	Category	χ^2 value	P value	Cramer's V	Category score	−	+	Range	Partial correlation coefficient
Age group	Early elderly	0.0739	0.9637	0.0447	−0.3055			0.9581	0.2023
	Late-stage elderly				0.5227				
	Super elderly				−0.4354				
Gender	Male	0.0474	0.8277	0.0358	−0.2180			0.2602	0.0440
	Female				0.0422				
Children	Presence	0.5543	0.4565	0.1224	0.3406			0.8402	0.1885
	Absence				−0.4996				
Health	Cheerful	0.0236	0.9883	0.0252	0.0916			0.2254	0.0363
	Average				−0.0553				
	Anxious				−0.1338				
Medical examination	Consult a doctor	0.4587	0.4982	0.1113	−0.1976			1.2184	0.2020
	Do not consult a doctor				1.0209				
Lifestyle	Affluent	3.2503	0.1969	0.2964	−0.9576			2.2830	0.3883
	Average				0.8287				
	Poor				1.3255				
Care, support	Receive services	0.1951	0.6587	0.0726	1.0990			1.5060	0.2762
	Do not receive services				−0.4070				

and money, while positive values indicate a tendency to not become uninterested in things and money.

With age group as the explanatory variable, a tendency for participants to become uninterested in things and money was exhibited by the early elderly and the super elderly in the living-alone group, and by the super elderly and the early elderly in the living-together group. A tendency to not become uninterested in things and money was exhibited by the late-stage elderly in the living-alone group and by the late-stage elderly and the super elderly in the living-together group. In the living-alone group, the early elderly also demonstrated a tendency to not become uninterested in things and money. Therefore, no transcendental tendency was demonstrated. In addition, a tendency opposite to transcendence was shown in the living-together group.

As shown in Tables 12.6 and 12.7, when responses to item D-2 were used as an external criterion, the explanatory variable with a higher range was lifestyle in both the living-alone (2.2830) and the living-together (2.7624) groups. The explanatory variable with a higher partial correlation coefficient was lifestyle in the living-alone group (0.3883), similar to the range results, and gender in the living-together group (0.6877). The analysis using Cramer's coefficient of association indicated relationships between responses to item D-2 and the following general characteristics:

Table 12.7 Association and impact of each factor (living-together group) in D-2

Item	Category	χ^2 value	P value	Cramer's V	Category score	−	+	Range	Partial correlation coefficient
Age group	Early elderly				−0.3871				
	Late-stage elderly	3.9583	0.1382	0.4564	0.0729			1.6071	0.4183
	Super elderly				1.2200				
Gender	Male				−1.0632				
	Female	0.8816	0.3478	0.2154	0.7733			1.8365	0.6877
Children	Presence				0.1438				
	Absence	2.0299	0.1542	0.3269	−0.3115			0.4553	0.2759
Health	Cheerful				0.9320				
	Average	2.5909	0.2738	0.3693	−0.3909			1.3230	0.5400
	Anxious				−0.1199				
Medical examination	Consult a doctor				0.2256				
	Do not consult a doctor	0.8247	0.3638	0.2083	−1.2032			1.4288	0.4888
Lifestyle	Affluent				−1.3532				
	Average	4.4955	0.1056	0.4864	0.2860			2.7624	0.5448
	Poor				1.4092				
Care, support	Receive services				0.5358				
	Do not receive services	0.3233	0.5697	0.1304	−0.1429			0.6787	0.2322

lifestyle (0.2964) in the living-alone group, and age group (0.4564), children (0.3269), health (0.3693), and lifestyle (0.4864) in the living-together group.

12.4.3 Changes in Happiness in Small Matters

To clarify cases where participants feel happiness even in a small matter or do not feel happiness even in a small matter, associations between the general characteristics, including age group, were examined according to the residence status, with responses to item D-3: I feel happiness even in small matters (yes or no) as an external criterion (target variable).

Table 12.8 Average score, correlation ratio, and predictive value in item D-3

	D-3	Average score	Correlation ratio	Predictive value (%)
Living alone	Yes	0.1629	0.4647	97.30
	No	−2.8517		
Living together	Yes	−0.2180	0.8552	100
	No	3.9236		

Table 12.9 Association and impact of each factor (living-alone group) in item D-3

Item	Category	χ^2 value	P value	Cramer's V	Category score	− +	Range	Partial correlation coefficient
Age group	Early elderly				−0.1437			
	Late-stage elderly	0.4118	0.8139	0.1055	0.2097		0.3758	0.1539
	Super elderly				−0.1662			
Gender	Male				1.8582			
	Female	10.9238	0.0009	0.5434	−0.3597		2.2179	0.5740
Children	Presence				0.1497			
	Absence	1.4416	0.2299	0.1974	−0.2195		0.3692	0.1569
Health	Cheerful				−0.5828			
	Average	1.7971	0.4072	0.2204	0.4081		0.9909	0.3869
	Anxious				0.2896			
Medical examination	Consult a doctor				0.0018			
	Do not consult a doctor	0.4092	0.5224	0.1052	−0.0093		0.0110	0.0035
Lifestyle	Affluent				−0.2950			
	Average	5.6271	0.0600	0.3900	0.2139		0.9246	0.2465
	Poor				0.6296			
Care, support	Receive services				0.0258			
	Do not receive services	0.7831	0.3762	0.1455	−0.0096		0.0354	0.0128

The results of the analysis are shown in Tables 12.8, 12.9, and 12.10. Negative values for the category scores indicate a tendency to feel happiness even in small matters, while positive values indicate a tendency to not feel happiness even in small matters.

With age group as an explanatory variable, a tendency to feels happiness even in small matters was demonstrated by the early elderly and the super elderly in the living-alone group, and by the late-stage elderly in the living-together group. A tendency to not feel happiness even in a small matter was exhibited by the late-stage elderly in the living-alone group, and by the late-stage elderly, the early elderly, and the super elderly in the living-together group.

In the living-alone group, the early elderly also showed a tendency to feel happiness even in small matters and subsequently, no transcendental tendency was observed. In the living-together group, the late-stage elderly exhibited a tendency to feel happiness even in small matters and no transcendental tendency was observed.

As shown in Tables 12.9 and 12.10, when responses to item D-3 were used as an external criterion, the explanatory variables with a higher range were gender in the living-alone group (2.2179) and lifestyle in the living-together group (1.7433). The explanatory variables with a higher partial correlation coefficient were age-group in the living-alone group (0.5740), similar to the range results, and care and support in the living-together group (0.7149). The analysis of responses to item D-3 using Cramer's coefficient of association indicated a very strong relationship with gender

Table 12.10 Association and impact of each factor (living-together group) in D-3

Item	Category	χ^2 value	P value	Cramer's V	Category score	−	+	Range	Partial correlation coefficient
Age group	Early elderly	8.9722	0.0113	0.6872	0.0888			0.8302	0.4135
					−0.2155				
	Super elderly				0.6146				
Gender	Male	1.4514	0.2283	0.2764	0.1602			0.2767	0.2894
	Female				−0.1165				
Children	Presence	2.2870	0.1305	0.3469	−0.2097			0.6641	0.5862
	Absence				0.4544				
Health	Cheerful	5.6296	0.0599	0.5443	−0.4646			1.5219	0.6860
	Average				−0.0772				
	Anxious				1.0574				
Medical examination	Consult a doctor	0.1979	0.6564	0.1021	−0.1354			0.8578	0.5262
	Do not consult a doctor				0.7224				
Lifestyle	Affluent	0.3770	0.8282	0.1409	−1.3253			1.7433	0.6786
	Average				0.4180				
	Poor				−0.5505				
Care, support	Receive services	3.9583	0.0466	0.4564	1.3545			1.7156	0.7149
	Do not receive services				−0.3612				

(0.5434) and a moderate relationship with lifestyle (0.3900) in the living-alone group. In the living-together group, very strong relationships were shown between responses to item D3 and age group (0.6872) and health (0.5443), and moderate relationships were shown with children (0.3469) and care and support (0.4564).

12.4.4 Changes in Feelings of Being Made to Live

To elucidate cases where participants increasingly feel that they were made to live or do not feel that they were made to live, associations between the general characteristic categories, including age group, were examined according to the residence status, with responses to item D-4: Increasingly I feel that I am made to live (yes or no) as the external criterion (target variable).

The results of this analysis are shown in Tables 12.11, 12.12, and 12.13. Negative category score values indicate a tendency to increasingly feel that participants were made to live, while positive values indicate a tendency to not feel that they were made to live.

With age group as an explanatory variable, a tendency to increasingly feel that they were made to live was shown by the super elderly in the living-alone group, and by the late-stage elderly in the living-together group. A tendency for

Table 12.11 Average score, correlation ratio, and predictive value in item D-4

	D-4	Average score	Correlation ratio	Predictive value (%)
Living alone	Yes	−0.2212	0.2529	75.68
	No	1.1430		
Living together	Yes	−0.5504	0.6563	89.47
	No	1.1925		

Table 12.12 Association and impact of each factor (living-alone group) in item D-4

Item	Category	χ^2 value	P value	Cramer's V	Category score	− +	Range	Partial correlation coefficient
Age group	Early elderly	1.5483	0.4611	0.2046	0.9649		1.4290	0.2581
	Late-stage elderly				0.0733			
	Super elderly				−0.4641			
Gender	Male	0.0011	0.9739	0.0054	−0.0447		0.0534	0.0103
	Female				0.0087			
Children	Presence	1.6933	0.1932	0.2139	0.4854		1.1974	0.3095
	Absence				−0.7119			
Health	Cheerful	0.5669	0.7532	0.1238	0.2485		1.2624	0.1562
	Average				−0.0850			
	Anxious				−1.0138			
Medical examination	Consult a doctor	1.5443	0.2140	0.2043	−0.2851		1.7583	0.3291
	Do not consult a doctor				1.4731			
Lifestyle	Affluent	1.0678	0.5863	0.1699	−0.3891		1.7331	0.2514
	Average				0.1857			
	Poor				1.3440			
Care, support	Receive services	0.1444	0.7039	0.0625	1.1132		1.5255	0.3199
	Do not receive services				−0.4123			

participants to not feel that they were made to live was exhibited by the early elderly and late-stage elderly in the living-alone group, and by the early elderly and the super elderly in the living-together group.

In the living-alone group, participants tended to increasingly feel that they were made to live according to their age group; thus, for changes in feelings of being made to live, a transcendental tendency was demonstrated. In the living-together group, the late-stage elderly showed a tendency to increasingly feel that they were made to live and no transcendental tendency was observed.

As shown in Tables 12.12 and 12.13, when responses to item D-4 were used as an external criterion, the explanatory variables with a higher range were medical examination in the living-alone group (1.7583) and lifestyle in the living-together group (2.5759). The explanatory variables with a higher partial correlation coefficient were medical examination in the living-alone group (0.3291), similar to the

Table 12.13 Association and impact of each factor (living-together group) in D-4

Item	Category	χ^2 value	P value	Cramer's V	Category score	−	+	Range	Partial correlation coefficient
Age group	Early elderly	4.8583	0.0881	0.5057	0.1789			1.3420	0.4906
	Late-stage elderly				-0.3741				
	Super elderly				0.9679				
Gender	Male	6.1147	0.0134	0.5673	0.7715			1.3325	0.6174
	Female				-0.5611				
Children	Presence	1.3772	0.2406	0.2692	-0.2237			0.7085	0.3941
	Absence				0.4847				
Health	Cheerful	0.2628	0.8769	0.1176	0.1956			0.8991	0.3152
	Average				-0.2625				
	Anxious				0.6366				
Medical examination	Consult a doctor	1.6442	0.1997	0.2942	0.1471			0.9314	0.3669
	Do not consult a doctor				-0.7843				
Lifestyle	Affluent	3.1319	0.2089	0.4060	0.4927			2.5759	0.4377
	Average				0.0080				
	Poor				-2.0832				
Care, support	Receive services	0.1015	0.7500	0.0731	-1.0775			1.3648	0.4753
	Do not receive services				0.2873				

range results, and gender in the living-together group (0.6174). The analysis of the associations between responses to item D-4 and general characteristics using Cramer's coefficient of association revealed no correlation of 0.25 or more in the living-alone group, while very strong associations were shown with age group (0.5673) and gender (0.5673). Moreover, associations were demonstrated with children (0.2692), medical examination (0.2942), and lifestyle (0.4060) in the living-together group.

12.4.5 Changes in Faith

To elucidate cases where participants have a deeper faith or do not have a deeper faith, the associations between the general characteristics, including age group, were examined according to the residence status, with responses to item D-5: I have deeper faith (yes or no) as an external criterion (target variable).

The results of the analysis are shown in Tables 12.14, 12.15, and 12.16. Negative values for the category scores indicate a tendency for participants to have deeper faith, while positive values indicate a tendency to not have deeper faith.

With age group as explanatory variable, a tendency of having deeper faith was demonstrated by the early elderly and the late-stage elderly in the living-alone

Table 12.14 Average score, correlation ratio, and predictive value in item D-5

	D-5	Average score	Correlation ratio	Predictive value (%)
Living alone	Yes	0.4965	0.2602	72.97
	No	−0.5241		
Living together	Yes	−0.9011	0.4736	73.68
	No	0.5256		

Table 12.15 Association and impact of each factor (living-alone group) in item D-5

Item	Category	χ^2 value	P value	Cramer's V	Category score	−	+	Range	Partial correlation coefficient
Age group	Early elderly				−0.9522				
	Late-stage elderly	1.4907	0.4746	0.2007	−0.2110			1.5582	0.2945
	Super elderly				0.6060				
Gender	Male				−0.9999				
	Female	0.0052	0.9423	0.0119	0.1935			1.1934	0.2205
Children	Presence				−0.0369				
	Absence	0.0397	0.8421	0.0327	0.0541			0.0909	0.0250
Health	Cheerful				0.1164				
	Average	2.2413	0.3261	0.2461	−0.3097			2.5335	0.2960
	Anxious				2.2238				
Medical examination	Consult a doctor				0.0975				
	Do not consult a doctor	0.6724	0.4122	0.1348	−0.5038			0.6013	0.1209
Lifestyle	Affluent				−0.2499				
	Average	3.4477	0.1784	0.3053	−0.2269			2.9595	0.3900
	Poor				2.7096				
Care, support	Receive services				−0.4568				
	Do not receive services	0.0100	0.9203	0.0165	0.1692			0.6260	0.1366

group, and by the late-stage elderly in the living-together group. A tendency of not have deeper faith was exhibited by the super elderly in the living-alone group, and by the early elderly and the super elderly in the living-together group.

In addition, a tendency opposite to transcendence was shown in the living-alone group. In the living-together group, the late-stage elderly tended to have deeper faith, and a transcendental tendency was not observed.

As shown in Tables 12.15 and 12.16, when responses to item D-5 were used as an external criterion, the explanatory variable with a higher range was lifestyle in both the living-alone (2.9595) and the living-together (6.3861) groups. The explanatory variables with a higher partial correlation coefficient were lifestyle in the living-alone group (0.3900), similar to the range results, and care and support in the living-together group (0.5943). The analysis using Cramer's coefficient of association indicated relationships between responses to item D-5 and the following

Table 12.16 Association and impact of each factor (living-together group) in D-5

Item	Category	χ^2 value	P value	Cramer's V	Category score	− +	Range	Partial correlation coefficient
Age group	Early elderly	1.3917	0.4987	0.2706	0.0576		0.7423	0.1979
	Late-stage elderly				−0.1769			
	Super elderly				0.5654			
Gender	Male	0.0026	0.9596	0.0116	0.1433		0.2474	0.0977
	Female				−0.1042			
Children	Presence	1.5340	0.2155	0.2841	−0.2589		0.8198	0.3229
	Absence				0.5609			
Health	Cheerful	0.0384	0.9810	0.0449	−0.6552		3.4319	0.5347
	Average				−0.4594			
	Anxious				2.7766			
Medical examination	Consult a doctor	0.0188	0.8908	0.0315	0.0632		0.4002	0.1233
	Do not consult a doctor				−0.3370			
Lifestyle	Affluent	2.4235	0.2977	0.3571	1.2128		6.3861	0.5679
	Average				0.0230			
	Poor				−5.1732			
Care, support	Receive services	0.3770	0.5392	0.1409	−2.7128		3.4362	0.5943
	Do not receive services				0.7234			

general characteristics: lifestyle (0.3053) in the living-alone group, and age group (0.2706), children (0.2841), and lifestyle (0.3571) in the living-together group.

12.4.6   Changes in the Way of Feeling About a Supernatural Being, Such as God

To clarify the cases of participants feeling something familiar about a supernatural being, such as God, and feeling nothing familiar about a supernatural being, such as God, the associations between the general characteristics, including age group, were examined according to residence status, with responses to item D-6: I feel

Table 12.17 Average score, correlation ratio, and predictive value in item D-6

	D-6	Average score	Correlation ratio	Predictive value (%)
Living alone	Yes	0.4717	0.2349	67.57
	No	−0.4979		
Living together	Yes	−0.8772	0.5596	89.47
	No	0.6380		

Table 12.18 Association and impact of each factor (living-alone group) in item D-6

Item	Category	χ^2 value	P value	Cramer's V	Category score	−	+	Range	Partial correlation coefficient
Age group	Early elderly	0.7068	0.7023	0.1382	−0.9760			1.2986	0.2261
	Early elderly				0.0636				
	Super elderly				0.3225				
Gender	Male	0.9306	0.3347	0.1586	−0.3479			0.4152	0.0733
	Female				0.0673				
Children	Presence	0.0397	0.8421	0.0327	−0.1043			0.2573	0.0669
	Absence				0.1530				
Health	Cheerful				0.2833				
	Average	0.2398	0.8870	0.0805	−0.2696			0.8408	0.1521
	Anxious				0.5712				
Medical examination	Consult a doctor	2.9321	0.0868	0.2815	0.2638			1.6268	0.2935
	Do not consult a doctor				−1.3630				
Lifestyle	Affluent				−0.4365				
	Average	3.8647	0.1448	0.3232	0.0909			2.5702	0.3145
	Poor				2.1337				
Care, support	Receive services	0.4103	0.5218	0.1053	−0.6275			0.8599	0.1834
	Do not receive services				0.2324				

Table 12.19 Association and impact of each factor (living-together group) in D-6

Item	Category	χ^2 value	P value	Cramer's V	Category score	−	+	Range	Partial correlation coefficient
Age group	Early elderly				0.6553				
	Late-stage elderly	2.1921	0.3342	0.3397	−0.3861			1.5394	0.3877
	Super elderly				−0.8841				
Gender	Male	1.6586	0.1978	0.2955	0.9489			1.6390	0.5528
	Female				−0.6901				
Children	Presence	0.2768	0.5988	0.1207	−0.1200			0.3800	0.1777
	Absence				0.2600				
Health	Cheerful				−0.9037				
	Average	0.1544	0.9257	0.0901	−0.3013			3.5149	0.6069
	Anxious				2.6111				
Medical examination	Consult a doctor	0.8816	0.3478	0.2154	0.0394			0.2495	0.0908
	Do not consult a doctor				−0.2101				
Lifestyle	Affluent				0.2102				
	Average	4.2025	0.1223	0.4703	0.2982			5.3134	0.6755
	Poor				−5.0152				
Care, support	Receive services	0.1295	0.7189	0.0826	−1.3291			1.6835	0.4775
	Do not receive services				0.3544				

something familiar about a supernatural being, like God (yes or no) as the external criterion (target variable).

The results of the analysis are shown in Tables 12.17, 12.18, and 12.19. Negative values for the category scores indicate a tendency to feel something familiar about a supernatural being, such as God, while positive values indicate a tendency to feel nothing familiar about a supernatural being.

With age group as an explanatory variable, a tendency to feel something familiar about a supernatural being was exhibited by the early elderly in the living-alone group, and by the late-stage elderly and the super elderly in the living-together group. A tendency to feel something familiar about a supernatural existence was demonstrated by the late-stage elderly and the super elderly in the living-alone group, and by the early elderly in the living-together group.

In addition, a tendency opposite to transcendence was shown in the living-alone group. Participants in the living-together group tended to feel something familiar with a supernatural being, such as God, as they aged. Therefore, a transcendental tendency was demonstrated by the changes in feelings about a supernatural being, such as God.

As shown in Tables 12.18 and 12.19, when responses to item D-6 were used as an external criterion, the explanatory variable with a higher range was lifestyle in both the living-alone (2.5702) and living-together (5.3134) groups. The explanatory variable with a higher partial correlation coefficient was lifestyle in both the living-alone (0.3145) and living-together (0.6755) groups, similar to the range results. The analysis using Cramer's coefficient of association indicated relationships between responses to item D-6 and several general characteristics: medical examination (0.2815) and lifestyle (0.3232) in the living-alone group, and age group (0.3397), gender (0.2955), and lifestyle (0.4703) in the living-together group.

12.4.7   Changes in a Sense of Fear Regarding Death

To clarify cases of participants feeling that a sense of fear of death has disappeared or not feeling that a sense of fear of the death has disappeared, associations between the categories of general characteristics, including age group, were examined according to residence status, with responses to item D-7: I feel that a sense of fear of death has disappeared (yes or no) as an external criterion (target variable).

Table 12.20 Average score, correlation ratio, and predictive value in item D-7

	D-7	Average score	Correlation ratio	Predictive value (%)
Living alone	Yes	0.3717	0.3731	81.08
	No	−1.0037		
Living together	Yes	−0.9469	0.6522	89.47
	No	0.6887		

Table 12.21 Association and impact of each factor (living-alone group) in item D-7

Item	Category	χ^2 value	P value	Cramer's V	Category score	−	+	Range	Partial correlation coefficient
Age group	Early elderly	0.4722	0.7897	0.1130	−0.1157			0.6731	0.2217
	Late-stage elderly				0.3481				
	Super elderly				−0.3250				
Gender	Male	0.1444	0.7039	0.0625	0.6636			0.7920	0.1897
	Female				−0.1284				
Children	Presence	0.5087	0.4757	0.1173	−0.0270			0.0666	0.0236
	Absence				0.0396				
Health	Cheerful	6.9158	0.0315	0.4323	−0.6258			2.2678	0.3971
	Average				0.3052				
	Anxious				1.6420				
Medical examination	Consult a doctor	0.3897	0.5324	0.1026	−0.0361			0.2226	0.0586
	Do not consult a doctor				0.1865				
Lifestyle	Affluent	0.9792	0.6129	0.1627	−0.0350			0.4003	0.0738
	Average				−0.0291				
	Poor				0.3653				
Care, support	Receive services	7.5543	0.0060	0.4519	1.2298			1.6853	0.4562
	Do not receive services				−0.4555				

Table 12.22 Association and impact of each factor (living-together group) in D-7

Item	Category	χ^2 value	P value	Cramer's V	Category score	−	+	Range	Partial correlation coefficient
Age group	Early elderly	2.1921	0.3342	0.3397	0.4350			0.9271	0.4911
	Late-stage elderly				−0.4849				
	Super elderly				0.4422				
Gender	Male	0.3533	0.5522	0.1364	0.1789			0.3090	0.1705
	Female				−0.1301				
Children	Presence	0.2242	0.6358	0.1086	0.0727			0.2302	0.1340
	Absence				−0.1575				
Health	Cheerful	2.8893	0.2358	0.3900	−1.4453			3.6591	0.7153
	Average				0.0532				
	Anxious				2.2137				
Medical examination	Consult a doctor	2.5909	0.1075	0.3693	−0.3171			2.0082	0.6579
	Do not consult a doctor				1.6911				
Lifestyle	Affluent	0.8328	0.6594	0.2094	0.2277			1.0785	0.1961
	Average				−0.0043				
	Poor				−0.8508				
Care, support	Receive services	0.1295	0.7189	0.0826	−0.9129			1.1564	0.3793
	Do not receive services				0.2435				

The results of the analysis are shown in Tables 12.20, 12.21, and 12.22. Negative values for the category scores indicate a tendency that participants feel that a sense of fear of death has disappeared, while positive values indicate a tendency that participants do not feel that a sense of fear of death has disappeared.

With age group as an explanatory variable, a tendency to feel that a sense of fear of death has disappeared was exhibited by the early elderly and the super elderly in the living-alone group, and by the late-stage elderly in the living-together group. A tendency to not feel that a sense of fear of death has disappeared was demonstrated by the late-stage elderly in the living-alone group, and by the early elderly and the super elderly in the living-together group.

In the living-alone group, the early elderly also showed a tendency to feel that a sense of fear of death has disappeared. Therefore, there was no transcendental tendency. In the living-together group, the late-stage elderly also showed a tendency to feel that a sense of fear of death has disappeared. No transcendental tendency was observed.

As shown in Tables 12.21 and 12.22, when responses to item D-7 were used as an external criterion, the explanatory variable with a higher range was health in both the living-alone (2.2678) and living-together (3.6591) groups. The explanatory variables with a higher partial correlation coefficient were care and support in the living-alone group (0.4562), similar to the range result, and medical examination in the living-together group (0.6579). The analysis using Cramer's coefficient of association indicated relationships between responses to item D-7 and the following general characteristics: health (0.323) in the living-alone group, and care and support (0.4519), age group (0.3397), health (0.3900), and medical examination (0.3693) in the living-together group.

12.4.8 Changes in the Desire to Live Out One's Own Life

To elucidate cases where participants strongly want to live out their lives or do not want to live out their lives, associations between the categories of general characteristics, including age group, were examined according to residence status, with responses to item D-8: I strongly want to live out my life (yes or no) as an external criterion (target variable).

The results of the analysis are shown in Tables 12.23, 12.24, and 12.25. Negative values for the category scores indicate a tendency to strongly desire to live

Table 12.23 Average score, correlation ratio, and predictive value in item D-8

	D-8	Average score	Correlation ratio	Predictive value (%)
Living alone	Yes	−0.2300	0.1918	78.38
	No	0.8337		
Living together	Yes	−0.2952	0.7405	100
	No	2.5089		

Table 12.24 Association and impact of each factor (living-alone group) in item D-8

Item	Category	χ^2 value	P value	Cramer's V	Category score	−	+	Range	Partial correlation coefficient
Age group	Early elderly	0.5868	0.7457	0.1259	0.2545			0.3395	0.0538
	Late-stage elderly				−0.0850				
	Super elderly				−0.0111				
Gender	Male	0.1037	0.7474	0.0530	−1.1650			1.3904	0.2091
	Female				0.2255				
Children	Presence	0.0391	0.8432	0.0325	0.0998			0.2463	0.0568
	Absence				−0.1464				
Health	Cheerful	1.0047	0.6051	0.1648	0.1184			2.2313	0.2181
	Average				−0.2836				
	Anxious				1.9477				
Medical examination	Consult a doctor	0.1037	0.7474	0.0530	0.0219			0.1353	0.0225
	Do not consult a doctor				−0.1134				
Lifestyle	Affluent	3.9306	0.1401	0.3259	−0.3722			3.5640	0.3754
	Average				−0.1797				
	Poor				3.1917				
Care, support	Receive services	0.0213	0.8841	0.0240	−0.1773			0.2430	0.0453
	Do not receive services				0.0657				

Table 12.25 Association and impact of each factor (living-together group) in D-8

Item	Category	χ^2 value	P value	Cramer's V	Category score	−	+	Range	Partial correlation coefficient
Age group	Early elderly	4.4007	0.1108	0.4813	0.6954			1.3118	0.5318
	Late-stage elderly				−0.6164				
	Super elderly				−0.0078				
Gender	Male	0.0572	0.8111	0.0548	0.1174			0.2028	0.1276
	Female				−0.0854				
Children	Presence	4.8431	0.0278	0.5049	−0.4826			1.5281	0.7427
	Absence				1.0456				
Health	Cheerful	2.2692	0.3216	0.3456	−0.7815			1.8256	0.6063
	Average				0.0705				
	Anxious				1.0441				
Medical examination	Consult a doctor	0.4191	0.5174	0.1485	−0.0470			0.2975	0.1472
	Do not consult a doctor				0.2505				
Lifestyle	Affluent	0.7983	0.6709	0.2050	−0.9466			1.2306	0.4127
	Average				0.2840				
	Poor				−0.1893				
Care, support	Receive services	1.1270	0.2884	0.2435	0.8434			1.0683	0.4251
	Do not receive services				−0.2249				

out their lives, while positive values indicate a tendency to not want to live out their lives.

With age group as an explanatory variable, a tendency to strongly want to live out their lives was exhibited by the late-stage elderly and the super elderly in both the living-alone group and the living-together group. A tendency to not want to live out their lives was demonstrated by the early elderly in both the living-alone and living-together groups.

In both the living-alone and living-together groups, the late-stage elderly showed a higher negative value category score compared to the super elderly, meaning that the late-stage elderly showed a tendency to strongly want to live out their lives, and no transcendental tendency was observed.

As shown in Tables 12.24 and 12.25, when responses to item D-8 were used as an external criterion, the explanatory variables with a higher range were lifestyle in the living-alone group (3.5640) and health in the living-together group (1.8256). The explanatory variables with a higher partial correlation coefficient were lifestyle in the living-alone group (0.3754) and children in the living-together group (0.7427). The analysis using Cramer's coefficient of association indicated relationships between responses to item D-8 and the following general characteristic categories: health (0.323) in the living-alone group and care and support (0.3259) in the living-together group. A very strong association was shown with children (0.5049), as well as moderate associations with age group (0.4813) and health (0.3456) in the living-together group.

12.4.9 Conclusions Regarding Question D

The questionnaire survey results for question D: Transcendental changes such as faith and separation from obsession clinging to self were classified into the following four categories and examined: (1) responses indicating a transcendental tendency, (2) responses indicating a tendency opposite to transcendence, (3) characteristics in the living-alone group, and (4) characteristics in the living-together group.

1. Responses indicating transcendental tendencies

The category scores indicating a transcendental tendency were observed for responses to items D-1 and D-4 in the living-alone group, and items D-1 and D-6 in the living-together group. In particular, for item D-1 in the living-alone group, the association between the external criterion and age group was indicated by Cramer's coefficient of association; there was a difference in the category score ranges between the early elderly and the super elderly and the partial correlation coefficient was higher. Hence, a transcendental tendency was clearly demonstrated.

2. Responses indicating tendencies opposite to transcendence

Responses of "yes" by the early elderly and "no" by the super elderly were found for items D-5 and D-6 in the living-alone group and for item D-2 in the living-together group. In particular, for items D-5 and D-6 in the living-alone group, significant associations with lifestyle were indicated by Cramer's coefficient of association, as well as by the range and partial correlation coefficient.

Participants who thought their lifestyle was "affluent" tended to respond "yes" while participants who thought their lifestyle was "poor" showed a tendency to respond "no." For five characteristics (gender, children, health, medical examination, and care and support), there was a greater tendency to respond "yes" by those participants who were male, had children, had average health, did not consult a doctor, and receive care services, respectively.

3. Characteristics in the living-alone group

In the living-alone group, for items D-2, D-3, and D-7, the early elderly and the super elderly tended to respond "yes." There was a common tendency in items D-2, D-3, and D-7: participants were more likely to respond "yes" if they did not have children, were affluent, and did not receive care services.

According to the results for Cramer's coefficient of association, range, and partial correlation coefficients, there was an association between lifestyle and responses for items D-2, D-5, D-6, and D-8, which indicated that participants who thought their lifestyle was affluent showed a tendency to respond "yes," while participants who thought their lifestyle was poor showed a tendency to respond "no". Male participants also demonstrated a tendency toward "yes" responses.

4. Characteristics in the living-together group

In the living-together group, for items D-3, D-4, D-5, and D-7, the late-stage elderly exhibited a tendency of "yes" responses, while the early elderly and the super elderly showed a tendency to respond "no." There was a common tendency demonstrated in items D-3, D-4, D-5, and D-7 in that females tended to respond "yes."

For the analysis using the range, items demonstrating an association with lifestyle were D-1, D-2, D-3, D-4, D-5, and D-6. Of these items, for D-1, D-4, D-5, and D-6, poor participants showed a tendency of "yes" responses, while affluent participants had a tendency to respond "no" to the items. For items D-2 and D-3, affluent participants showed a tendency of "yes" responses while poor participants showed a tendency of responding "no".

In six of the items where the range results indicated an association, there was a large difference in the tendency between "yes" and "no" responses. For the general characteristic of lifestyle, there was a difference the responses to each item between affluent and poor participants in the living-together group.

12.5   Conclusion

In this study, we used quantification theory II to analyze the results of a survey conducted with the elderly using a questionnaire that was developed based on *Gerontranscendence: A Developmental Theory of Positive Aging* by L. Tornstam. The data were compared by classifying the participants according to their residence status: single-elderly households and elderly households with only a spouse. The results confirmed the impact of seven general characteristics and their associations with the eight external criterion regarding gerotranscendental changes such as religiosity and separation from clinging to self. Further examination focusing on the characteristics of each group based on these results will be performed in the future to explore support measures for the elderly.

The questionnaire included items in addition to the eight items regarding transcendental changes such as faith and separation from clinging to self [13]. Going forward, the responses to these items will also be comprehensively analyzed to further elucidate characteristics of the elderly.

References

1. The Statistics Bureau of the Ministry of Internal Affairs and Communications: The population statistics released in July 2016 (2016)
2. Director General for Policy on Cohesive Society: The white paper on aging society in 2016 (2016)
3. Ministry of Health, Labour and Welfare: Comprehensive survey of living conditions in 2014 (2014)
4. Hasehe, M.: A study of autobiographical memory functions in elderly people who live alone. Study Soc. Work **48**, 111–114 (2009)
5. Mizuho Information & Research Institute: Business report on surveys and research of community infrastructure and its plan to support in elderly people who live alone and aged household (2012)
6. Tomizawa, K.: Gerotranscendence as an adjustment outcome in the ninth stage of life cycle: a case study of the oldest old in the Amami Archipelago, Bulletin of Graduate School of Human Development and Environment, Kobe University, **2**(2), 111–119 (2009)
7. Tsukuda, A.: One consideration for the re-formulation of 'Successful Aging', Ritsumeikan review of industrial society. Ritsumeikan Soc. Sci. rev. **43**(4), 133–154 (2008)
8. Tornstam, L.: Gerotranscendence. Springer, New York (2005)
9. Kan, T., Yasunori, F.: Quantification Theory Type II, Gendai-Sugakusha (2011)
10. Fujii, Y.: Vol. 1 Categorical data analysis, "Data Science Learning Through R", Kyoritsu Shuppan (2010)
11. Watada, J., Tanaka, H., Asai, K.: Analysis of purchasing factors by using fuzzy quantification theory type II. J. Japan Ind. Manage. Assoc. **32**(5), 385–392 (1981)
12. Kusunoki, M.: The sensibility to cold in female students: the quantification method entitled "II". J. Yasuda Women's Univ. **39**, 193–200 (2011)
13. Kawaura, T., Sugatani, Y.: A study of older people's socializing form with others: comparative analysis of "Living alone" and "Living with a spouse" using quantification theory type II. Innov. Med. Healthc. **2015**, 115–126 (2015)